Nevanlinna Theory and Its Relation to Diophantine Approximation

Nevanlinna Theory and Its Relation to Diophantine Approximation

Min Ru
Department of Mathematics
University of Houston
USA

World Scientific
Singapore • New Jersey • London • Hong Kong

Published by

World Scientific Publishing Co. Pte. Ltd.

P O Box 128, Farrer Road, Singapore 912805

USA office: Suite 1B, 1060 Main Street, River Edge, NJ 07661

UK office: 57 Shelton Street, Covent Garden, London WC2H 9HE

Library of Congress Cataloging-in-Publication Data

Ru, Min.

Nevanlinna theory and its relation to Diophantine approximation / Min Ru.

p. cm.

Includes bibliographical references and index.

ISBN 9810244029 (alk. paper)

1. Nevanlinna theory. 2. Diophantine approximation. I. Title.

QA331 .R82 2001

515--dc21 2001017930

British Library Cataloguing-in-Publication Data

A catalogue record for this book is available from the British Library.

Printed in Singapore.

to My Wife Yu Shen, My Son Aaron and My Daughter Christina

Preface

Diophantine equations are systems of polynomial equations to be solved in integers, in rational numbers, or in various generalizations, such as finitely generated rings over **Z** or finitely generated fields over **Q**. Diophantine approximation is the study of Diophantine equations using the method of approximations. The Nevanlinna theory, on the other hand, studies holomorphic solutions of the systems of polynomial equations. More precisely, since the complex solutions to a system of polynomial equations form an algebraic variety, Diophantine approximation studies the rational points in algebraic varieties defined over **Q** and Nevanlinna theory investigates the properties of holomorphic curves in algebraic varieties over **C**. Nevanlinna theory and Diophantine approximation have developed independently of one another for several decades. It has been, however, discovered by C.F. Osgood, P. Vojta, Serge Lang and others that a number of striking similarities exist between these two subjects. Generally speaking, a non-constant holomorphic curve in an algebraic variety corresponds to an infinite set of rational points, so, in this way, any theorem in Nevanlinna theory should translate into a true statement in Diophantine approximation. A growing understanding of these connections over the last 15 years has led to significant advances in both fields. Outstanding conjectures from decades ago are being solved.

This book presents, in a systematic and almost self-contained way, the analogy between the Nevanlinna theory and Diophantine approximation. Although the emphasis is on Nevanlinna theory, both theories are presented in this book, including some results of recent research. We divide each chapter into Part A and Part B. Part A deals with Nevanlinna theory and Part B covers Diophantine approximation. At the end of each chapter, a table is provided to indicate the correspondence of the theorems.

Let us first review the historical development of Nevanlinna theory. Nevanlinna theory begins with the study of the distribution of values of meromorphic functions. In 1929, R. Nevanlinna extended the classical little Picard's theorem by proving the two elegant theorems (we call them the *First* and the *Second Main Theorem*). The work of Nevanlinna evoked a very strong interest of research in his theory. Right after Nevanlinna, a number of important papers by researchers such as Bloch [Blo], Cartan[Car1] [Car2], and Weyls [We-We] were published. Cartan extended Nevanlinna's result to holomorphic curves in projective spaces and Bloch considered holomorphic curves in Abelian varieties. In 1941, Ahlfors[Ahl], following Weyls' work, gave a geometric approach to the theory of holomorphic curves in projective spaces. Stoll [Sto1] generalized the work of Weyl-Ahlfors to the case of several complex variables, and gave a foundation of the Nevanlinna theory in several complex variables. The task of generalizing the Nevanlinna theory to higher dimensional complex manifolds is, in general, very difficult. Griffiths et al. [Ca-G], in the 1970's, successfully proved the Second Main Theorem for equi-dimensional holomorphic mappings. Their results also gave a new insight to the theory in terms of Chern invariants after the work of Bott-Chern [Bo-C]. A recent discovery of the relationship to Diophantine approximation has generated greatly renewed interest in Nevanlinna theory. Great progress has been made recently: Siu-Yeung [Siu-Y2] settled Lang's conjecture for Abelian varieties and significant progress towards solving Griffiths' conjecture concerning holomorphic curves in algebraic varieties has been made (see Siu-Yeung [Siu-Y1], Dethloff, Schumacher and Wong [D-S-W1] [D-S-W2], Wong[Wong6], McQuillan [McQ3], Demailly and J.El Goul [Dem-G2] etc.). McQuillan's work of translating Faltings' proof in Diophantine approximation to Nevanlinna theory is also worthy of mention.

Diophantine problems have also had a long history. In the first half of the 20th century, Thue and Siegel developed the method of Diophantine approximation to prove finiteness for integer solutions of certain polynomial equations. To give an example, if (x, y) is a large integral solution of the equation $x^3 - 2y^3 = 1$, then

$$|\frac{x}{y} - 2^{1/3}| = \frac{1}{|y||x^2 + 2^{1/3}xy + 4^{1/3}y^2|} << \frac{1}{|y|^3}.$$

By continued fractions, for any irrational real number α there exists a constant $c = c(\alpha)$ and infinitely many rational numbers x/y ($x, y \in \mathbf{Z}$) with $|x/y - \alpha| < c/|y|^2$. Thue and Siegel used a weak converse of this fact to obtain their finiteness statements. Roth [Rot] in 1955 proved the celebrated Roth's theorem, which provides a "sharp approximation" to algebraic numbers. W. Schmidt extended Roth's result for simultaneous approximation to algebraic numbers. Faltings [Fal1] in 1983 solved Mordell's conjecture: *A curve of genus $g \geq 2$ has only finitely many rational points.* P. Vojta [Voj2] obtained an alternative proof of Faltings' theorem, using Diophantine approximation techniques similar to those used in the proof of Roth's theorem. In the same year, G. Faltings, using an adaptation of Vojta's method, extended the theory of Diophantine approximation to Abelian varieties.

This book begins with Nevanlinna theory of meromorphic functions. In Chapter I, we prove Nevanlinna's first and second main theorem for meromorphic functions, as well as Roth's theorem. We also carefully examine the "error terms" appearing in Nevanlinna's Second Main Theorem. The precise error term gives a better analogy. Chapter II presents the general one dimensional Nevanlinna theory and Diophantine approximation. Both theories depend on the genus. The genus 0 case has been included in chapter I. *If the genus is 1, then any affine Riemann surface does not carry a non-constant holomorphic curve*, while in Diophantine approximation, *a curve of genus $g \geq 1$ has only finitely many integer points.* The first is referred to as Picard's theorem and the latter is referred to as Siegel's theorem. We also know that *the set of rational points on a curve of genus 1 forms a group and this group is of finite rank.* This is the Mordell-Weil theorem. *A curve of genus $g \geq 2$ has only finitely many rational points.* This was known as Mordell's conjecture, but is now Faltings' theorem. On the other hand, the classical Picard's theorem states that *every holomorphic map from $\mathbf{C}$ to a compact Riemann surface of genus greater than or equal to two must be constant.* Chapter III introduces Cartan's theory for holomorphic curves in projective spaces and Schmidt's subspace theorem. We then use these theorems to study the hyperbolicity of the complement of hyperplanes in Nevanlinna theory and to study the finiteness of the number of integer solutions of decomposable polynomial equations in Diophantine approximation. Chapter IV extends the theory of Chapter III to the moving target case. Chapter V covers the equi-dimensional Nevanlinna theory developed by Griffiths and his school in the late 1970's. Chapter VI studies holomorphic curves in Abelian varieties, as well as rational points of Abelian varieties. We also include the proof by McQuillan of Bloch's conjecture in Nevanlinna theory. This method is parallel to Faltings' proof regarding the Diophantine problems of the Abelian variety. The last chapter includes the theory of complex hyperbolicity, focusing on the method of negative curvature.

Because this book covers subjects in complex analysis and number theory, I hope it will be useful to many types of mathematicians: complex analysts, differential geometers, algebraic geometers, and number theorists at the very least. It is also my hope that this book will help you, the reader, to appreciate some of deep and elegant results currently known in both fields and will inspire you to do further research in these two beautiful subjects.

I wish to thank my Ph.D. advisor Wilhelm Stoll for introducing me to this field, and for his constant encouragements. I am very much indebted to many friends and colleagues, without whose help and encouragement this text would never have been written. I am especially grateful to Zhihua Chen, S.S. Chern, D. Drasin, S. Lang, R. Osserman, Y.T. Siu and P.M. Wong for their constant encouragements. I also want to express my appreciation to the many people from whom I learned these two subjects, including (but certainly not limited to) W. Cherry, J.P. Demailly, A. Eremenko, S. Lang, M. McQuillan, J. Noguchi, W.M. Schmidt, B. Shiffman, Y.T. Siu, P. Vojta, and P. M. Wong. I wish to thank LeeAnn Chastain and E H Chionh for editorial assistance. I would like to acknowledge the generous financial support of the U.S. National Science Foundation (through grants DMS-9596181, DMS-9800361) and U.S. Security Agency (through grants MDA904-99-1-0034, MDA904-01-1-0051). Finally, but most importantly, I want to thank my wife Yu Shen, my son Aaron, and my daughter Christina for their love, support, understanding and patience.

Contents

Chapter 1

Nevanlinna Theory for Meromorphic Functions and Roth's Theorem

In this chapter we introduce Nevanlinna's theory of meromorphic functions and Roth's theorem in Diophantine approximation. Nevanlinna theory and Diophantine approximation are counterpart to each other. As throughout this book, we divide each chapter into two parts. Part A covers Nevanlinna theory and Part B covers Diophantine approximation. At the end of the chapter, we provide a table describing the correspondence of the theorems.

Part A: Nevanlinna Theory

Nevanlinna theory studies holomorphic maps $f : \mathbf{C} \to M$ where $\mathbf{C}$ is the complex number plane and M is a complex manifold. In particular, it asks when f is forced to be constant. A simple example is taking M as a disc of finite radius. In this case, the classical Liouville theorem says that every holomorphic map sending $\mathbf{C}$ into the disc must be constant. This chapter deals with the case that $M = \mathbf{P}^1$, the complex projective space of dimension 1.

The fundamental tool of this subject is the measurement of the growth of f. Hadamard made the first discovery in this direction. Given an entire function, there are two different ways of measuring its rate of growth–its maximum modulus on the disc of radius r (viewed as a function of r) and the maximum number of times at the value in the image is taken on this disc. The insight is that these two rates of growth are essentially the same, the former being roughly the exponential of the latter. As an example, consider the function $e^{z^n} - 1$. Its maximum modulus on the disc of radius r grows like e^{r^n} while the number of zeros in this disc grows like

r^n. The maximum modulus itself clearly does not work for meromorphic functions since it may become infinite for finite values of r. R. Nevanlinna [Nev], in 1929, found the right substitute for the maximum modulus. He introduced the characteristic function $T_f(r)$ to measure the growth of the meromorphic function f. Starting from the Poisson-Jensen formula, he was able to derive a more subtle growth estimate for meromorphic functions in what he called the *Second Main Theorem.* It gives a quantitative version of the classical Picard's theorem for meromorphic functions. This chapter introduces Nevanlinna's *First* and *Second Main Theorem* for meromorphic functions. We note that while we present the *Second Main Theorem*, we also, following W. Cherry and Zhuan Ye [Ch-Y], carefully examine the "error term". The Second Main Theorem with a "good" error term provides a more precise analogy to Roth's theorem in Diophantine approximation.

A1.1 The First Main Theorem

We begin by recalling the following well-known Poisson-Jensen formula in the classical complex analysis.

Theorem A1.1.1 (Poisson-Jensen Formula) Let $f \not\equiv 0$ be meromorphic on the closed disc $\overline{\mathbf{D}}(R)$, $R < \infty$. Let $a_1, \ldots, a_p$ denote the zeros of f in $\overline{\mathbf{D}}(R)$, counting multiplicities, and let $b_1, \ldots, b_q$ denote the poles of f in $\overline{\mathbf{D}}(R)$, also counting multiplicities. Then for any z in $|z| < R$ which is not a zero or pole, we have

$$\begin{aligned}\log|f(z)| &= \int_0^{2\pi} \frac{R^2 - |z|^2}{|Re^{i\theta} - z|^2} \log|f(Re^{i\theta})| \frac{d\theta}{2\pi} \\ &\quad - \sum_{i=1}^{p} \log\left|\frac{R^2 - \bar{a}_i z}{R(z - a_i)}\right| + \sum_{j=1}^{q} \log\left|\frac{R^2 - \bar{b}_j z}{R(z - b_j)}\right|.\end{aligned}$$

Proof. We note that it suffices to prove the theorem when f has no zeros or poles on the circle $|z| = R$. Otherwise, we consider the function $f(\rho z)$ and let $\rho \to 1$.

We first consider the case when f is analytic and has no zeros in the closed disc $|z| \leq R$. Then $\log|f|$ is harmonic. For a given z in $\mathbf{D}(R)$, we consider the linear transformation $L(w) = \dfrac{R^2(z - w)}{R^2 - \bar{z}w}$. L sends z to zero and satisfies $|L(w)| = R$ if $|w| = R$. Let $F(w) = \log f(L(w))$. Applying the

Mean Value Theorem for harmonic functions to $F(w)$, we have

$$\log f(z) = F(0) = \int_0^{2\pi} F(Re^{i\theta}) \frac{d\theta}{2\pi} = \int_{|w|=R} F(w) \frac{dw}{2\pi i w}. \qquad (1.1)$$

We let $\zeta = L(w)$, then

$$w = L^{-1}(\zeta) = \frac{R^2(z-\zeta)}{R^2 - \bar{z}\zeta}.$$

So, for $|\zeta| = R$,

$$\begin{aligned} \frac{dw}{2\pi i w} &= \frac{1}{2\pi i}\left(\frac{-1}{z-\zeta} + \frac{\bar{z}}{R^2 - \bar{z}\zeta}\right) d\zeta = \left(\frac{-1}{z-\zeta} + \frac{\bar{z}}{\bar{\zeta}\zeta - \bar{z}\zeta}\right)\frac{d\zeta}{2\pi i} \\ &= \left(\frac{-\zeta}{z-\zeta} + \frac{\bar{z}}{\bar{\zeta} - \bar{z}}\right)\frac{d\zeta}{2\pi i \zeta} = \frac{R^2 - |z|^2}{|\zeta - z|^2}\frac{d\zeta}{2\pi i \zeta}. \end{aligned} \qquad (1.2)$$

Note that when $|w| = R$, $|\zeta| = R$, and $\frac{d\zeta}{i\zeta} = d\theta$, so by combining (1.1) and (1.2)

$$\log f(z) = \int_0^{2\pi} \log f(Re^{i\theta}) \frac{R^2 - |z|^2}{|Re^{i\theta} - z|^2} \frac{d\theta}{2\pi}.$$

Thus

$$\log |f(z)| = \int_0^{2\pi} \log |f(Re^{i\theta})| \frac{R^2 - |z|^2}{|Re^{i\theta} - z|^2} \frac{d\theta}{2\pi}. \qquad (1.3)$$

The Theorem is proved in this case.

For the general case, we consider the function

$$g(z) = f(z) \frac{\prod_{\mu=1}^{p} \frac{R^2 - \bar{a}_\mu z}{R(z-a_\mu)}}{\prod_{\nu=1}^{q} \frac{R^2 - \bar{b}_\nu z}{R(z-b_\nu)}}.$$

Then g has no zeros or poles in $|z| \le R$. Note that when $|z| = R$, $|g(z)| = |f(z)|$. Applying (1.3) to g yields the theorem. □

Applying Theorem A1.1.1 with $f(z) \equiv e$, we have

Corollary A1.1.2 *Let $|z| < R$, then*

$$\int_0^{2\pi} \frac{1}{|Re^{i\theta} - z|^2} \frac{d\theta}{2\pi} = \frac{1}{R^2 - |z|^2}.$$

Let $z_0 \in \mathbf{D}(R)$. If $f(z) = c(z-z_0)^m + \cdots$, where c is the leading nonzero coefficient, then m is called the order of f at z_0 and is denoted by $\mathrm{ord}_{z_0} f$.

Corollary A1.1.3 (Jensen's Formula) *Let $f \not\equiv 0$ be meromorphic on $\overline{\mathbf{D}}(R)$, $R < \infty$. Let $a_1, \ldots, a_p$ denote the zeros of f in $\overline{\mathbf{D}}(R) - \{0\}$, counting multiplicities, and let $b_1, \ldots, b_q$ denote the poles of f in $\overline{\mathbf{D}}(R) - \{0\}$, also counting multiplicities. Then*

$$\log |c_f| = \int_0^{2\pi} \log |f(Re^{i\theta})| \frac{d\theta}{2\pi} - \sum_{\mu=1}^{p} \log \left| \frac{R}{a_\mu} \right| + \sum_{\nu=1}^{q} \log \left| \frac{R}{b_\nu} \right| - (\mathrm{ord}_0 f) \log R,$$

where $f(z) = c_f z^{\mathrm{ord}_0 f} + \cdots$, $\mathrm{ord}_0 f \in \mathbf{Z}$, and c_f is the leading nonzero coefficient.

Proof. Applying Theorem A1.1.1 with $z = 0$ to the function

$$f(z) z^{-\mathrm{ord}_0 f}.$$

□

We now proceed to define Nevanlinna functions. Let f be a meromorphic function on $\mathbf{D}(R)$, where $0 < R \leq \infty$ and let $r < R$. Denote the number of poles of f on the closed disc $\overline{\mathbf{D}}(r)$ by $n_f(r, \infty)$, counting multiplicity. We then define the **counting function** $N_f(r, \infty)$ to be

$$N_f(r, \infty) = n_f(0, \infty) \log r + \int_0^r [n_f(t, \infty) - n_f(0, \infty)] \frac{dt}{t},$$

here $n_f(0, \infty)$ is the multiplicity if f has a pole at $z = 0$. For each complex number a, we then define the **counting function** $N_f(r, a)$ to be

$$N_f(r, a) = N_{1/(f-a)}(r, \infty). \tag{1.4}$$

So, in particular, by the definition of the Lebesgue-Stieltjes integral,

$$N_f(r, 0) = (\mathrm{ord}_0^+ f) \log r + \sum_{z \in \mathbf{D}(r), z \neq 0} (\mathrm{ord}_z^+ f) \log \left| \frac{r}{z} \right| \tag{1.5}$$

where $\mathrm{ord}_z^+ f = \max\{0, \mathrm{ord}_z f\}$ is just the multiplicity of the zero at z. We note that $N_f(r, a)$ measures how many times f takes value a. With this notation, we can rewrite Corollary A1.1.3 as

Corollary A1.1.4 *Let $f \not\equiv 0$ be meromorphic on $\overline{\mathbf{D}}(r)$. Then*

$$\log|c_f| = \int_0^{2\pi} \log|f(re^{i\theta})|\frac{d\theta}{2\pi} - \sum_{z\in\mathbf{D}(r),z\neq 0} (\mathrm{ord}_z f)\log\left|\frac{r}{z}\right| - (ord_0 f)\log r,$$

or equivalently,

$$\log|c_f| = \int_0^{2\pi} \log|f(re^{i\theta})|\frac{d\theta}{2\pi} + N_f(r,\infty) - N_f(r,0).$$

The Nevanlinna's **proximity function** $m_f(r,\infty)$ is defined by

$$m_f(r,\infty) = \int_0^{2\pi} \log^+|f(re^{i\theta})|\frac{d\theta}{2\pi}, \tag{1.6}$$

where $\log^+ x = \max\{0, \log x\}$. For any complex number a, the **proximity function** $m_f(r,a)$ of f with respect to a is then defined by

$$m_f(r,a) = m_{1/(f-a)}(r,\infty). \tag{1.7}$$

We note that $m_f(r,a)$ measures how close f is, on average, to a on the circle of radius r. Finally, the **Nevanlinna's characteristic function** of f is defined by

$$T_f(r) = m_f(r,\infty) + N_f(r,\infty). \tag{1.8}$$

$T_f(r)$ measures the growth of f. For example: $T_f(r) = O(1)$ if and only if f is constant; $T_f(r) = O(\log r)$ if and only if f is a rational function.

The characteristic function T, the proximity function m and the counting function N are the three main **Nevanlinna functions**. Nevanlinna theory can be described as the study of how the growth of these three functions is interrelated. The First Main Theorem is a reformulation of Corollary A1.1.4.

Theorem A1.1.5 (First Main Theorem) *Let $f \not\equiv 0$ be meromorphic on $\overline{\mathbf{D}}(R)$, $R \leq \infty$. Then, for any $0 \leq r < R$,*

(i) $T_f(r) = m_f(r,0) + N_f(r,0) + \log|c_f|$.

(ii) Given a complex number a,

$$|T_f(r) - m_f(r,a) - N_f(r,a)| \leq \left|\log|c_{1/(f-a)}|\right| + \log^+|a| + \log 2,$$

where $c_{1/(f-a)}$ *is the leading non-zero coefficient in the Taylor's expansion of* $1/(f-a)$ *around* 0.

Proof. (i) is derived directly from Corollary A1.1.4. To prove (ii), applying Corollary A1.1.4 to $1/(f-a)$ yields

$$\log|c_{1/(f-a)}| = \int_0^{2\pi} \log\frac{1}{|f(re^{i\theta})-a|}\frac{d\theta}{2\pi} + N_{1/(f-a)}(r,\infty) - N_{1/(f-a)}(r,0).$$

Since $\log x = \log^+ x - \log^+(1/x)$,

$$\begin{aligned}\log|c_{1/(f-a)}| &= \int_0^{2\pi} \log^+\frac{1}{|f(re^{i\theta})-a|}\frac{d\theta}{2\pi} - \int_0^{2\pi} \log^+|f(re^{i\theta})-a|\frac{d\theta}{2\pi}\\ &\quad +N_f(r,a) - N_f(r,\infty).\end{aligned}$$

Thus,

$$\int_0^{2\pi} \log^+|f(re^{i\theta})-a|\frac{d\theta}{2\pi} = -N_f(r,\infty)+m_f(r,a)+N_f(r,a)-\log|c_{1/(f-a)}|.$$

Note that if x and y are positive real numbers, then

$$\log^+(x+y) \le \log^+ 2\max\{x,y\} \le \log^+ x + \log^+ y + \log 2.$$

So

$$|\log^+|x-y| - \log^+|x|| \le \log^+|y| + \log 2.$$

Thus

$$|T_f(r) - m_f(r,a) - N_f(r,a) + \log|c_{1/(f-a)}|| \le \log^+|a| + \log 2.$$

□

Theorem A1.1.6 *Let* f *and* $a_j, 0 \le j \le m$, *be meromorphic functions on* **C** *satisfying*

$$\sum_{j=0}^m a_j f^j \equiv 0$$

on **C**. *Then*

$$T_f(r) \le \sum_{j=0}^m T_{a_j}(r) + O(1).$$

Proof. Without loss of generality, we assume that a_m and f are not identically zero. We have

$$\begin{aligned} m\log^+|f| &= \log^+(|f|^m) = \log^+\left(\left|\sum_{j=0}^{m-1}\frac{a_j}{a_m}f^j\right|\right) \\ &\leq \log^+\left(\max_{0\leq j\leq m-1}\left|\frac{a_j}{a_m}\right|\right) + \log^+\left(\sum_{j=0}^{m-1}|f|^j\right) \\ &\leq \log^+\left(\max_{0\leq j\leq m-1}\left|\frac{a_j}{a_m}\right|\right) + \log^+\left(m\max(1,|f|^{m-1})\right) \\ &\leq \log^+\frac{1}{|a_m|} + \sum_{j=0}^{m-1}\log^+|a_j| + (m-1)\log^+|f| + \log m. \end{aligned}$$

It follows that

$$m_f(r,\infty) \leq m_{a_m}(r,0) + \sum_{j=0}^{m-1} m_{a_j}(r,\infty) + O(1).$$

We now estimate $N_f(r,\infty)$. Let b be an entire function on $\mathbf{C}$ whose zeros divisor is the maximum of the pole divisors of $a_0,\ldots,a_{m-1}$ so that $ba_0,\ldots,ba_{m-1}$ are entire functions. Clearly we have

$$N_b(r,0) \leq \sum_{j=0}^{m-1} N_{a_j}(r,\infty).$$

It follows from $ba_m = -\sum_{j=0}^{m-1} ba_j f^{-m+j}$ that $N_f(r,\infty) \leq N_{ba_m}(r,0)$. Thus

$$N_f(r,\infty) \leq N_{a_m}(r,0) + N_b(r,0) \leq N_{a_m}(r,0) + \sum_{j=0}^{m-1} N_{a_j}(r,\infty).$$

So

$$T_f(r) \leq T_{1/a_m}(r) + \sum_{j=0}^{m-1} T_{a_j}(r) + O(1) = \sum_{j=0}^{m} T_{a_j}(r) + O(1).$$

□

The First Main Theorem states that $T_f(r) = m_f(r,a) + N_f(r,a) + O(1)$. It gives us an upper bound on $N_f(r,a)$ in terms of $T_f(r)$, hence on the number of times f takes on the value a. It can be regarded as the generalization of the statement that the number of solutions of any polynomial equation $P = a$, counting multiplicity, is at most $d = \deg P$. Our goal is to prove the Second Main Theorem which concerns the lower bounds of $N_f(r, a)$ in terms of $T_f(r)$. To do that, we need the "logarithmic derivative lemma", which appears in the next section.

A1.2 The Logarithmic Derivative Lemma

In this section, we derive the Logarithmic Derivative Lemma. We closely follow the presentation by W. Cherry and Zhuan Ye (cf. [Ch-Y]), focusing on developing a good error term. We note that if one ignores the sharpness of the error term, then the proof presented below can be significantly simplified (see [Hay] or [Lang2]).

Lemma A1.2.1 (Smirnov's inequality) *Let $R < \infty$, and let $F(z)$ be analytic in the disc $\mathbf{D}(R)$. If either of the functions $ReF(z)$ or $ImF(z)$ has constant sign in the disc $\mathbf{D}(R)$ where $ReF(z)$(resp. $ImF(z)$) is the real part (resp. imaginary part) of $F(z)$, then for any α with $0 < \alpha < 1$ and $0 < r < R$, we have*

$$\int_0^{2\pi} |F(re^{i\theta})|^\alpha \frac{d\theta}{2\pi} \le \sec(\pi\alpha/2)|F(0)|^\alpha.$$

Proof. Assume, without loss of generality, that $\mathrm{Re}F(z) > 0$ for $|z| < R$. The function F is non-zero in $\mathbf{D}(R)$, and we can fix a choice of $\arg F(z)$ so that $|\arg F(z)| < \pi/2$ for $|z| < R$. Thus the function

$$F^\alpha(z) = |F(z)|^\alpha e^{i\alpha \arg F(z)}$$

is analytic in $\mathbf{D}(R)$, and it follows that $\mathrm{Re}\{F^\alpha(z)\}$ is harmonic. Since

$$\mathrm{Re}F^\alpha(z) = |F(z)|^\alpha \cos(\alpha \arg F(z)) \ge |F(z)|^\alpha \cos(\alpha\pi/2),$$

then

$$\int_0^{2\pi} |F(re^{i\theta})|^\alpha \frac{d\theta}{2\pi} \le \sec(\alpha\pi/2) \int_0^{2\pi} \mathrm{Re}F^\alpha(re^{i\theta}) \frac{d\theta}{2\pi} = \sec(\alpha\pi/2)\mathrm{Re}F^\alpha(0),$$

and so

$$\int_0^{2\pi} |F(re^{i\theta})|^\alpha \frac{d\theta}{2\pi} \le \sec(\alpha\pi/2)|F(0)|^\alpha.$$

The lemma is thus proven. □

Theorem A1.2.2 (Gol'dberg-Grinshtein) *Let f be a meromorphic function in $\mathbf{D}(R)(0 < R \leq \infty)$, and let $0 < \alpha < 1$, then, for $r < s < R$, we have*

$$\begin{aligned}\int_0^{2\pi} \left|\frac{f'(re^{i\theta})}{f(re^{i\theta})}\right|^{\alpha} \frac{d\theta}{2\pi} \leq & \left(\frac{s}{r(s-r)}\right)^{\alpha} (m_f(s,0)+m_f(s,\infty))^{\alpha} \\ & +9\sec(\alpha\pi/2)r^{-\alpha}\left(n_f^{\alpha}(s,0)+n_f^{\alpha}(s,\infty)\right).\end{aligned}$$

Proof. Let $a_1, \ldots, a_p$ (resp. $b_1, \ldots, b_q$) denote the zeros (resp. poles) of f in $\mathbf{D}(s)$, counting multiplicities. We start from the Poisson-Jensen Formula,

$$\begin{aligned}\log|f(z)| = & \int_0^{2\pi} \frac{s^2-|z|^2}{|se^{i\phi}-z|^2}\log|f(se^{i\phi})|\frac{d\phi}{2\pi} - \sum_{\mu=1}^{p}\log\left|\frac{s^2-\bar{a}_\mu z}{s(z-a_\mu)}\right| \\ & +\sum_{\nu=1}^{q}\log\left|\frac{s^2-\bar{b}_\nu z}{s(z-b_\nu)}\right|. \qquad (1.9)\end{aligned}$$

Since

$$\log|f(z)| = \frac{1}{2}[\log f(z) + \log\overline{f}(z)],$$

and $(\log\overline{f}(z))' = 0$,

$$\frac{f'(z)}{f(z)} = (\log f(z))' = 2(\log|f(z)|)'.$$

Differentiating (1.9) yields

$$\begin{aligned}\frac{f'(z)}{f(z)} = & \int_0^{2\pi} \frac{2se^{i\phi}}{(se^{i\phi}-z)^2}\log|f(se^{i\phi})|\frac{d\phi}{2\pi} \\ & +\sum_{\mu=1}^{p}\left(\frac{\bar{a}_\mu}{s^2-\bar{a}_\mu z}+\frac{1}{z-a_\mu}\right) - \sum_{\nu=1}^{q}\left(\frac{\bar{b}_\nu}{s^2-\bar{b}_\nu z}+\frac{1}{z-b_\nu}\right).\end{aligned}$$

Consequently,

$$\left|\frac{f'(z)}{f(z)}\right| \leq \int_0^{2\pi} \frac{2s}{|se^{i\phi}-z|^2}\left|\log|f(se^{i\phi})|\right|\frac{d\phi}{2\pi} + \left|\sum_{\mu=1}^{p}\frac{\bar{a}_\mu}{s^2-\bar{a}_\mu z}\right|$$

$$+\left|\sum_{\nu=1}^{q}\frac{\bar{b}_\nu}{s^2-\bar{b}_\nu z}\right|+\left|\sum_{\mu=1}^{p}\frac{1}{z-a_\mu}\right|+\left|\sum_{\nu=1}^{q}\frac{1}{z-b_\nu}\right|.$$

For $0<\alpha<1$ and for positive real numbers d_j, we know that

$$(\sum d_j)^\alpha \le \sum d_j^\alpha.$$

So the above inequality becomes

$$\begin{aligned}\left|\frac{f'(z)}{f(z)}\right|^\alpha &\le \left(\int_0^{2\pi}\frac{2s}{|se^{i\phi}-z|^2}\left|\log|f(se^{i\phi})|\right|\frac{d\phi}{2\pi}\right)^\alpha+\left|\sum_{\mu=1}^{p}\frac{\bar{a}_\mu}{s^2-\bar{a}_\mu z}\right|^\alpha\\ &\quad+\left|\sum_{\nu=1}^{q}\frac{\bar{b}_\nu}{s^2-\bar{b}_\nu z}\right|^\alpha+\left|\sum_{\mu=1}^{p}\frac{1}{z-a_\mu}\right|^\alpha+\left|\sum_{\nu=1}^{q}\frac{1}{z-b_\nu}\right|^\alpha.\end{aligned}$$

Now, set $z=re^{i\theta}$ and integrate with respect to θ to arrive at

$$\begin{aligned}\int_0^{2\pi}\left|\frac{f'(re^{i\theta})}{f(re^{i\theta})}\right|^\alpha\frac{d\theta}{2\pi} &\le \int_0^{2\pi}\left(\int_0^{2\pi}\frac{2s}{|se^{i\phi}-re^{i\theta}|^2}\left|\log|f(se^{i\phi})|\right|\frac{d\phi}{2\pi}\right)^\alpha\frac{d\theta}{2\pi}\\ &\quad+\int_0^{2\pi}\left|\sum_{\mu=1}^{p}\frac{\bar{a}_\mu}{s^2-\bar{a}_\mu re^{i\theta}}\right|^\alpha\frac{d\theta}{2\pi}+\int_0^{2\pi}\left|\sum_{\nu=1}^{q}\frac{\bar{b}_\nu}{s^2-\bar{b}_\nu re^{i\theta}}\right|^\alpha\frac{d\theta}{2\pi}\\ &\quad+\int_0^{2\pi}\left|\sum_{\mu=1}^{p}\frac{1}{re^{i\theta}-a_\mu}\right|^\alpha\frac{d\theta}{2\pi}+\int_0^{2\pi}\left|\sum_{\nu=1}^{q}\frac{1}{re^{i\theta}-b_\nu}\right|^\alpha\frac{d\theta}{2\pi}\\ &= I_1+I_2+I_3+I_4+I_5.\end{aligned}\tag{1.10}$$

By applying the Hölder inequality and interchanging the order of integration, we see that

$$\begin{aligned}I_1^{1/\alpha} &\le \int_0^{2\pi}\left(\int_0^{2\pi}\frac{2s}{|se^{i\phi}-re^{i\theta}|^2}\left|\log|f(se^{i\phi})|\right|\frac{d\phi}{2\pi}\right)\frac{d\theta}{2\pi}\\ &= 2s\int_0^{2\pi}\left(\int_0^{2\pi}\frac{1}{|se^{i\phi}-re^{i\theta}|^2}\frac{d\theta}{2\pi}\right)\left|\log|f(se^{i\phi})|\right|\frac{d\phi}{2\pi}.\end{aligned}$$

By Corollary A1.1.2,

$$\int_0^{2\pi}\frac{1}{|se^{i\phi}-re^{i\theta}|^2}\frac{d\theta}{2\pi}=\frac{1}{s^2-r^2}.$$

Also

$$\int_0^{2\pi} \left|\log|f(se^{i\phi})|\right| \frac{d\phi}{2\pi} = m_f(s,\infty) + m_f(s,0).$$

So

$$\begin{aligned} I_1 &\leq \left(\frac{2s}{s^2-r^2}(m_f(s,\infty)+m_f(s,0))\right)^\alpha \\ &\leq \left(\frac{s}{r(s-r)}\right)^\alpha (m_f(s,\infty)+m_f(s,0))^\alpha \qquad (1.11) \end{aligned}$$

where the last inequality follows directly from $s+r \geq 2r$.

Our next step is to estimate I_2, where

$$I_2 = \int_0^{2\pi} \left|\sum_{\mu=1}^{p} \frac{\bar{a}_\mu}{s^2 - \bar{a}_\mu re^{i\theta}}\right|^\alpha \frac{d\theta}{2\pi}.$$

Since $\bar{a}_\mu = \text{Re}a_\mu - i\text{Im}a_\mu$,

$$\begin{aligned} \sum_{\mu=1}^{p} \frac{\bar{a}_\mu}{s^2-\bar{a}_\mu z} &= \sum_{Rea_\mu>0} \frac{\text{Re}a_\mu}{s^2-\bar{a}_\mu z} - \sum_{Rea_\mu\leq 0} \frac{-\text{Re}a_\mu}{s^2-\bar{a}_\mu z} \\ &\quad -i \sum_{Ima_\mu>0} \frac{\text{Im}a_\mu}{s^2-\bar{a}_\mu z} + i \sum_{Ima_\mu\leq 0} \frac{-\text{Im}a_\mu}{s^2-\bar{a}_\mu z}. \end{aligned}$$

Use the inequality $(\sum d_j)^\alpha \leq \sum d_j^\alpha$ again,

$$\begin{aligned} I_2 \leq & \int_0^{2\pi} \left|\sum_{Rea_\mu>0} \frac{\text{Re}a_\mu}{s^2-\bar{a}_\mu re^{i\theta}}\right|^\alpha \frac{d\theta}{2\pi} + \int_0^{2\pi} \left|\sum_{Rea_\mu\leq 0} \frac{-\text{Re}a_\mu}{s^2-\bar{a}_\mu re^{i\theta}}\right|^\alpha \frac{d\theta}{2\pi} \\ & + \int_0^{2\pi} \left|\sum_{Ima_\mu>0} \frac{\text{Im}a_\mu}{s^2-\bar{a}_\mu re^{i\theta}}\right|^\alpha \frac{d\theta}{2\pi} + \int_0^{2\pi} \left|\sum_{Ima_\mu\leq 0} \frac{-\text{Im}a_\mu}{s^2-\bar{a}_\mu re^{i\theta}}\right|^\alpha \frac{d\theta}{2\pi}. \qquad (1.12) \end{aligned}$$

Applying Lemma A1.2.1 to the analytic function

$$F(z) = \sum_{\text{Re}a_\mu>0} \frac{\text{Re}a_\mu}{s^2-\bar{a}_\mu z},$$

note that $\mathrm{Re}(s^2 - \bar{a}_\mu z)$ is positive for $|z| < s$ so $\mathrm{Re}F(z) > 0$, yields

$$\int_0^{2\pi} \left| \sum_{Rea_\mu > 0} \frac{\mathrm{Re}a_\mu}{s^2 - \bar{a}_\mu re^{i\theta}} \right|^\alpha \frac{d\theta}{2\pi} \leq \sec(\alpha\pi/2) \left(\sum_{Rea_\mu > 0} \frac{\mathrm{Re}a_\mu}{s^2} \right)^\alpha .$$

Since $|a_\mu| < s$, we have

$$\begin{aligned} \int_0^{2\pi} \left| \sum_{Rea_\mu > 0} \frac{Rea_\mu}{s^2 - \bar{a}_\mu re^{i\theta}} \right|^\alpha \frac{d\theta}{2\pi} &\leq \sec(\alpha\pi/2) s^{-\alpha} \left(\sum_{Rea_\mu > 0} 1 \right)^\alpha \\ &\leq \sec(\alpha\pi/2) \left(\frac{n_f(s,0)}{s} \right)^\alpha \\ &\leq \sec(\alpha\pi/2) \left(\frac{n_f(s,0)}{r} \right)^\alpha , \end{aligned} \tag{1.13}$$

where the last inequality is simply $s > r$. The rest of the terms in (1.12) can also be estimated in the same way, so

$$I_2 \leq 4 \sec(\alpha\pi/2) \left(\frac{n_f(s,0)}{r} \right)^\alpha . \tag{1.14}$$

By the same argument, replacing a_μ by b_ν, we have

$$I_3 \leq 4 \sec(\alpha\pi/2) \left(\frac{n_f(s,\infty)}{r} \right)^\alpha . \tag{1.15}$$

Finally we estimate I_4 and I_5. Recall that

$$I_4 = \int_0^{2\pi} \left| \sum_{\mu=1}^p \frac{1}{re^{i\theta} - a_\mu} \right|^\alpha \frac{d\theta}{2\pi}. \tag{1.16}$$

Denote by $\phi_\mu = \arg a_\mu$. Because $e^{i\phi_\mu}(\cos\phi_\mu - i \sin\phi_\mu) = 1$, we can write

$$\begin{aligned} \sum_{\mu=1}^p \frac{1}{z - a_\mu} &= \sum_{|a_\mu| > r} \frac{1}{z - a_\mu} + \sum_{|a_\mu| \leq r} \frac{1}{z - a_\mu} \\ &= \sum_{|a_\mu| > r} \frac{e^{i\phi_\mu} \cos\phi_\mu}{z - a_\mu} - i \sum_{|a_\mu| > r} \frac{e^{i\phi_\mu} \sin\phi_\mu}{z - a_\mu} + \sum_{|a_\mu| \leq r} \frac{1}{z - a_\mu} \\ &= \sum_{\cos\phi_\mu > 0, |a_\mu| > r} \frac{e^{i\phi_\mu} \cos\phi_\mu}{z - a_\mu} + \sum_{\cos\phi_\mu < 0, |a_\mu| > r} \frac{e^{i\phi_\mu} \cos\phi_\mu}{z - a_\mu} \end{aligned}$$

$$-i\sum_{|a_\mu|>r,\sin\phi_\mu>0}\frac{e^{i\phi_\mu}\sin\phi_\mu}{z-a_\mu}-i\sum_{|a_\mu|>r,\sin\phi_\mu<0}\frac{e^{i\phi_\mu}\sin\phi_\mu}{z-a_\mu}$$
$$+\sum_{|a_\mu|\le r}\frac{1}{z-a_\mu}.$$

Denote by $p_\mu=\cos\phi_\mu$ and $q_\mu=\sin\phi_\mu$, then the above equation becomes

$$\begin{aligned}\sum_{\mu=1}^{p}\frac{1}{z-a_\mu} &= \sum_{p_\mu>0,|a_\mu|>r}\frac{e^{i\phi_\mu}p_\mu}{z-a_\mu}+\sum_{p_\mu<0,|a_\mu|>r}\frac{e^{i\phi_\mu}p_\mu}{z-a_\mu}\\ &\quad -i\sum_{|a_\mu|>r,q_\mu>0}\frac{e^{i\phi_\mu}q_\mu}{z-a_\mu}-i\sum_{|a_\mu|>r,q_\mu<0}\frac{e^{i\phi_\mu}q_\mu}{z-a_\mu}\\ &\quad +\sum_{|a_\mu|\le r}\frac{1}{z-a_\mu}. \end{aligned}\tag{1.17}$$

Note that when $|a_\mu|>r$ and $|z|<r$ then $\mathrm{Re}\{e^{i\phi_\mu}/(z-a_\mu)\}<0$. So by applying Lemma A1.2.1 to

$$F(z)=\sum_{p_\mu>0,|a_\mu|>r}\frac{e^{i\phi_\mu}p_\mu}{z-a_\mu}$$

we conclude

$$\begin{aligned}\int_0^{2\pi}\left|\sum_{\cos\phi_\mu>0,|a_\mu|>r}\frac{e^{i\phi_\mu}\cos\phi_\mu}{re^{i\theta}-a_\mu}\right|^\alpha\frac{d\theta}{2\pi} &= \int_0^{2\pi}\left|\sum_{p_\mu>0,|a_\mu|>r}\frac{e^{i\phi_\mu}p_\mu}{re^{i\theta}-a_\mu}\right|^\alpha\frac{d\theta}{2\pi}\\ &\le \sec(\alpha\pi/2)\left(\sum_{|a_\mu|>r}\frac{1}{|a_\mu|}\right)^\alpha\\ &\le \sec(\alpha\pi/2)\left(\frac{n_f(s,0)-n_f(r,0)}{r}\right)^\alpha. \end{aligned}\tag{1.18}$$

Similarly,

$$\int_0^{2\pi}\left|\sum_{\cos\phi_\mu<0,|a_\mu|>r}\frac{e^{i\phi_\mu}\cos\phi_\mu}{re^{i\theta}-a_\mu}\right|^\alpha\frac{d\theta}{2\pi}\le\sec(\alpha\pi/2)\left(\frac{n_f(s,0)-n_f(r,0)}{r}\right)^\alpha,\tag{1.19}$$

$$\int_0^{2\pi} \left| \sum_{\sin\phi_\mu > 0, |a_\mu| > r} \frac{e^{i\phi_\mu} \sin\phi_\mu}{re^{i\theta} - a_\mu} \right|^\alpha \frac{d\theta}{2\pi} \leq \sec(\alpha\pi/2) \left(\frac{n_f(s,0) - n_f(r,0)}{r} \right)^\alpha, \tag{1.20}$$

and

$$\int_0^{2\pi} \left| \sum_{\sin\phi_\mu < 0, |a_\mu| > r} \frac{e^{i\phi_\mu} \sin\phi_\mu}{re^{i\theta} - a_\mu} \right|^\alpha \frac{d\theta}{2\pi} \leq \sec(\alpha\pi/2) \left(\frac{n_f(s,0) - n_f(r,0)}{r} \right)^\alpha. \tag{1.21}$$

When $|z| = r$, we write

$$\left| \sum_{|a_\mu| \leq r} \frac{1}{z - a_\mu} \right| = \left| \sum_{|a_\mu| \leq r} \frac{r}{r^2 - \bar{a}_\mu z} \right|.$$

Note that, when $|a_\mu| \leq r$ and $|z| < r$, $Re\{r^2 - \bar{a}_\mu z\} > 0$. So by applying Lemma A1.2.1 to

$$F(z) = \sum_{|a_\mu| \leq r} \frac{r}{r^2 - \bar{a}_\mu z}$$

we arrive at

$$\begin{aligned} \int_0^{2\pi} \left| \sum_{|a_\mu| \leq r} \frac{1}{re^{i\theta} - a_\mu} \right|^\alpha \frac{d\theta}{2\pi} &\leq \sec(\alpha\pi/2) \left(\sum_{|a_\mu| \leq r} \frac{1}{|a_\mu|} \right)^\alpha \\ &\leq \sec(\alpha\pi/2) \left(\frac{n_f(s,0)}{r} \right)^\alpha. \end{aligned} \tag{1.22}$$

Combining (1.16) to (1.22) and using the fact that if d_j are non-negative real numbers and $0 < \alpha < 1$ then $(\sum_j d_j)^\alpha \leq \sum_j d_j^\alpha$, we have

$$I_4 \leq 5 \sec(\alpha\pi/2) \left(\frac{n_f(s,0)}{r} \right)^\alpha. \tag{1.23}$$

By replacing a_μ by b_ν and repeating the above argument again, we can prove

$$I_5 \leq 5 \sec(\alpha\pi/2) \left(\frac{n_f(s,\infty)}{r} \right)^\alpha. \tag{1.24}$$

The theorem follows by combining (1.10), (1.11), (1.14), (1.15), (1.23), and (1.24). □

Theorem A1.2.3 (Gol'dberg-Grinshtein Estimate) *Let f be a meromorphic function in $\mathbf{D}(R)(0 < R \leq \infty)$, and let $0 < \alpha < 1$, then, for $r_0 < r < \rho < R$, we have*

$$\int_0^{2\pi} \left|\frac{f'(re^{i\theta})}{f(re^{i\theta})}\right|^{\alpha} \frac{d\theta}{2\pi} \leq 2^{\alpha}\left(\frac{\rho}{r(\rho-r)}\right)^{\alpha}(2T_f(\rho) - \log|c_f|)^{\alpha}$$
$$+2^{\alpha+4}\sec(\alpha\pi/2)\left(\frac{\rho}{r(\rho-r)}\right)^{\alpha}\left(T_f(\rho) + |ord_0 f|\log^+\frac{1}{r_0}\right)^{\alpha},$$

where $f(z) = c_f z^{\mathrm{ord}_0 f} + \ldots$, $\mathrm{ord}_0 f \in \mathbf{Z}$, and c_f is the leading nonzero coefficient.

Proof. Let $s = (r+\rho)/2$. Then $s - r = \rho - s = (\rho - r)/2$, so

$$\frac{s}{r(s-r)} = \frac{r+\rho}{r(\rho-r)} \leq \frac{2\rho}{(\rho-r)}. \tag{1.25}$$

Also, by definition,

$$\begin{aligned}(N_f(\rho,0) + |\mathrm{ord}_0 f|\log^+\frac{1}{r_0}) &\geq \int_s^{\rho} n_f(t,0)\frac{dt}{t}\\ &\geq n_f(s,0)\int_s^{\rho}\frac{dt}{\rho}\\ &= \frac{\rho-s}{\rho}n_f(s,0).\end{aligned}$$

So

$$\begin{aligned}n_f(s,0) &\leq \frac{\rho}{\rho-s}\left(N_f(\rho,0) + |\mathrm{ord}_0 f|\log^+\frac{1}{r_0}\right)\\ &\leq \frac{\rho}{\rho-s}\left(T_f(\rho) + |\mathrm{ord}_0 f|\log^+\frac{1}{r_0}\right).\end{aligned} \tag{1.26}$$

Similarly,

$$n_f(s,\infty) \leq \frac{\rho}{\rho-s}\left(T_f(\rho) + |\mathrm{ord}_0 f|\log^+\frac{1}{r_0}\right). \tag{1.27}$$

By the First Main Theorem,

$$m_f(s,0) + m_f(s,\infty) \leq 2T_f(s) - \log|c_f| \leq 2T_f(\rho) - \log|c_f|.$$

The Theorem is proved by applying Theorem A1.2.2 with s as chosen above, using (1.25), (1.26) and (1.27). □

Before we derive the Lemma of the logarithmic derivative, we need the following lemma.

Lemma A1.2.4 (Borel's Growth Lemma) *Let $F(r)$ be a positive, non-decreasing, continuous function defined on $[r_0, \infty)$ with $r_0 \geq e$ such that $F(r) \geq e$ on $[r_0, \infty)$. Then, for every $\epsilon > 0$, there exists a closed set $E \subset [r_0, \infty)$ (called the "exceptional set") of finite Lebesgue measure such that if we set $\rho = r + 1/\log^{1+\epsilon} F(r)$ for all $r \geq r_0$ and not in E, we have*

$$\log F(\rho) \leq \log F(r) + 1 \tag{1.28}$$

and

$$\log^+ \frac{\rho}{r(\rho - r)} \leq (1+\epsilon)\log^+ \log F(r) + \log 2. \tag{1.29}$$

Proof. Let

$$E = \left\{ r \in [r_0, \infty) : \, F\left(r + \frac{1}{\log^{1+\epsilon} F(r)}\right) \geq eF(r) \right\}.$$

We may assume that E is non-empty, otherwise, the lemma is trivial. We claim that E is of finite Lebesgue measure.

Let r_1 be the smallest $r \in E$ with $r \geq r_0$. Now assume that we have found numbers $r_1, \ldots, r_n, s_1, \ldots, s_{n-1}$. We describe here how to inductively extend this set, and we continue this process as long as possible. If there is no number s with $F(s) \geq eF(r_n)$, then we stop here. Otherwise, by continuity of F, there exists an s with $F(s) = eF(r_n)$. Let s_n be the smallest such s. Then, if there is an $r \in E$ with $r \geq s_n$, let r_{n+1} be the smallest such r. Otherwise, we stop here.

For each pair r_j, s_j, clearly $s_j > r_j$, and since $r_j \in E$,

$$F\left(r_j + \frac{1}{\log^{1+\epsilon} F(r_j)}\right) \geq eF(r_j) = F(s_j).$$

Since F is nondecreasing, this implies

$$r_j + \frac{1}{\log^{1+\epsilon} F(r_j)} \geq s_j,$$

and so

$$s_j - r_j \leq \frac{1}{\log^{1+\epsilon} F(r_j)}. \tag{1.30}$$

Moreover, $F(r_{j+1}) \geq F(s_j) = eF(r_j)$ since $s_{j+1} \geq s_j$. Hence,

$$F(r_{n+1}) \geq eF(r_n) \geq e^2 F(r_{n-1}) \geq \cdots \geq e^n F(r_1) \geq e^{n+1}. \tag{1.31}$$

It follows that either we can only find finitely many r_n or else the sequence r_n goes to the infinity as n goes to the infinity. Since the set E is contained in the union of $[r_n, s_n]$, if we can only find finitely many r_n, then E is of finite Lebesgue measure. Now consider the case where n goes to ∞. Let $m(E)$ be the Lebesgue measure of E, then

$$m(E) \leq \sum_{n=1}^{\infty} (s_n - r_n).$$

By (1.30) and (1.31),

$$\sum_{n=1}^{\infty} (s_n - r_n) \leq \sum_{n=1}^{\infty} \frac{1}{\log^{1+\epsilon} F(r_n)} \leq \sum_{n=1}^{\infty} \frac{1}{n^{1+\epsilon}} < +\infty.$$

Thus the claim is proved.

To verify (1.28), let $r \geq r_0$ where r is not contained in E, then, by the construction of E,

$$F(\rho) = F\left(r + \frac{1}{\log^{1+\epsilon} F(r)}\right) \leq eF(r).$$

Thus $\log F(\rho) \leq \log F(r) + 1$. So (1.28) holds. Finally, we verify (1.29).

$$\frac{\rho}{r(\rho - r)} = \frac{1}{\rho - r} + \frac{1}{r} \leq \log^{1+\epsilon} F(r) + 1 \leq 2\log^{1+\epsilon} F(r).$$

Hence

$$\log^+ \frac{\rho}{r(\rho - r)} \leq (1+\epsilon)\log^+ \log F(r) + \log 2.$$

□

The following Lemma on the Logarithmic Derivative with a good error term is due to Z. Ye.

Theorem A1.2.5 (Lemma on the Logarithmic Derivative) *Let f be a non-constant meromorphic function on $\mathbf{C}$. Assume that $T_f(r_0) \geq e$ for some r_0. Then for any $\epsilon > 0$, the inequality*

$$m_{f'/f}(r, \infty) \leq \log T_f(r) + (1+\epsilon)\log^+ \log T_f(r) + C$$

holds for all $r \geq r_0$ outside a set $E \subset (0, +\infty)$ with finite Lebesgue measure, where C is a constant which depends only on f.

Proof. Using the con-cavity of $\log^+$ to pull the $\log^+$ outside the integral to get

$$m_{f'/f}(r, \infty) = \frac{1}{\alpha} \int_0^{2\pi} \log^+ \left| \frac{f'(re^{i\theta})}{f(re^{i\theta})} \right|^\alpha \frac{d\theta}{2\pi} \leq \frac{1}{\alpha} \log^+ \int_0^{2\pi} \left| \frac{f'(re^{i\theta})}{f(re^{i\theta})} \right|^\alpha \frac{d\theta}{2\pi}. \tag{1.32}$$

Take $\rho = r + \frac{1}{\log^{1+\epsilon} T_f(r)}$. Applying Lemma A1.2.4 with $F(r) = T_f(r)$, we have, for $r \geq r_0$ where r is not contained in E,

$$\log T(\rho) \leq \log T_f(r) + 1 \tag{1.33}$$

and

$$\log^+ \frac{\rho}{r(\rho - r)} \leq (1 + \epsilon) \log^+ \log T_f(r) + \log 2. \tag{1.34}$$

The theorem follows from (1.32), (1.33), (1.34), Theorem A1.2.3 and the inequality

$$\log^+(x + y) \leq \log^+ x + \log^+ y + \log 2.$$

□

A1.3 The Second Main Theorem for Meromorphic Functions

Before we state the Second Main Theorem, we introduce the **ramification term** $N_{\text{ram},f}(r)$ of f, which is defined by

$$N_{\text{ram},f}(r) = N_{f'}(r, 0) + 2N_f(r, \infty) - N_{f'}(r, \infty).$$

We note that $N_{\text{ram},f}(r) \geq 0$.

The following Second Main Theorem with a good error term is due to P.M. Wong, and Y. Zhuan.

Theorem A1.3.1 (The Second Main Theorem) *Let $a_1, \ldots, a_q$ be a set of distinct complex numbers. Let f be a non-constant meromorphic function on* **C**. *Then for any $\epsilon > 0$, the inequality*

$$(q - 1)T_f(r) + N_{\text{ram},f}(r)$$

$$\leq \sum_{j=1}^{q} N_f(r, a_j) + N_f(r, \infty) + \log T_f(r) + (1+\epsilon)\log^+ \log T_f(r) + O(1),$$

holds for all $r \geq r_0$ outside a set $E \subset (0, +\infty)$ with finite Lebesgue measure.

Proof. Let $\delta = \min_{i \neq j}\{|a_i - a_j|, 1\}$. For each z with $f(z) \neq \infty$ and $f(z) \neq a_j$ for $1 \leq j \leq q$, let j_0 be the index among $\{1, 2, \ldots, q\}$, such that

$$|f(z) - a_{j_0}| \leq |f(z) - a_j| \quad \text{forall } 1 \leq j \leq q.$$

Then, for $j \neq j_0$, by the triangle inequality, $|f(z) - a_j| \geq \delta/2$. Thus, for $j \neq j_0$,

$$\begin{aligned}\log^+ |f(z)| &\leq \log^+ |f(z) - a_j| + \log^+ |a_j| + \log 2 \\ &\leq \log |f(z) - a_j| + \log^+ \frac{2}{\delta} + \log^+ |a_j| + \log 2.\end{aligned}$$

Therefore,

$$(q-1)\log^+ |f(z)| \leq \sum_{j \neq j_0} \log |f(z)-a_j| + \sum_{j=1}^{q} \log^+ |a_j| + (q-1)(\log^+ \frac{2}{\delta} + \log 2).$$

Now

$$\begin{aligned}\sum_{j \neq j_0} \log |f(z) - a_j| &= \sum_{j=1}^{q} \log |f(z) - a_j| - \log |f'(z)| + \log \frac{|f'(z)|}{|f(z) - a_{j_0}|} \\ &\leq \sum_{j=1}^{q} \log |f(z) - a_j| - \log |f'(z)| + \log \left(\sum_{j=1}^{q} \frac{|f'(z)|}{|f(z) - a_j|} \right).\end{aligned}$$

Thus

$$\begin{aligned}(q-1)\log^+ |f(z)| &\leq \sum_{j=1}^{q} \log |f(z) - a_j| - \log |f'(z)| + \log \left(\sum_{j=1}^{q} \frac{|f'(z)|}{|f(z) - a_j|} \right) \\ &\quad + \sum_{j=1}^{q} \log^+ |a_j| + (q-1)(\log^+ \frac{2}{\delta} + \log 2).\end{aligned}$$

Now, set $z = re^{i\theta}$ and integrate with respect to θ to get

$$(q-1)m_f(r, \infty) = (q-1)\int_0^{2\pi} \log^+ |f(re^{i\theta})| \frac{d\theta}{2\pi}$$

$$\le \sum_{j=1}^{q}\int_0^{2\pi}\log|f(re^{i\theta})-a_j|\frac{d\theta}{2\pi}-\int_0^{2\pi}\log|f'(re^{i\theta})|\frac{d\theta}{2\pi}$$
$$+\int_0^{2\pi}\log\left(\sum_{j=1}^{q}\frac{|f'(re^{i\theta})|}{|f(re^{i\theta})-a_j|}\right)\frac{d\theta}{2\pi}+O(1).$$

From Corollary A1.1.4, we have

$$\int_0^{2\pi}\log|f(re^{i\theta})-a_j|\frac{d\theta}{2\pi}=N_f(r,a_j)-N_f(r,\infty)+\log|c_{f-a_j}|,$$

and

$$\int_0^{2\pi}\log|f'(re^{i\theta})|\frac{d\theta}{2\pi}=N_{f'}(r,0)-N_{f'}(r,\infty)+\log|c_{f'}|.$$

Thus, the above inequality becomes

$$(q-1)m_f(r,\infty)-\sum_{j=1}^{q}N_f(r,a_j)+qN_f(r,\infty)+N_{f'}(r,0)-N_{f'}(r,\infty)$$
$$\le \int_0^{2\pi}\log\left(\sum_{j=1}^{q}\frac{|f'(re^{i\theta})|}{|f(re^{i\theta})-a_j|}\right)\frac{d\theta}{2\pi}+\sum_{j=1}^{q}\log^+|a_j|$$
$$+(q-1)(\log^+\frac{2}{\delta}+\log 2)+\sum_{j=1}^{q}\log|c_{f-a_j}|-\log|c_{f'}|.$$

However, by the First Main Theorem and the definition of $N_{\mathrm{ram},f}(r)$, the left-hand side of the above inequality is

$$(q-1)T_f(r)-\sum_{j=1}^{q}N_f(r,a_j)-N_f(r,\infty)+N_{\mathrm{ram},f}(r).$$

To complete the proof, we still need to estimate

$$\int_0^{2\pi}\log\left(\sum_{j=1}^{q}\frac{|f'(re^{i\theta})|}{|f(re^{i\theta})-a_j|}\right)\frac{d\theta}{2\pi}.$$

Let α be a real number between 0 and 1, then

$$\int_0^{2\pi}\log\left(\sum_{j=1}^{q}\frac{|f'(re^{i\theta})|}{|f(re^{i\theta})-a_j|}\right)\frac{d\theta}{2\pi} = \frac{1}{\alpha}\int_0^{2\pi}\log\left(\sum_{j=1}^{q}\frac{|f'(re^{i\theta})|}{|f(re^{i\theta})-a_j|}\right)^{\alpha}\frac{d\theta}{2\pi}$$

$$\leq \frac{1}{\alpha}\int_0^{2\pi} \log\left(\sum_{j=1}^q \left|\frac{f'(re^{i\theta})}{f(re^{i\theta})-a_j}\right|^\alpha\right)\frac{d\theta}{2\pi}$$

$$\leq \frac{1}{\alpha}\log\left(\sum_{j=1}^q \int_0^{2\pi} \left|\frac{f'(re^{i\theta})}{f(re^{i\theta})-a_j}\right|^\alpha\right)\frac{d\theta}{2\pi},$$

where in the last inequality, we used the con-cavity property of the logarithm and the inequality $(\sum d_j)^\alpha \leq \sum d_j^\alpha$ for positive real numbers d_j and $0 < \alpha < 1$. Now applying Theorem A1.2.3 and using $\log^+(x+y) \leq \log^+ x + \log^+ y$, to get that the above expression is

$$\leq \log^+ \frac{\rho}{r(\rho - r)} + \log^+ \sum_{j=1}^q 2T_{f-a_j}(\rho) + C(\alpha)$$

where C_α is a constant depends only α. By Lemma A1.2.4, by taking $\rho = r + 1/\log^{1+\epsilon} T_f(r)$, we have, for $r \geq r_0$ and not in E, $\log T_f(\rho) \leq \log T_f(r) + 1$ and

$$\log^+ \frac{\rho}{r(\rho - r)} \leq (1+\epsilon)\log^+ \log T_f(r) + \log 2.$$

Thus, for $r \geq r_0$ and not in E,

$$\begin{aligned}
&\log^+ \frac{\rho}{r(\rho - r)} + \log^+ \left(\sum_{j=1}^q 2T_{f-a_j}(\rho)\right) + C(\alpha) \\
\leq\ & (1+\epsilon)\log^+ \log T_f(r) + \log^+ \max_{1\leq j\leq q}\{2T_{f-a_j}(\rho)\} + C(\alpha) \\
\leq\ & (1+\epsilon)\log^+ \log T_f(r) + \log(2T_f(\rho)) + C(\alpha) \\
\leq\ & (1+\epsilon)\log^+ \log T_f(r) + \log T_f(r) + C(\alpha).
\end{aligned}$$

Combining these various estimates completes the proof of the theorem. □

Define the **truncated counting function** by

$$N_f^{(1)}(r,0) = \min\{\mathrm{ord}_0^+ f, 1\}\log r + \sum_{z\in\mathbf{D}(r), z\neq 0} \min\{\mathrm{ord}_z^+ f, 1\}\log\left|\frac{r}{z}\right|. \quad (1.35)$$

Theorem A1.3.1 can be restated as the following corollary.

Corollary A1.3.2 *Let $a_1, \ldots, a_q$ be a set of distinct complex numbers. Let f be a non-constant meromorphic function on $\mathbf{C}$. Then for any $\epsilon > 0$, the inequality*

$$\begin{aligned}(q-1)T_f(r) \quad &\leq \quad \sum_{j=1}^{q} N_f^{(1)}(r, a_j) + N_f^{(1)}(r, \infty) + \log T_f(r) \\ &\quad + (1+\epsilon)\log^+ \log T_f(r) + O(1),\end{aligned}$$

holds for all $r \geq r_0$ outside a set $E \subset (0, +\infty)$ with finite Lebesgue measure.

Proof. It is easy to check that

$$\sum_{j=1}^{q} N_f(r, a_j) + N_f(r, \infty) - N_{\mathrm{ram}, f}(r) \leq \sum_{j=1}^{q} N_f^{(1)}(r, a_j) + N_f^{(1)}(r, \infty).$$

Combining the above inequality and Theorem A1.3.1, we can imply this Corollary. □

Corollary A1.3.3 (Picard's theorem) *If a meromorphic function f on $\mathbf{C}$ omits three distinct points $a_1, a_2, a_3 \in \mathbf{C} \cup \{\infty\}$, then f must be constant.*

Proof. Assume that f is not constant. Applying Theorem A1.3.1, we have

$$\sum_{j=1}^{3} m_f(r, a_j) \leq 2T_f(r) + \log T_f(r) + (1+\epsilon)\log^+ \log T_f(r) + O(1),$$

which holds for all $r \geq r_0$ outside a set $E \subset (0, +\infty)$ with a finite Lebesgue measure. However, since f omits a_j, $N_f(r, a_j) = 0$, for $1 \leq i \leq 3$. So $m_f(r, a_j) = T_f(r) + O(1)$. Thus

$$3T_f(r) \leq 2T_f(r) + \log T_f(r) + (1+\epsilon)\log^+ \log T_f(r) + O(1)$$

holds for all $r \geq r_0$ outside a set $E \subset (0, +\infty)$ with a finite Lebesgue measure. This is a contradiction. □

Part B: Diophantine Approximation

B1.1 Introduction to Diophantine Approximation

The fundamental problem in the subject of Diophantine approximation is the question of how closely an irrational number can be approximated by a rational number. For example, for irrational number $\sqrt{2}$, since $\mathbf{Q}$ is dense in $\mathbf{R}$, an appropriate choice of p/q will make $|(p/q) - \sqrt{2}|$ as small as desired. The problem is: can we make it small without taking p, q too large? The next two elementary propositions answer this question in different directions.

Proposition B1.1.1 (Dirichlet) *Let $\alpha \in \mathbf{R}$ with $\alpha \notin \mathbf{Q}$. Then there are infinitely many rational numbers $p/q \in \mathbf{Q}$ satisfying*

$$|\frac{p}{q} - \alpha| \le \frac{1}{q^2}.$$

Proof. Let N be a large integer, and look at the set

$$\{q\alpha - [q\alpha] : q = 0, 1, \ldots, N\}$$

(here [ ] means greatest integer). Since α is irrational, this set consists of $N+1$ distinct numbers in the interval between 0 and 1. If we divide the unit interval $[0,1]$ into N line segments of equal length, the pigeonhole principle tells us that one of the segments must contain two of the numbers. Hence there are two of these numbers whose difference has absolute value at most $1/N$. In other words, there are integers $0 \le q_1 < q_2 \le N$ satisfying

$$|(q_1\alpha - [q_1\alpha]) - (q_2\alpha - [q_2\alpha])| \le 1/N.$$

Hence

$$|\frac{[q_2\alpha] - [q_1\alpha]}{q_2 - q_1} - \alpha| \le \frac{1}{(q_2 - q_1)N} \le \frac{1}{(q_2 - q_1)^2}.$$

This provides one rational approximation to α with the desired property, and by increasing N one can obtain infinitely many approximations. □

Proposition B1.1.2 (Liouville) *Let α be an algebraic number of degree $d \ge 2$. There is a constant $C > 0$, depending on α, such that for all rational* *numbers* p/q

$$|\frac{p}{q} - \alpha| \geq \frac{C}{q^d}.$$

Proof. Let $P(X) = a_0X^d + a_1X^{d-1} + \cdots + a_d \in \mathbf{Z}[X]$ be the minimal polynomial of α. Suppose now that $|\frac{p}{q} - \alpha| \leq 1$, then Taylor's formula yields

$$|P(\frac{p}{q})| = |\sum_{i=1}^{d}(\frac{p}{q} - \alpha)^i \frac{1}{i!}P^{(i)}(\alpha)| < |\frac{p}{q} - \alpha| \cdot d \cdot \max_{1\leq i\leq d}\left|\frac{1}{i!}P^{(i)}(\alpha)\right| = C|\frac{p}{q} - \alpha|.$$

On the other hand, $q^dP(\frac{p}{q}) \in \mathbf{Z}$ and $P(p/q) \neq 0$ since the minimal polynomial P does not have rational roots. Hence

$$|q^dP(\frac{p}{q})| \geq 1.$$

Combining the last two inequalities gives Liouville's Theorem if $|\frac{p}{q} - \alpha| \leq 1$. The theorem is obvious if $|\frac{p}{q} - \alpha| > 1$. □

Proposition B1.1.1 says that every real number can be approximated by rational numbers to within $1/q^2$, while Proposition B1.1.2 says that an algebraic number of degree d can be approximated no closer than C/q^d. The natural question is: for a given algebraic number α, what is the best exponent κ, such that there are only finitely many $p/q \in \mathbf{Q}$ (written in lowest terms) satisfying an inequality of the form

$$|\frac{p}{q} - \alpha| < \frac{c}{|q|^\kappa}$$

for some constant $c > 0$. It took many decades to obtain the best value for κ: letting $d = [\mathbf{Q}(\alpha), \mathbf{Q}]$, the progress is as follows:

Liouville 1844 $\kappa = d$ and c is computable
Thue 1909 $\kappa = (d+1)/2 + \epsilon$
Siegel 1921 $\kappa = 2\sqrt{d} + 1 + \epsilon$
Gelfond, Dyson 1947 $\kappa = \sqrt{2}d + \epsilon$
Roth 1955 $\kappa = 2 + \epsilon$.

How do theorems on Diophantine approximation lead to results concerning Diophantine equations? Consider the simple example of solving the equation

$$x^3 - 2y^3 = 1, x, y \in \mathbf{Z}.$$

Suppose (x, y) is a solution with $y \neq 0$, Let ζ be a primitive cube root of unity, and factor the equation as

$$(\frac{x}{y} - 2^{1/3})(\frac{x}{y} - \zeta 2^{1/3})(\frac{x}{y} - \zeta^2 2^{1/3}) = \frac{1}{y^3}.$$

The second and third terms in the product are bounded away from 0 so we obtain an estimate

$$|\frac{x}{y} - 2^{1/3}| \leq \frac{C}{|y|^3}$$

for some constant C independent of x and y. Then the result of Thue implies that there are only finitely many possibilities for x and y. So the equation $x^3 - 2y^3 = 1$ has only finitely many solutions in integers.

B1.2 Roth's Theorem and Vojta's Dictionary

Definition B1.2.1 *Let F be a field. By an* **absolute value** *on F, we mean a real-valued function $|\ |$ on F satisfying the following three conditions:*

(i) $|a| \geq 0$, *and* $|a| = 0$ *if and only if* $a = 0$.

(ii) $|ab| = |a||b|$.

(iii) $|a + b| \leq |a| + |b|$.

Two absolute values $|\ |_1$ and $|\ |_2$ are called **equivalent** if there is a positive constant λ such that $|\ |_1 = |\ |_2^{\lambda}$. Over the field of rational numbers $\mathbf{Q}$ we have the following absolute values: **the standard Archimedean absolute value** $|\ |$ (we also denote it by $|\ |_\infty$), which is defined by $|x| = x$ if $x \geq 0$, and $|x| = -x$ if $x < 0$; **p-adic absolute value** $|\ |_p$, for each prime number p, defined by $|x|_p = p^{-r}$, if $x = p^r a/b$, for some integer r, where a and b are integers relatively prime to p. For $x = 0$, $|x|_p = 0$. The p-adic absolute value $|\ |_p$ satisfies (i) and (ii), and a property stronger than (iii) in Definition B1.2.1, namely

$$(iii)' \qquad |a + b|_p \leq \max\{|a|_p, |b|_p\}.$$

An absolute value that satisfies $(iii)'$ is called a **non-Archimedean absolute value**. Every nonzero rational number has a factorization into prime factors. So for every $x \in \mathbf{Q}$ with $x \neq 0$, we have

$$|x|_\infty \cdot \prod_p |x|_p = 1, \tag{1.36}$$

where in the product, p runs for all prime numbers. (1.36) is called the **product formula**.

Theorem B1.2.2 (A. Ostrowski) *Any absolute value on* $\mathbf{Q}$ *is equivalent to one of the following: a* p*-adic absolute value for some prime number* p*, the standard Archimedean absolute value* $|\ |_\infty$*, or the trivial absolute value* $|\ |_0$ *defined by* $|x|_0 = 1$ *for all* $x \neq 0$.

To clearly see how Roth's theorem connects to Nevanlinna theory, we have to consider the fields more general than $\mathbf{Q}$, namely the number fields. Let us first consider the extension of an absolute value to $\mathbf{Q}(\alpha)$ where α is an algebraic number. We know that an algebraic number is usually viewed as a complex roots of its minimal polynomial. Then $|\alpha|$ is just the modulus of this complex number, and extends $|\ |_\infty$ to an absolute value of $\mathbf{Q}(\alpha)$. To extend a p-adic absolute value is less easy. But if one is willing to accept the p-adic closure $\mathbf{Q}_p$ of $\mathbf{Q}$ and the algebraic closure $\mathbf{C}_p$ of $\mathbf{Q}_p$, with the corresponding extension of $|\ |_p$ to $\mathbf{C}_p$, this becomes just as easy as for $|\ |_\infty$. Namely, every embedding $\sigma : \mathbf{Q}(\alpha) \to \mathbf{C}_p$ gives an extension of $|\ |_p$ defined by $|\beta|_p = |\sigma(\beta)|_p$, for $\beta \in \mathbf{Q}(\alpha)$. More precisely, we present, in the following, the theory of the extension of absolute values to a number field k. A number field k is a finite extension of the rationals $\mathbf{Q}$. Absolute values on $\mathbf{Q}$ extend to absolute values on k. The absolute values on k are divided into Archimedeans and non-Archimedeans. The Archimedean absolute values arise in the following ways: Let $n = [k : \mathbf{Q}]$. It is a standard fact from the field theory that k admits exactly n distinct embeddings $\sigma : k \hookrightarrow \mathbf{C}$. Each such embedding is used to define an absolute value on k according to the rule

$$|x|_\sigma = |\sigma(x)|_\infty$$

where $|\ |_\infty$ is the usual absolute value on $\mathbf{C}$. Recall that the embeddings $\sigma : k \hookrightarrow \mathbf{C}$ come in two flavors, the real embeddings (i.e., $\sigma(k) \subset \mathbf{R}$) and complex embeddings (i.e. $\sigma(k) \not\subset \mathbf{R}$). The complex embeddings come in pairs that differ by complex conjugation. The usual notation is that there are r_1 real embeddings and $2r_2$ pairs of complex embeddings, so $n = r_1 + 2r_2$. The normalized almost absolute value corresponding to σ is then defined by

$$||x||_\sigma = |x|_\sigma, \tag{1.37}$$

if σ is a real embedding, and

$$||x||_\sigma = |x|_\sigma^2, \tag{1.38}$$

if σ is a complex embedding. We note that the normalized almost-absolute values arising from the complex embedding do not satisfy the triangle inequality. This is why they are called almost-absolute values.

The non-Archimedean absolute values on k arise in much the same way as they do on $\mathbf{Q}$. However, one may not be able to uniquely factor elements of k into primes. A key idea in number theory is to look at prime ideals instead. To be more precise, let $\mathbf{R}_k$ be the ring of algebraic integers of k. Recall that $x \in k$ is called an **algebraic integer** if x is a root of a monic polynomial with coefficients in $\mathbf{Z}$. Note that, although $\mathbf{R}_k$ is not a principle ideal domain, for every $x \in \mathbf{R}_k$, the principal ideal (x) in $\mathbf{R}_k$ generated by x does factor uniquely into a product of prime ideals. For every prime ideal $\mathcal{P}$ of $\mathbf{R}_k$, we denote by $\mathrm{ord}_\mathcal{P} x$ the number of times the prime ideal $\mathcal{P}$ appears in this ideal factorization. Every prime ideal $\mathcal{P}$ lies above some prime p in $\mathbf{Q}$. For every element $x \in \mathbf{R}_k$, we define

$$|x|_\mathcal{P} = p^{-\mathrm{ord}_\mathcal{P} x/\mathrm{ord}_\mathcal{P} p}.$$

Of course, we always understand that $\mathrm{ord}_\mathcal{P} 0 = \infty$. The absolute value $|\ |_\mathcal{P}$ extends to k by writing any $x \in k$ as the quotient of two elements in $\mathbf{R}_k$. Note that the $\mathrm{ord}_\mathcal{P} p$ is needed to ensure that $|p|_\mathcal{P} = p^{-1}$. To get the normalized non-Archimedean absolute values, let $\mathbf{Q}_p$ be the completion of $\mathbf{Q}$ with respect to the p-adic absolute value $|\ |_p$ on $\mathbf{Q}$. Write $k = \mathbf{Q}(x_1, \ldots, x_q)$ so that $x_j, 1 \leq j \leq q$, generate k. Let $k_\mathcal{P} = \mathbf{Q}_p(x_1, \ldots, x_q)$. For every element $x \in \mathbf{R}_k$, we define

$$||x||_\mathcal{P} = p^{-[k_\mathcal{P}:\mathbf{Q}_p]\mathrm{ord}_\mathcal{P} x}. \tag{1.39}$$

The absolute value $||\ ||_\mathcal{P}$ extends to k by writing any $x \in k$ as the quotient of two elements in $\mathbf{R}_k$. Note the definition in (1.39) can also be written as

$$\|x\|_{\mathcal{P}} = \left(N_{k/\mathbf{Q}}\mathcal{P}\right)^{-\mathrm{ord}_{\mathcal{P}} x},$$

where $N_{k/\mathbf{Q}}\mathcal{P}$ is the norm of the ideal $\mathcal{P}$.

Theorem B1.2.2 is then extended to the following theorem.

Theorem B1.2.3 (A. Ostrowski) *Let k be a number field. Any almost-absolute value on k is equivalent to one of the following: the Archimedean absolute values which come from the real embeddings $\sigma : k \to \mathbf{R}$ defined by (1.37); the Archimedean almost-absolute values which come from the complex embeddings $\sigma : k \to \mathbf{C}$ defined by (1.38); and the non-Archimedean absolute value $\| \ \|_{\mathcal{P}}$ for some prime number $p \in \mathbf{Q}$, defined by (1.39).*

We refer to the real embeddings $\sigma : k \to \mathbf{R}$, the complex conjugate pairs $\{\sigma, \bar{\sigma}\}$ of the complex embeddings $\sigma : k \to \mathbf{C}$, and the nonzero prime ideals $\mathcal{P}$ in the ring $\mathbf{R}_k$ as **real places**, **complex places** and **non-Archimedean places**. We denote by M_k the canonical set of all the non-equivalent places. The set of non-equivalent Archimedean places of k is denoted by M_k^∞, the set of non-equivalent non-Archimedean places of k is denoted by M_k^0. For every place $v \in M_k$, v has **almost-absolute values** $\| \cdot \|_v$ defined by

$$\|x\|_v = \begin{cases} |\sigma(x)| & \text{if } v \text{ is real, corresponding to } \sigma : k \to \mathbf{R} \\ |\sigma(x)|^2 & \text{if } v \text{ is complex, corresponding to } \sigma, \bar{\sigma} : k \to \mathbf{C} \\ p^{-[k_{\mathcal{P}}:\mathbf{Q}_p]\mathrm{ord}_{\mathcal{P}} x} & \text{if } v \text{ is non-Arch., corresponding to } \mathcal{P} \subset \mathbf{R}_k \end{cases} \tag{1.40}$$

for $x \neq 0 \in k$. We also define $\|0\|_v = 0$. As we noted, these are not necessarily genuine absolute values. However, instead of having the triangle inequality, we have a value such that if $a_1, \ldots, a_n \in k$, then

$$\left\| \sum_{i=1}^n a_i \right\|_v \leq n^{N_v} \max_{1 \leq i \leq n} \|a_i\|_v, \tag{1.41}$$

where

$$N_v = \begin{cases} 1 & \text{if } v \text{ is real} \\ 2 & \text{if } v \text{ is complex} \\ 0 & \text{if } v \text{ is non-Archimedean.} \end{cases}$$

If L is a finite extension of k, $v \in M_k$, and $x \in k$, then

$$\prod_{w \in M_L, w|v} \|x\|_w = \|x\|_v^{[L:k]}. \tag{1.42}$$

Artin-Whaples extended the product formula (1.36) on $\mathbf{Q}$ to the number fields.

Theorem B1.2.4 (Product Formula) *Let k be a number field. Let M_k be the canonical set of non-equivalent places on k. Then, for every $x \in k$ with $x \neq 0$,*

$$\prod_{v \in M_k} \|x\|_v = 1. \tag{1.43}$$

Roth's theorem was extended by Mahler to number field k as follows:

Theorem B1.2.5 (Roth) *Given $\epsilon > 0$, a finite set of places S of k containing M_k^∞, and $\alpha_v \in \overline{\mathbf{Q}}$ for each $v \in S$. Then for all, except for finitely many, $x \in k$,*

$$\frac{1}{[k:\mathbf{Q}]} \sum_{v \in S} -\log \min(\|x - \alpha_v\|_v, 1) \leq (2+\epsilon) h(x), \tag{1.44}$$

where $h(x)$ is the absolute logarithmic height defined by

$$h(x) = \frac{1}{[k:\mathbf{Q}]} \sum_{v \in M_k} \log^+ \|x\|_v. \tag{1.45}$$

Fix a finite set S containing M_k^∞, we define, for $a, x \in k$,

$$m(x,a) = \frac{1}{[k:\mathbf{Q}]} \sum_{v \in S} \log^+ \frac{1}{\|x-a\|_v}, \tag{1.46}$$

$$N(x,a) = \frac{1}{[k:\mathbf{Q}]} \sum_{v \notin S} \log^+ \frac{1}{\|x-a\|_v}. \tag{1.47}$$

Then the product formula (Theorem B1.2.4) reads

Theorem B1.2.6 *For all $x \in k^*$, $a \in k$*

$$m(x,a) + N(x,a) = h(x) + O(1).$$

Theorem B1.2.5 can be restated as

Theorem B1.2.7 (Roth) *Given $\epsilon > 0$, a finite set $S \subset M_k$ containing M_k^∞, and distinct points $a_1, \ldots, a_q \in k$. Then the inequality*

$$\sum_{j=1}^{q} m(x, a_j) \leq (2+\epsilon)h(x)$$

holds for all, except for finitely many, $x \in k$.

Lang made the following conjecture with a more precise error term.

Conjecture B1.2.8 (Lang) *Given $\epsilon > 0$, a finite set $S \subset M_k$ containing M_k^∞, and distinct points $a_1, \ldots, a_q \in k$, the inequality*

$$(q-2)h(x) \leq \sum_{j=1}^{q} N^{(1)}(x, a_j) + (1+\epsilon)\log h(x)$$

holds for all, except for finitely many, $x \in k$.

Roth's theorem implies the following analogy of Picard's Theorem.

Theorem B1.2.9 *Let k be a number field, and let $a_1, \ldots, a_q$ be distinct numbers in $k \cup \{\infty\}$. If $q \geq 3$, then there are only finitely many elements $x \in k$ such that $1/(x - a_j)$ (or x itself if $a_j = \infty$) is an algebraic integer for all $1 \leq j \leq q$.*

To further explore the analogy, we introduce more notation. Recall the Nevanlinna counting function for a meromorphic function f is defined by

$$N_f(r, a) = \sum_{z \in \mathbf{D}(r), z \neq 0} \mathrm{ord}_z^+(f-a)\log\frac{r}{|z|} + \mathrm{ord}_0^+(f-a)\log r.$$

On the other hand, take $S = M_k^\infty$, then the number theoretic counting function $N(x, a)$ defined by (1.47) can be rewritten as, using (1.39),

$$\begin{aligned} N(x,a) &= \frac{1}{[k:\mathbf{Q}]}\sum_{v \notin M_k^\infty} \log^+\frac{1}{\|x-a\|_v} \\ &= \frac{1}{[k:\mathbf{Q}]}\sum_{\mathcal{P} \subset \mathbf{R}_k} \mathrm{ord}_{\mathcal{P}}^+(x-a)[k_{\mathcal{P}}:\mathbf{Q}_p]\log p, \qquad (1.48) \end{aligned}$$

where $\mathrm{ord}_{\mathcal{P}}^{+}x = \max\{0, \mathrm{ord}_{\mathcal{P}}x\}$. So $N_f(r,a)$ and $N(x,a)$ can be compared by replacing $\log(r/|z|)$ in the definition of $N_f(r,a)$ with $[k_{\mathcal{P}} : \mathbf{Q}_p]\log p$ in the definition of $N(x,a)$. From this point of view, Paul Vojta has compiled a dictionary to translate the terms in Nevanlinna theory to the terms in Diophantine approximation. It is provided on p. 32.

Vojta's Dictionary

Nevanlinna Theory	**Diophantine Approximation**
non-constant meromorphic function f	infinite $\{x\}$ in a number field k
A radius r	An element of k
A finite measure set E of radii	A finite subset of $\{x\}$
An angle θ	An embedding $\sigma : k \to \mathbf{C}$
$\|f(re^{i\theta})\|$	$\|x\|_\sigma$
$(\mathrm{ord}_z f)\log\frac{r}{\|z\|}$	$(\mathrm{ord}_{\mathcal{P}} x)[k_{\mathcal{P}} : \mathbf{Q}_p]\log p$
Proximity function	Proximity function
$m_f(r,a) = \int_0^{2\pi}\log^+\left\|\frac{1}{f(re^{i\theta})-a}\right\|\frac{d\theta}{2\pi}$	$m(x,a) = \sum_{\sigma:k\to\mathbf{C}}\log^+\left\Vert\frac{1}{x-a}\right\Vert_\sigma$
Counting function:	Counting function:
$N_f(r,a) = \mathrm{ord}_0^+(f-a)\log r$ $+\sum_{0<\|z\|<r}\mathrm{ord}_z^+(f-a)\log\frac{r}{\|z\|}$	$N(x,a) =$ $\frac{1}{[k:\mathbf{Q}]}\sum_{\mathcal{P}\subset\mathbf{R}_k}\mathrm{ord}_{\mathcal{P}}^+(x-a)[k_{\mathcal{P}} : \mathbf{Q}_p]\log p$
Characteristic function	Logarithmic height
$T_f(r) = \int_0^{2\pi}\log^+\|f(re^{i\theta})\|\frac{d\theta}{2\pi}$ $+N_f(r,\infty)$	$h(x) = \dfrac{1}{[k:\mathbf{Q}]}\sum_{\sigma:k\to\mathbf{C}}\log^+\Vert x\Vert_\sigma$ $+N(x,\infty)$
Jensen's formula:	Atin-Whaples Product Formula:
$\int_0^{2\pi}\log\|f(re^{i\theta})\|\frac{d\theta}{2\pi}$ $= N_f(r,0) - N_f(r,\infty) + O(1)$	$\sum_{\sigma:k\to\mathbf{C}}\log\Vert x\Vert_\sigma$ $= N(x,0) - N(x,\infty)$
First Main Theorem:	Height Property:
$m_f(r,a) + N_f(r,a) = T_f(r) + O(1)$	$m(x,a) + N(x,a) = h(x) + O(1)$
Weaker Second Main Theorem:	Roth's Theorem:
$(q-2)T_f(r) - \sum_{j=1}^{q} N_f(r,a_j) .\le .\epsilon T_f(r)$	$(q-2)h(x) - \sum_{j=1}^{q} N(x,a_j) .\le .\epsilon h(x)$
Second Main Theorem:	Lang's conjecture:
$(q-2)T_f(r) - \sum_{j=1}^{q} N_f^{(1)}(r,a_j)$ $.\le .(1+\epsilon)\log T_f(r)$	$(q-2)h(x) - \sum_{j=1}^{q} N^{(1)}(x,a_j)$ $.\le .(1+\epsilon)\log h(x)$

Note that, in above, we use the notation $.\le.$ to denote that the inequality holds for all r except a set $E \subset (0,+\infty)$ with finite Lebesgue measure in Nevanlinna theory and the inequality holds for all, except for finitely many, $x \in k$ in Diophantine approximation.

B1.3 Proof of Roth's Theorem

The goal of this section is to prove Theorem B1.2.5. Write, for $x \in k$,

$$H_k(x) = \prod_{v \in M_k} \max\{1, \|x\|_v\},$$

then (1.44) is equivalent to

$$\prod_{v \in S} \min(\|x - \alpha_v\|_v, 1) \geq \frac{1}{H_k(x)^{2+\epsilon}}. \tag{1.49}$$

Note that in this section, it will be convenient to use both multiplicative and logarithmic heights, where the logarithmic height $h_k(x) = \log H_k(x)$.

To prove Roth's theorem, we first state several lemmas. The first one is the so-called Siegel's lemma. Siegel's lemma is a corollary of the "pigeonhole principle."

Lemma B1.3.1(Siegel's Lemma) *Let A be an $M \times N$ matrix with $M < N$ and having entries in $\mathbf{Z}$ of absolute value at most Q, where $\mathbf{Z}$ is the set of integers. Then there exists a nonzero vector $\mathbf{x} = (x_1, \ldots, x_N) \in \mathbf{Z}^N$ with $A\mathbf{x} = 0$, such that*

$$|x_i| \leq [(NQ)^{M/(N-M)}] =: Z, \qquad i = 1, \ldots, N.$$

Proof. The number of integer points in the box

$$0 \leq x_i \leq Z, \quad i = 1, \ldots, N$$

is $(Z+1)^N$. On the other hand, for all $j = 1, \ldots, N$ and for each such $\mathbf{x}$, the jth coordinate y_j of the vector $\mathbf{y} := A\mathbf{x}$ lies in the interval $[-n_j QZ, (N - n_j)QZ]$, where n_j is the number of negative entries in the jth row of A. Therefore, there are at most $(NQZ + 1)^M < (Z + 1)^N$ possible values of $A\mathbf{x}$. Hence, there must exist vectors $\mathbf{x}_1 \neq \mathbf{x}_2$ in the box and such that $A\mathbf{x}_1 = A\mathbf{x}_2$. Then $\mathbf{x} = \mathbf{x}_1 - \mathbf{x}_2$ satisfies the conditions of the lemma. □

For a number field k, we will now apply the same sort of pigeonhole principle argument to solve linear equations with algebraic coefficients. If the coefficients lie in k of degree d, and if we have M equations in N unknowns, then choosing a basis for the number fields allows us to translate the problem into dM equations with coefficients in $\mathbf{Q}$. Thus the relevant linear algebra condition is now $dM < N$. The generalization of Siegel's lemma to number field k is the following.

Lemma B1.3.2 *Let k be a number field with $d = [k : \mathbf{Q}]$, let $a_{ij} \in k$ be elements not all zero, and let $A := H(\ldots, a_{ij}, \ldots)$ be the height of the vector formed by the a_{ij}'s. Assume that $dM < N$. Then there exists a nonzero vector $\mathbf{x} \in \mathbf{Z}^N$ such that*

$$\sum_{i=1}^{N} a_{ij} x_i = 0 \quad \text{for all } 1 \leq j \leq M$$

and

$$\max_{1 \leq i \leq N} |x_i| \leq (NA)^{dM/(N-dM)}.$$

Proof. We begin by computing how many algebraic numbers are contained in various boxes.

Claim. *Let k be a number field of degree d, fix an element $a_v \in k$ for each $v \in M_k$, and let $c = \{c_v\}_v$ be a multiplicative M_k-constant. That is, $c_v \geq 1$ for all $v \in M_k$, and $c_v = 1$ for all but finitely many $v \in M_k$. Set $C := \prod_v c_v$. Then*

$$\#\left\{ x \in k \;\middle|\; |x - \alpha_v|_v \leq c_v \text{ for all } v \in M_k \right\} \leq \left(2C^{1/d} + 1\right)^d.$$

We now prove the claim. Call $\mathcal{T}$ the set whose cardinality we are trying to bound. Each $v \in M_k^\infty$ is associated with an embedding $\sigma_v : k \hookrightarrow \mathbf{C}$. So let

$$E = \prod_{v \in M_k^\infty} k_v = \mathbf{R}^{r_1} \times \mathbf{C}^{r_2}.$$

We write every $x \in E$ as $x = (x_v)$, where v in M_k^∞. For $\alpha \in k$ and $\epsilon > 0$, consider the box

$$B(\alpha, \epsilon) = \left\{ x \in E \;\middle|\; |x_v - \sigma_v(\alpha)| < \epsilon c_v \text{ for all } v \in M_k^\infty \right\}.$$

We first observe that if $\alpha, \beta \in \mathcal{T}$ and if we take $\epsilon = \frac{1}{2} C^{-1/d}$, then the intersection $B(\alpha, \epsilon) \cap B(\beta, \epsilon)$ is empty. To verify this, suppose that x sits in both boxes. If v is Archimedean, then

$$|\alpha-\beta|_v = |\sigma_v(\alpha)-\sigma_v(\beta)| \le |x_v-\sigma_v(\alpha)| + |x_v-\sigma_v(\beta)| < 2\epsilon c_v;$$

and if v is non-Archimedean, then

$$|\alpha-\beta|_v \le \max\{|\alpha-\alpha_v|_v, |\beta-\alpha_v|_v\} \le c_v.$$

It follows that $\prod|\alpha-\beta|_v < (2\epsilon)^d C = 1$, and then the product formula tells us that $\alpha=\beta$.

Now the disjointness of the $B(\alpha,\epsilon)$'s for $\alpha\in\mathcal{T}$ implies that

$$\mathrm{Vol}\left(\cup_{\alpha\in\mathcal{T}}B(\alpha,\epsilon)\right) = \#(\mathcal{T})\mathrm{Vol}\left(B(0,\epsilon)\right) = \#(\mathcal{T})\epsilon^d\mathrm{Vol}\left(B(0,1)\right).$$

Next, if $x\in B(\alpha,\epsilon)$ with $\alpha\in\mathcal{T}$, then

$$|x_v-\sigma_v(\alpha_v)| \le |x_v-\sigma_v(\alpha)| + |\alpha-\alpha_v|_v \le (1+\epsilon)c_v.$$

These inequalities define a box with volume equal to $(1+\epsilon)^d Vol(B(0,1))$; hence

$$\#\mathcal{T} \le \left(\frac{1+\epsilon}{\epsilon}\right)^d = \left(2C^{1/d}+1\right)^d.$$

This proves the Claim.

We now proceed with the proof of Lemma B1.3.2. To ease notation, put $\delta = dM/(N-dM)$ and $X=[(NA)^\delta]$. We define linear forms $L_j(\mathbf{t}) = \sum_{i=1}^N a_{ij}t_i$ for $1\le j\le M$. We apply the Claim with

$$\alpha_v = \begin{cases} L_j(X/2,\ldots,X/2) & \text{if } v \text{ is Archimedean} \\ 0 & \text{otherwise;} \end{cases}$$

$$c_v = \begin{cases} NX\max|a_{ij}|_v/2 & \text{if } v \text{ is Archimedean} \\ \max|a_{ij}|_v & \text{otherwise.} \end{cases}$$

We then compute the associated "constant"

$$C = (NX/2)^d\prod_v \max|a_{ij}|_v \le (NXA/2)^d.$$

We conclude that the linear forms $L_j(x_1,\ldots,x_N)$ takes at most $(1+NXA)^d$ values, and hence that $L=L_1\cdots L_M$ takes at most $(1+NXA)^{dM}$ values. But $X+1>(NA)^\delta$, which implies that

$$(X+1)^N = (X+1)^{N-dM}(X+1)^{dM} > (NA)^{dM}(X+1)^{dM} \ge (NAX+1)^{dM}.$$

The pigeonhole principle says that there are distinct N-tuples of integers $\mathbf{x}'$ and $\mathbf{x}''$ satisfying $L(\mathbf{x}') = L(\mathbf{x}'')$. Hence

$$L(\mathbf{x}' - \mathbf{x}'') = 0 \text{ and } |\mathbf{x}' - \mathbf{x}''| \leq X \leq (NA)^{\delta}$$

as required. □

We also need Roth's lemma. It roughly says: Suppose that the degrees $d_1, \ldots, d_n$ are fairly rapidly decreasing (the rate of decreasing depending on n). Given any points $x_1, \ldots, x_n \in k$ with heights fairly rapidly increasing (the rate of increase depending on n and $d_1, \ldots, d_n$), then any nonzero homogeneous polynomial $P(X_1, \ldots, X_n)$ with degree d_i in X_i and coefficients in k whose heights are bounded in terms of d_1 and $h_k(x_1)$, vanishes at $(x_1, \ldots, x_n)$ to only a fairly low order. To be more precise, we introduce the concept of "index," which measures how high the vanishing order is.

Definition B1.3.3 *Let*

$$Q(X_1, \ldots, X_n) = \sum_{l_1, \ldots, l_n \geq 0} a_{l_1, \ldots, l_n} X_1^{l_1} \cdots X_n^{l_n}$$

be a nonzero polynomial in n variables, and let $d_1, \ldots, d_n$ be positive real numbers. Then the **index of Q at 0 with weights** $d_1, \ldots, d_n$ *is*

$$t(Q, (0, \ldots, 0), d_1, \ldots, d_n) = \min \left\{ \sum_{i=1}^{n} \frac{l_i}{d_i} \;\middle|\; a_{l_1, \ldots, l_n} \neq 0 \right\}.$$

The precise statement of Roth's lemma is as follows:

Lemma B1.3.4 (Roth's Lemma) *Let k be a number field. Let $S \subset M_k$ be a finite set containing M_k^{∞}. Let R_S be the set of S-integers, i.e., those $x \in k$ such that $\|x\|_v \leq 1$ for $v \notin S$. Let*

$$0 < \delta < \frac{1}{16^{n+\#S}}.$$

Let $d_j (j = 1, \ldots, n)$ be integers δ-decreasing, that is

$$10 < \delta d_n, \qquad \frac{d_{i+1}}{d_i} < \delta, \qquad i = 1, \ldots, n-1. \tag{1.50}$$

Let $Q(X_1, \ldots, X_n) \not\equiv 0$ be a polynomial in $R_S[X_1, \ldots, X_n]$ of degree at most d_i in X_i. Denote by $B_{\infty}(Q) = \max_{v \in S} \|Q\|_v$, where $\|Q\|_v$ is the maximum *of the v-absolute values of the coefficients of Q. Let $x_1, \ldots, x_n$ be elements in k, and let*

$$t = t(Q, (x_1, \ldots, x_n), d_1, \ldots, d_n)$$

be the index of Q at $(x_1, \ldots, x_n)$. Suppose that

$$\#S \log 4 + 4n \leq \delta h_k(x_1), \qquad d_1 h_k(x_1) \leq d_i h_k(x_i), \qquad i = 1, \ldots, n \tag{1.51}$$

and

$$\log B_\infty(Q) \leq \delta d_1 h_k(x_1), \tag{1.52}$$

then

$$t = t(Q, (x_1, \ldots, x_n), d_q, \ldots, d_n) \leq (20)^n \delta^{2^{-n}}.$$

The proof of Roth's lemma can be found in [Lang1].

We now prove Roth's Theorem.

Proof. First, we may assume that all $\alpha_v \in k$. Otherwise, let k' be some finite extension field of k containing all α_v, let S' be the set of places w of k' lying over $v \in S$, and for each $w|v$ let α_w be a certain conjugate of α_v. (In order to write $\|x - \alpha_v\|_v$ when $\alpha_v \notin k$, some extension of $\|\cdot\|_v$ to $k(\alpha_v)$ must be chosen, then the α_w should be chosen accordingly.) With proper choices of α_w, the left-hand side of (1.44) will remain unchanged when k is replaced by k', as will the right-hand side.

The basic idea is to assume that there are infinitely many counterexamples to (1.44), and we then derive a contradiction. We derive a contradiction as follows: We choose n good approximations $(x_1, \ldots, x_n)$ which satisfy certain additional constraints that appeared in Roth's lemma. We then use Siegel's lemma to construct a nonzero polynomial Q, of degree d_i in X_i, which vanishes to a fairly high order at the points $(\alpha_v, \ldots, \alpha_v)$. Thus $\prod \|Q(x_1, \ldots, x_n)\|_v$ vanishes at a fairly high order, by the selection of good approximation $(x_1, \ldots, x_n)$. This will contradict Roth's lemma, which says that the vanishing order of Q at $(x_1, \ldots, x_n)$ cannot exceed a certain number. The detail of the proof is provided below:

Step 1: Let $\alpha_1, \ldots, \alpha_m$ be the distinct values taken on by all $\alpha_v, v \in S$. Let $d_1, \ldots, d_n$ be positive integers. We wish to construct a nonzero auxiliary polynomial Q in $X_1, \ldots, X_n$ variables of degree d_i in X_i for each i and where Q vanishes to fairly high order $\geq n(\frac{1}{2} - \epsilon_1)$ at each point $(\alpha_j, \ldots, \alpha_j), j = 1, 2, \ldots, m$. The precise statement is as follows.

Lemma B1.3.5 Given $\epsilon_1 > 0$, and let $n > (2m[k : \mathbf{Q}]/\epsilon_1)^2$. Then there exists a nonzero polynomial Q in $X_1, \ldots, X_n$ variables, of degree d_i in each X_i, and

$$t(Q, (\alpha_j, \ldots, \alpha_j), d_1, \ldots, d_n) \geq n(1/2 - \epsilon_1), \qquad j = 1, \ldots, m.$$

Proof. The polynomial Q will be constructed by using the simple linear algebraic fact that a system of linear equations always has a non-trivial solution if the number of unknowns exceed the number of linear equations. Let

$$Q(X_1, \ldots, X_n) = \sum_{j_1=0}^{d_1} \cdots \sum_{j_n=0}^{d_n} a_{j_1,\ldots,j_n} X_1^{j_1} \cdots X_n^{j_n}$$

where the integers $a_{j_1,\ldots,j_n}$ are unknowns to be determined. Clearly, the number of $a_{j_1,\ldots,j_n}$ is

$$N = (d_1 + 1) \cdots (d_n + 1).$$

The index

$$t(Q, (\alpha_j, \ldots, \alpha_j), d_1, \ldots, d_n) \geq n(1/2 - \epsilon_1)$$

means that

$$\frac{\partial^{l_1 + \ldots + l_n}}{\partial X_1^{l_1} \cdots \partial X_n^{l_n}} Q(\alpha_j, \ldots, \alpha_j) = 0 \tag{1.53}$$

wherever

$$\sum_{i=1}^{n} \frac{l_i}{d_i} \leq n(1/2 - \epsilon_1). \tag{1.54}$$

Since (1.53) is always true if $l_i \geq d_i$ for some i, the number of non-trivial equations (1.53) is the number of points

$$(\frac{l_1}{d_1}, \ldots, \frac{l_n}{d_n})$$

in the unit cube I^n where $I = [0,1]$ with (1.54). We denote this number by N_1. It turns out that $N_1/N \to 0$ as $n \to \infty$. More precisely, we have the following lemma.

Lemma B1.3.6 (A Combinatorial Lemma) Let $d_1, \ldots, d_n$ be integers greater than or equal to 1 and let $\epsilon_1 > 0$. The number of sets of integers $(l_1, \ldots, l_n)$ satisfying

$$0 \le l_1 \le d_1, \ldots, 0 \le l_n \le d_n$$

and

$$\frac{l_1}{d_1} + \cdots + \frac{l_n}{d_n} \le n(\frac{1}{2} - \epsilon_1)$$

does not exceed

$$\frac{1}{\epsilon_1 n^{1/2}}(d_1+1)\ldots(d_n+1).$$

Proof. We prove by induction on n. The assertion is trivial if $n = 1$. Take $n > 1$. We write

$$n(\frac{1}{2} - \epsilon_1) = \frac{1}{2}(n - \lambda)$$

so that $n\epsilon_1 = \lambda/2$. In terms of λ our upper bound reads

$$\frac{2n^{1/2}}{\lambda}(d_1+1)\ldots(d_n+1).$$

Our assertion is trivial if $\lambda \le 2n^{1/2}$. We may therefore assume that $\lambda > 2n^{1/2}$. For each l_n and d_n fixed, we consider the solutions of

$$\begin{aligned} \frac{l_1}{d_1} + \cdots + \frac{l_{n-1}}{d_{n-1}} &\le \frac{1}{2}(n-\lambda) - \frac{l_n}{d_n} \\ &= \frac{1}{2}(n - 1 - (\lambda - 1 + 2l_n/d_n)). \end{aligned}$$

By the induction hypothesis, we have that the number of sets of integers $(l_1, \ldots, l_{n-1})$ satisfying $0 \le l_1 \le d_1, \ldots, 0 \le l_{n-1} \le d_{n-1}$ and

$$\frac{l_1}{d_1} + \cdots + \frac{l_{n-1}}{d_{n-1}} \le \frac{1}{2}(n-\lambda) - \frac{l_n}{d_n}$$

does not exceed

$$\frac{2(n-1)^{1/2}}{\lambda - 1 + 2l_n/d_n}(d_1+1)\dots(d_{n-1}+1).$$

Take the sum over l_n with $l_n \leq d_n$, we have that the number of sets of integers $(l_1, \dots, l_n)$ satisfying

$$0 \leq l_1 \leq d_1, \dots, 0 \leq l_n \leq d_n$$

and

$$\frac{l_1}{d_1} + \dots + \frac{l_n}{d_n} \leq n(\frac{1}{2} - \epsilon_1) = \frac{1}{2}(n - \lambda)$$

does not exceed

$$\sum_{l_n=1}^{d_n} \frac{2(n-1)^{1/2}}{\lambda - 1 + 2l_n/d_n}(d_1+1)\dots(d_{n-1}+1).$$

So it remains to prove that

$$\sum_{l_n=1}^{d_n} \frac{2(n-1)^{1/2}}{\lambda - 1 + 2l_n/d_n} \leq \frac{2n^{1/2}}{\lambda}(d_n+1).$$

To do this, we first make a computation:

$$\begin{aligned}
\sum_{i=1}^{r} \frac{2}{\lambda - 1 + 2i/r} &= \sum_{i=1}^{r} \left[\frac{1}{\lambda - 1 + 2i/r} + \frac{1}{\lambda + 1 - 2i/r}\right] \\
&= \sum_{i=1}^{r} \frac{2\lambda}{\lambda^2 - (1 - 2i/r)^2} \\
&\leq \sum_{i=1}^{r} \frac{2\lambda}{\lambda^2 - 1} = (r+1)\frac{2\lambda}{\lambda^2 - 1}.
\end{aligned}$$

Let $i = l_n$ and $r = d_n$, then the above inequality becomes

$$\sum_{l_n=1}^{d_n} \frac{2}{\lambda - 1 + 2l_n/d_n} \leq (d_n+1)\frac{2\lambda}{\lambda^2 - 1}.$$

Therefore, the proof is reduced to check whether

$$\frac{2\lambda(n-1)^{1/2}}{\lambda^2 - 1} \leq \frac{2n^{1/2}}{\lambda},$$

or equivalently, whether

$$[(n-1)/n]^{1/2} \leq (\lambda^2 - 1)/\lambda^2$$

holds. Observe that $(1 - 1/n)^{1/2} \leq 1 - \frac{1}{2}n$. It suffices therefore that $1 - \frac{1}{2}n \leq 1 - 1/\lambda^2$. But this is true in view of our original hypothesis on λ, and our lemma is proved. $\square$

Back to Lemma B1.3.5. Since $n > (2m[k : \mathbf{Q}]/\epsilon_1)^2$, Lemma B1.3.6 tells us that $N_1 < N/(2m[k : \mathbf{Q}])$, where $N = (d_1 + 1) \dots (d_n + 1)$. That is the number of nontrivial conditions (1.53) is at most $N/(2m[k : \mathbf{Q}])$. Each condition (1.53) is a homogeneous linear equation in the unknowns $a_{j_1,\dots,j_n}$. Hence, each condition follows from m linear homogeneous equations whose coefficients are rational integers, and altogether our unknown integers $a_{j_1,\dots,j_n}$ have to satisfy at most $N/(2[k : \mathbf{Q}])$ linear homogeneous equations with rational coefficients. A nonzero solution exists since the total number of unknowns (i.e. the coefficients of Q) is N. This finishes the proof of Lemma B1.3.5. $\square$

Step 2: Note that in Lemma B1.3.5, we only use the fact that $mN_1 < N$. However, what we derive from Lemma B1.3.6 is $mN_1 < N/(2[k : \mathbf{Q}])$. This allows us to apply Siegel's lemma to derive the following more precise lemma.

Lemma B1.3.7 *Let ϵ_1, m, n be defined as stated in Lemma B1.3.5. Then there is a nonzero polynomial $Q \in k[X_1, \dots, X_n]$ with integer coefficients which is bounded by*

$$B_\infty(Q) \leq c_1^{d_1 + \dots + d_n},$$

where c_1 is a constant dependent on k, S, n and $\alpha_1, \dots, \alpha_m$, and

$$t(Q, (\alpha_j, \dots, \alpha_j), d_1, \dots, d_n) \geq n(1/2 - \epsilon_1), \qquad j = 1, \dots, m.$$

Here $B_\infty(Q) = \max_{v \in S} \|Q\|_v$, where $\|Q\|_v$ is the maximum of the v-absolute values of the coefficients of Q.

Proof. For each $1 \leq j \leq m$, the number of non-trivial equations (1.53) is N_1. So we have $M = mN_1$ linear homogeneous equations for the N variables $a_{j_1,\dots,j_n}$ and we know, from above that $M \leq N/(2[k : \mathbf{Q}])$. In order to apply Siegel's lemma, we need to estimate the size of the coefficients of these equations. Note that, for each n-tuple $(l_1, \ldots, l_n)$,

$$\begin{aligned}&\frac{\partial^{l_1+\ldots+l_n}}{\partial X_1^{l_1}\cdots\partial X_n^{l_n}}Q(\alpha_j,\ldots,\alpha_j)\\&= \sum_{j_1=l_1}^{d_1}\cdots\sum_{j_n=l_n}^{d_n} a_{j_1,\ldots,j_n}\binom{j_1}{l_1}\cdots\binom{j_n}{l_n}\alpha_j^{j_1-l_1}\cdots\alpha_j^{j_n-l_n}.\end{aligned}$$

So we can get estimate of the coefficients of our linear homogeneous equations (1.53) by

$$\begin{aligned}\left|\binom{j_1}{l_1}\cdots\binom{j_n}{l_n}C^{j_1+\cdots+j_n-l_1-\cdots-l_n}\right| &\leq 2^{j_1+\cdots+j_n}C^{j_1+\cdots+j_n}\\ &\leq (2C)^{d_1+\cdots+d_n},\end{aligned}$$

where C is a constant depends only on $\alpha_j, 1 \leq j \leq m$, k and S. Now applying Lemma B1.3.2(Siegel's Lemma), we find that there is a polynomial Q with $t(Q, (\alpha_j, \ldots, \alpha_j), d_1, \ldots, d_n) \geq n(1/2 - \epsilon_1)$ for $j = 1, \ldots, m$ and its coefficients $a_{j_1,\ldots,j_n}$ are bounded by

$$|Q| \leq \left(N(2C)^{d_1+\cdots+d_n}\right)^{[k:\mathbf{Q}]M/(N-[k:\mathbf{Q}]M)} \leq (4C)^{d_1+\cdots+d_n},$$

using the facts that $M \leq \frac{1}{2[k:\mathbf{Q}]}N$ and $N \leq 2^{d_1+\cdots+d_n}$. This finishes the proof of Lemma B1.3.7. □

Step 3: We now proceed with the proof of Roth's theorem. Assume that Roth's theorem is not true. Then there exists infinitely many $x \in k$ such that

$$\prod_{v\in S}\min(\|x-\alpha_v\|_v, 1) \leq \frac{1}{H_k(x)^{2+\epsilon}}.$$

We claim that there must exist non-negative real numbers $k_v, v \in S$, with $\sum_{v\in S}\kappa_v = 2 + \frac{\epsilon}{2}$ such that

$$\min(\|x-\alpha_v\|_v, 1) \leq \frac{1}{H_k(x)^{\kappa_v}}, \qquad v \in S \tag{1.55}$$

holds for infinitely many $x \in k$. In fact, we write

$$\min(\|x-\alpha_v\|_v, 1) = \frac{1}{H_k(x)^{(2+\epsilon)\xi_v}}.$$

Then, for those infinitely many x with

$$\prod_{v\in S}\min(\|x-\alpha_v\|_v,1)\le\frac{1}{H_k(x)^{2+\epsilon}},$$

we have

$$\sum_{v\in S}\xi_v(x)\ge 1.$$

Choose a positive integer A such that

$$A\left(\frac{2+\epsilon}{2+\epsilon/2}-1\right)>s$$

where $s=\#S$. Using induction and the obvious fact that $[x+y]\le[x]+[y]+1$(the bracket being the largest integer $\le$) we get

$$\begin{aligned}A+s &\le A\frac{2+\epsilon}{2+\epsilon/2}\le\left[\sum_{v\in S}A\frac{2+\epsilon}{2+\epsilon/2}\xi_v(x)\right]+1\\ &\le\sum_{v\in S}\left[A\frac{2+\epsilon}{2+\epsilon/2}\xi_v(x)\right]+s,\end{aligned}$$

whence

$$A\le\sum_{v\in S}\left[A\frac{2+\epsilon}{2+\epsilon/2}\xi_v(x)\right].$$

Consequently, there exist integers $a_v(x)\ge 0$ such that

$$a_v(x)\le\left[A\frac{2+\epsilon}{2+\epsilon/2}\xi_v(x)\right]\le A\frac{2+\epsilon}{2+\epsilon/2}\xi_v(x)$$

and $\sum_{v\in S}a_v(x)=A$. From this we see that there is only a finite number of possible distributions of such integers $a_v(x)$, and hence, restricting our attention to a subsequence of x, if necessary, we can assume that the $a_v(x)$ are the same for all x. We write them a_v. We then put

$$\kappa_v=\frac{(2+\frac{\epsilon}{2})a_v}{A}$$

so that $0\le\kappa_v\le(2+\frac{\epsilon}{2})$ and $\sum_{v\in S}\kappa_v=2+\frac{\epsilon}{2}$. For each x in our subsequence we have $\kappa_v\le(2+\epsilon)\xi_v(x)$, and hence x satisfies the simultaneous system of the inequality

$$\min(\|x-\alpha_v\|_v, 1) \leq \frac{1}{H_k(x)^{\kappa_v}}, \qquad v \in S.$$

This verifies our claim.

Now choose $\epsilon_1 < \epsilon/(20+4\epsilon)$. We choose n, δ, points $x_1, \ldots, x_n$ and integers $d_1, \ldots, d_n$ subject to the following conditions: We first select $n > (2m[k:\mathbf{Q}]/\epsilon_1)^2$ so that we can apply Lemma B1.3.7; We then select $\delta < \epsilon_1$ so that we also have

$$(20)^n \delta^{2^{-n}} < \epsilon_1, \tag{1.56}$$

and

$$6\delta \# S + (5+\epsilon)\epsilon_1 - \epsilon/4 < 0; \tag{1.57}$$

We choose $x_1 \in k$ satisfying (1.55), with $\# S \log 4 + 4n \leq \delta h_k(x_1)$ and

$$\delta h_k(x_1) > \max\{n \log c_1, \log 2\} \tag{1.58}$$

where c_1 is the constant that appeared in Lemma B1.3.7; We then choose $x_2, \ldots, x_n \in k$, satisfying (1.55) and with

$$h_k(x_{i+1})/h_k(x_i) > 2/\delta \qquad 1 \leq i \leq n-1; \tag{1.59}$$

we choose d_1 so large that

$$d_1 > \frac{10 h_k(x_i)}{\delta h_k(x_1)}, \quad 2 \leq i \leq n; \tag{1.60}$$

and finally, we choose decreasing integers $d_2, \ldots, d_n$ such that

$$\frac{d_1 h_k(x_1)}{h_k(x_j)} \leq d_j < 1 + \frac{d_1 h_k(x_1)}{h_k(x_j)}, \quad 2 \leq j \leq n. \tag{1.61}$$

It follows immediately from the above inequality that d_j satisfies the condition

$$d_1 h_k(x_1) \leq d_j h_k(x_j) \leq (1+\epsilon_1) d_1 h_k(x_1). \tag{1.62}$$

These conditions are easily satisfied, since by assumption there are infinitely many $x \in k$ satisfying (1.55) and the heights of these x go to infinity.

Step 4: Having chosen $d_1, \ldots, d_n$, and points $x_1, \ldots, x_n \in k$, we choose a polynomial Q according to Lemma B1.3.7. We want to obtain a lower bound for the index of Q at the point $(x_1, \ldots, x_n)$. We claim that $t = t(Q, (x_1, \ldots, x_n), d_1, \ldots, d_n) \geq \epsilon_1$. That is, for every $i_1, \ldots, i_n$ with

$$\frac{i_1}{d_1} + \ldots + \frac{i_n}{d_n} \leq \epsilon_1, \tag{1.63}$$

$$\frac{\partial^{i_1+\ldots+i_n}}{\partial X_1^{i_1} \ldots X_n^{i_n}} Q(x_1, \ldots, x_n) = 0. \tag{1.64}$$

To verify (1.64), fix $i_1, \ldots, i_n$ satisfying (1.63), we let $\frac{\partial^{i_1+\ldots+i_n}}{\partial X_1^{i_1}\ldots X_n^{i_n}} Q = F$. It is clear, from the definition, that we only need to prove that $F(x_1, \ldots, x_n) = 0$. Let $\alpha_1, \ldots, \alpha_m$ be the distinct values taken on by all $\alpha_v, v \in S$. Since by the construction of Q,

$$t(Q, (\alpha_j, \ldots, \alpha_j), d_1, \ldots, d_n) \geq n(1/2 - \epsilon_1), \quad j = 1, \ldots, m,$$

we have,

$$F^{(l)}(\alpha_v, \ldots, \alpha_v) = 0, \quad v \in S$$

whenever $(l) = (l_1, \ldots, l_n)$ satisfies $l_1/d_1 + \ldots l_n/d_n \leq n(1/2 - \epsilon_1) - \epsilon_1$. We first estimate $\|F(x_1, \ldots, x_n)\|_v$ for $v \in S$. By Taylor's expansion,

$$F(X_1, \ldots, X_n) = \sum_{(l) \geq 0} F^{(l)}(\alpha_v, \ldots, \alpha_v)(X - \alpha_v)^{(l)}.$$

Here, as we have indicated above, all the terms will be 0 except those belonging to (l) with

$$\frac{l_1}{d_1} + \cdots + \frac{l_n}{d_n} \geq n(\frac{1}{2} - 2\epsilon_1).$$

The total number of terms in the sum is bounded by $(d_1 + 1) \cdots (d_n + 1)$ which is also bounded by, using (1.58),

$$2^{d_1+\cdots+d_n} \leq 2^{nd_1} \leq H_k(x_1)^{n\delta d_1},$$

since the d_j are decreasing. Because

$$B_\infty(Q) \leq c_1^{d_1+\ldots+d_n} \leq c_1^{nd_1}, \tag{1.65}$$

and $F = \frac{\partial^{i_1+\ldots+i_n}}{\partial X_1^{i_1}\ldots X_n^{i_n}} Q$, an upper bound for $\|F^{(l)}(\alpha_v, \ldots, \alpha_v)\|_v$ is certainly

$$\leq 2^{nd_1} 2^{nd_1} c_1^{nd_1} B_\infty(Q) \leq 2^{2nd_1} c_1^{2nd_1} \leq H_k(x_1)^{2\delta d_1 n},$$

using (1.58) in the last step. Thus, putting everything together, we have,

$$||F(x_1,\ldots,x_n)||_v \leq H_k(x_1)^{3n\delta d_1} \cdot \sup(||x_1-\alpha_v||_v^{l_1}\cdots||x_n-\alpha_v||_v^{l_n}),$$

where the supremum is taken for those (l) with $l_1/d_1+\cdots+l_n/d_n \geq n(1/2-2\epsilon_1)$. Using (1.55), we have

$$\begin{aligned}\log||F(x_1,\ldots,x_n)||_v &\leq 3n\delta d_1 h_k(x_1) - \kappa_v(l_1h_k(x_1)+\cdots+l_nh_k(x_n))\\ &\leq 3n\delta d_1 h_k(x_1) - \kappa_v(\frac{l_1}{d_1}h(x_1)d_1+\cdots+\frac{l_n}{d_n}h(x_n)d_n)\\ &\leq 3n\delta d_1 h_k(x_1) - \kappa_v h_k(x_1)d_1(\frac{l_1}{d_1}+\cdots+\frac{l_n}{d_n}),\end{aligned}$$

using $d_ih_k(x_i) \geq d_1h_k(x_1)$. Since $l_1/d_1+\cdots+l_n/d_n \geq n(\frac{1}{2}-2\epsilon_1)$,

$$\log||F(x_1,\ldots,x_n)||_v \leq 3n\delta d_1 h_k(x_1) - \kappa_v h_k(x_1)d_1 n(\frac{1}{2}-2\epsilon_1).$$

If $v \in S$ and $\kappa_v = 0$, we estimate $||F(x_1,\ldots,x_n)||_v$ naively, and get

$$||F(x_1,\ldots,x_n)||_v \leq H_k(x_1)^{3n\delta d_1}\max(1,||x_1||_v)^{d_1}\cdots\max(1,||x_n||_v^{d_n}).$$

At $v \notin S$, v is non-Archimedean, then we simply get

$$\log||F(x_1,\ldots,x_n)||_v \leq \log\left((\max(1,||x_1||_v)^{d_1})\cdots(\max(1,||x_n||_v)^{d_n})\right).$$

Hence

$$\begin{aligned}&\sum_{v\in M_k}\log||F(x_1,\ldots,x_n)||_v\\ &\leq 6\delta n d_1 h_k(x_1)\#S - \sum_{v\in S}\kappa_v d_1 h_k(x_1) n(\frac{1}{2}-2\epsilon_1)\\ &\quad +d_1h_k(x_1)+\ldots+d_nh_k(x_n)\\ &\leq 6\delta n d_1 h_k(x_1)\#S - h_k(x_1)d_1 n(\frac{1}{2}-2\epsilon_1)(2+\epsilon/2)\\ &\quad +nd_1h_k(x_1)(1+\epsilon_1)\\ &= d_1h_k(x_1)n(6\delta\#S+(1+\epsilon_1)-(\frac{1}{2}-2\epsilon_1)(2+\epsilon/2)),\end{aligned}$$

using $d_jh_k(x_j) \leq d_1h_k(x_1)(1+\epsilon_1)$ by (1.62). By (1.57),

$$6\delta\#S+(1+\epsilon_1)-(\frac{1}{2}-2\epsilon_1)(2+\epsilon/2)) < 0,$$

which contradicts the product formula, unless $F(x_1, \ldots, x_n) = 0$. So $F(x_1, \ldots, x_n) = 0$, which means

$$t = t(Q, (x_1, \ldots, x_n), d_1, \ldots, d_n) \geq \epsilon_1. \tag{1.66}$$

Step 5: We now apply Roth's lemma to Q. We verify the conditions in Roth's lemma. We first verify (1.50). In fact,

$$\frac{d_{i+1}}{d_i} = \frac{d_i}{d_1} \cdot \frac{d_1}{d_{i+1}} \geq \frac{r_1 h(x_1)}{h(x_i) d_1} \cdot \frac{r_1 h(x_{i+1})}{d_1 h(x_1)(1+\epsilon_1)} = \frac{h(x_{i+1})}{h(x_i)} \frac{1}{1+\epsilon_1} > \frac{1}{\delta},$$

using (1.62), and (1.60). So (1.50) is satisfied. (1.51) in Roth's lemma is satisfied because of (1.61). Finally, (1.58) and (1.65) imply that (1.52). Thus Roth's lemma implies that

$$t < (20)^n \delta^{2^{-n}} \leq \epsilon_1$$

using (1.56) in the last step. This contradicts (1.66). So Roth's theorem is proven. □

The Correspondence Table

Nevanlinna Theory	**Diophantine Approximation**
non-constant meromorphic function f	infinite $\{x\}$ in a number field k
Jensen's Formula(Corollary A1.1.3)	Product Formula(Theorem B1.2.4)
(1.5)	(1.48)
(1.6)	(1.46)
(1.8)	(1.45)
Theorem A1.1.5	Theorem B1.2.6
Theorem A1.3.1	Theorem B1.2.7
Corollary A1.3.2	Conjecture B1.2.8
Corollary A1.3.3	Theorem B1.2.9

Chapter 2

Holomorphic Curves into Compact Riemann Surfaces and Theorems of Siegel, Roth and Faltings

Part A: Nevanlinna Theory

In chapter 1, we introduced Nevanlinna theory for meromorphic functions. A meromorphic function can be regarded as a holomorphic map from $\mathbf{C}$ into $\mathbf{P}^1(\mathbf{C})$, while we identify $\mathbf{P}^1(\mathbf{C})$ with $\mathbf{C} \cup \{\infty\}$. In this chapter, we will extend the theory to the holomorphic map $f : \mathbf{C} \to M$ where M is a compact Riemann surface. We closely follow Ahlfors' negative curvature method. We note that there is a simpler approach of using the Logarithmic Derivative Lemma, which will be presented in chapter 5.

A2.1 Some Lemmas

We first introduce some notation. Let $z = x + iy$ be an analytic coordinate. Let

$$\frac{\partial}{\partial \bar{z}} = \frac{1}{2}(\frac{\partial}{\partial x} + \sqrt{-1}\frac{\partial}{\partial y}), \text{ and } \quad \frac{\partial}{\partial z} = \frac{1}{2}(\frac{\partial}{\partial x} - \sqrt{-1}\frac{\partial}{\partial y}).$$

For a function f, we define

$$\partial f = \frac{\partial f}{\partial z} dz, \quad \text{and } \bar{\partial} f = \frac{\partial f}{\partial \bar{z}} d\bar{z},$$

so ∂ and $\bar{\partial}$ send functions to 1-forms. Note that

$$\partial + \bar{\partial} = d,$$

where d is the ordinary exterior derivative of differential forms. We define the real operator d^c by

$$d^c = \frac{\sqrt{-1}}{4\pi}(\bar{\partial} - \partial) = \frac{1}{4\pi}(r\frac{\partial}{\partial r} \otimes d\theta - r^{-1}\frac{\partial}{\partial \theta} \otimes dr), \tag{2.1}$$

so that

$$dd^c = \frac{\sqrt{-1}}{2\pi}\partial\bar{\partial}. \tag{2.2}$$

Before we state the Green-Jensen's formula, we introduce the concept of currents. Let g be a sub-harmonic function (resp. super-harmonic function). Let Z be the set of singularities of g. Denote by $S(Z,\epsilon)(t)$ the union of small circles around the singularities in $\mathbf{D}(t)$. The $(1,1)$ **current** $dd^c[g]$ is the functional such that, for any fixed number $r_0 > 0$,

$$\int_{r_0}^{r} \frac{dt}{t} \int_{|\zeta|<t} dd^c[g] = \int_{r_0}^{r} \frac{dt}{t} \int_{|\zeta|<t} dd^c g + \int_{r_0}^{r} \frac{dt}{t} \lim_{\epsilon \to 0} \int_{S(Z,\epsilon)(t)} d^c g.$$

In particular, taking $g = \log|f|^2$ for some holomorphic function f, we have the following Lemma.

Lemma A2.1.1 *Let f be a holomorphic function on $\mathbf{D}(r)$. Let Z denote the zeros of f inside $\mathbf{D}(r)$. Then, for any t with $0 < t < r$,*

$$\lim_{\epsilon \to 0} \int_{S(Z,\epsilon)(t)} d^c \log|f|^2 = n_f(t,0). \tag{2.3}$$

Proof. Let p be a zero of f. It suffices to prove that

$$\lim_{\epsilon \to 0} \int_{S(p,\epsilon)} d^c \log|f|^2 = \operatorname{ord}_p(f),$$

where $S(p,\epsilon)$ is the circle centered at p with radius ϵ. Without loss of generality, we may assume that $p = 0$. We then have to prove

$$\lim_{\epsilon \to 0} \int_{S(\epsilon)} d^c \log|f|^2 = \operatorname{ord}_0(f),$$

where $S(\epsilon)$ is the circle centered at 0 with radius ϵ. Let $k = \operatorname{ord}_0 f$. We can write $f\bar{f} = r^{2k}h(r,\theta)$ where h is smooth and positive. So

$$\lim_{\epsilon \to 0} \int_{S(p,\epsilon)} d^c \log|f|^2 = \lim_{\epsilon \to 0} \int_{S(p,\epsilon)} d^c \log r^{2k},$$

since $\log h(z)$ is smooth. By (2.1),

$$d^c \log r^{2k} = \frac{1}{4\pi} r \frac{\partial(\log r^{2k})}{\partial r} d\theta = \frac{1}{2\pi} k d\theta.$$

Thus

$$\lim_{\epsilon\to 0} \int_{S(p,\epsilon)} d^c \log |f|^2 = \lim_{\epsilon\to 0} \int_0^{2\pi} k \frac{d\theta}{2\pi} = k.$$

This finishes the proof. □

So we have the following theorem.

Theorem A2.1.2 (Poincaré-Lelong Formula) *Let f be a holomorphic function on an open neighborhood of $\mathbf{D}(r)$. Then, for every $0 < r_0 < r$,*

$$\int_{r_0}^{r} \frac{dt}{t} \int_{|\zeta|<t} dd^c[\log |f|^2] = \int_{r_0}^{r} \frac{n_f(t,0)}{t} dt,$$

or we simply write, in the sense of current,

$$dd^c[\log |f|^2] = [D_f],$$

where $D_f = \sum_p (\mathrm{ord}_p f) \cdot p$ is the divisor associated with f.

Proof. By the definition,

$$\int_{r_0}^{r} \frac{dt}{t} \int_{|\zeta|<t} dd^c[\log |f|^2] = \int_{r_0}^{r} \frac{dt}{t} \int_{|\zeta|<t} dd^c \log |f|^2 + \int_{r_0}^{r} \frac{dt}{t} \lim_{\epsilon\to 0} \int_{S(p,\epsilon)(t)} d^c \log |f|^2.$$

Since f is holomorphic, $dd^c \log |f|^2 = 0$ in the sense of the differential form, and by Lemma A2.1.1,

$$\lim_{\epsilon\to 0} \int_{S(p,\epsilon)(t)} d^c \log |f|^2 = n_f(t,0).$$

So the theorem is verified. □

We use the concept of current because of the following theorem.

Theorem A2.1.3 (Green-Jensen's Formula) *Let g be a function of class C^2 on $\overline{\mathbf{D}}(R)$ or a sub-harmonic (resp. super-harmonic) function on $\overline{\mathbf{D}}(R)$. Then, for any $0 \le r < R$,*

$$\int_r^R \frac{dt}{t} \int_{|\zeta|<t} dd^c[g] = \frac{1}{2}\int_0^{2\pi} g(Re^{i\theta})\frac{d\theta}{2\pi} - \frac{1}{2}\int_0^{2\pi} g(re^{i\theta})\frac{d\theta}{2\pi}.$$

Proof. Denote by Z the set of singularities of g, and $S(Z,\epsilon)(t)$ the union of small circles around singularities in $\mathbf{D}(t)$. Then Stokes formula implies that, using (2.1),

$$\begin{aligned}\int_{|\zeta|<t} dd^c g &= \int_{|\zeta|=t} d^c g - \lim_{\epsilon\to 0}\int_{S(Z,\epsilon)(t)} d^c g \\ &= \frac{1}{2}\int_{|\zeta|=t} t\frac{\partial g}{\partial t}\frac{d\theta}{2\pi} - \lim_{\epsilon\to 0}\int_{S(Z,\epsilon)(t)} d^c g.\end{aligned}$$

Integrating the above with respect to $1/t$, we get

$$\begin{aligned}&\int_r^R \frac{dt}{t}\int_{|\zeta|<t} dd^c g \\ &= \int_r^R \frac{dt}{t}\int_{|\zeta|=t}\frac{1}{2}t\frac{\partial g}{\partial t}\frac{d\theta}{2\pi} - \int_r^R\frac{dt}{t}\lim_{\epsilon\to 0}\int_{S(Z,\epsilon)(t)} d^c g \\ &= \frac{1}{2}\int_0^{2\pi} g(Re^{i\theta})\frac{d\theta}{2\pi} - \frac{1}{2}\int_0^{2\pi} g(re^{i\theta})\frac{d\theta}{2\pi} \\ &\quad -\int_r^R\frac{dt}{t}\lim_{\epsilon\to 0}\int_{S(Z,\epsilon)(t)} d^c g.\end{aligned}$$

So, by the definition,

$$\begin{aligned}\int_r^R \frac{dt}{t}\int_{|\zeta|<t} dd^c[g] &= \int_r^R\frac{dt}{t}\int_{|\zeta|<t} dd^c g + \int_r^R\frac{dt}{t}\lim_{\epsilon\to 0}\int_{S(Z,\epsilon)(t)} d^c g \\ &= \frac{1}{2}\int_0^{2\pi} g(Re^{i\theta})\frac{d\theta}{2\pi} - \frac{1}{2}\int_0^{2\pi} g(re^{i\theta})\frac{d\theta}{2\pi}.\end{aligned}$$

The theorem is proven. □

We shall need the following form of the Calculus Lemma.

Lemma A2.1.4 (Calculus Lemma) *Let T be a strictly nondecreasing function of class C^1 defined on $(0,\infty)$. Let $\gamma > 0$ be a number such that $T(\gamma) \geq e$. Let ϕ be a strictly positive nondecreasing function such that*

$$\int_e^\infty \frac{1}{t\phi(t)}dt = c_0(\phi) < \infty.$$

Then the inequality

$$T'(r) \leq T(r)\phi(T(r))$$

holds for all $r \geq \gamma$ outside a set of Lebesgue measure $\leq c_0(\phi)$.

Proof. Let $A \subset [\gamma, \infty)$ be the set of r such that $T'(r) \geq T(r)\phi(T(r))$. Then

$$\text{meas}(A) = \int_A dr \leq \int_\gamma^\infty \frac{T'(r)}{T(r)\phi(T(r))}dr = \int_e^\infty \frac{dt}{t\phi(t)} = c_0(\phi),$$

which proves the lemma. □

Lemma A2.1.5 *Let T be a function of class C^2 defined on $(0,\infty)$. Assume that (i) there exists $\gamma \geq 1$ such that $T(r) \geq e$ for all $r \geq \gamma$ and (ii) that both $T(r)$ and $T'(r)$ are strictly nondecreasing functions of r. Let $b \geq 1$ be a number such that $brT'(r) \geq e$ for all $r \geq 1$ (such number clearly exists). Then*

$$\frac{1}{r}\frac{d}{dr}(r\frac{dT}{dr}) \leq T(r)\phi(T(r))\phi[brT(r)\phi(T(r))]$$

for all $r \geq \gamma$ outside a set of measure $\leq 2c_0(\phi)$.

Proof. The assumptions guarantee that we may apply the Calculus lemma twice, first to the function $brT'(r)$ and then to the function $T(r)$. □

Lemma A2.1.6 *Let τ be a non-negative function of class C^2 when $\tau > 0$. Then*

$$dd^c \log\left(\frac{1}{\log\tau}\right)^2 = 2\{\frac{1}{\tau(\log\tau)^2}dd^c\tau - \frac{1+\log\tau}{(\log\tau)^2}dd^c\log\tau\}.$$

Proof. By direct computation. See [Wong3]. □

A2.2 Divisors, Line Bundles, and the First Main Theorem

Given a compact Riemann surface M. Let $f : \mathbf{C} \to M$ be a holomorphic map. Given a point $a \in M$, to introduce the proximity function of $m_f(r, a)$, we need to measure the distance between $f(z)$ and a. So we borrow some terminologies from algebraic geometry. Instead of one point, we consider a more general case: a formal finite sum of distinct points $\sum_j n_j a_j$, where $n_j \in \mathbf{Z}, a_j \in M$.

Definition A2.2.1 *Let M be a compact Riemann surface. A* **divisor** *on M, denoted by D, is a formal finite sum $D = \sum_j n_j a_j$, where $n_j \in \mathbf{Z}$ and $a_j \in M$ are distinct points on M. If $n_j \geq 0$ for all j, then we call D an* **effective divisor**.

Let $D = \sum_j n_j a_j$ be a divisor. We choose a local coordinate z_j for a coordinate chart U_j centered at a_j so that all U_j are disjoint. Let U_0 be an open subset of M so that no a_j belongs to the topological closure of U_0 and the set $U - \cup_{j=1}^k U_j$ is contained in U_0. Take $g_j = z_j^{n_j}$ on U_j $(1 \leq j \leq k)$ and $g_0 = 1$ on U_0. Let $g_{ij} = g_i/g_j$ on $U_i \cap U_j$, then g_{ij} are nowhere-zero holomorphic functions, and g_{ij} satisfies $g_{it} = g_{ij}g_{jt}$, on $U_i \cap U_j \cap U_t$. We call the collection $\{U_j, g_{ij}\}$ the **line bundle** associated with D, and denote it by $\mathcal{O}(D)$. The collection $\{g_j\}$ is called the **canonical section** of $\mathcal{O}(D)$, and is denoted by s_D. We have the following general definition of line bundles.

Definition A2.2.2 By a **line bundle** L *over a compact Riemann surface M, we mean a collection $\{U_\alpha, g_{\alpha\beta}\}$ where $\{U_\alpha\}$ is a finite open cover of M and $g_{\alpha\beta}$ is a nowhere-zero holomorphic function on $U_\alpha \cap U_\beta$ satisfying the compatibility condition $g_{\alpha\gamma} = g_{\alpha\beta}g_{\beta\gamma}$ on $U_\alpha \cap U_\beta \cap U_\gamma$. The functions $\{g_{\alpha\beta}\}$ are called* **transition functions**.

Definition A2.2.3 *Let $L = \{U_\alpha, g_{\alpha\beta}\}$ be a line bundle. A* **holomorphic section** *s of L is a collection $\{s_\alpha\}$ where each s_α is a holomorphic function defined on U_α and satisfying $s_\alpha = g_{\alpha\beta}s_\beta$ on $U_\alpha \cap U_\beta$. We note that, for any $p \in U_\alpha \cap U_\beta, ord_p s_\alpha = ord_p s_\beta$ by the transition property. So we define $D = \sum_{p \in M}(ord_p s_\alpha)p$, where for each $p \in M$ we pick U_α with $p \in U_\alpha$. D is called the* **divisor associated with the section** *s.*

Definition A2.2.4 *Let $L = \{U_\alpha, g_{\alpha\beta}\}$ be a line bundle. A* **metric** *on L is* *a collection of positive smooth functions*

$$h_\alpha : U_\alpha \to \mathbf{R}_{>0}$$

such that on $U_\alpha \cap U_\beta$ we have

$$h_\beta = |g_{\alpha\beta}|^2 h_\alpha.$$

Example A2.2.5 *We consider the canonical line bundle K_M. Cover M by coordinate charts $\{U_\alpha\}$ so that the coordinate for U_α is z_α. The transition functions $g_{\alpha\beta}$ of K_M on $U_\alpha \cap U_\beta$ is given by $g_{\alpha\beta} = \dfrac{dz_\beta}{dz_\alpha}$. A section of K_M is precisely a $(1,0)$ form on M. The reason is as follows. A section of K_M is given by a collection $\{s_\alpha\}$ so that $s_\alpha = g_{\alpha\beta}s_\beta$ on $U_\alpha \cap U_\beta$. In other words, $s_\alpha = \dfrac{dz_\beta}{dz_\alpha}s_\beta$ on $U_\alpha \cap U_\beta$ which is the same as saying that $\omega = s_\alpha dz_\alpha$ on U_α is well-defined. The divisor of any $(1,0)$ form ω on M is called the* **canonical divisor** *on M. To give a metric for K_M is the same as giving a volume form $\Omega = h\sqrt{-1}dz \wedge d\bar{z}$. The reason is as follows. A metric for K_M is a collection $\{h_\alpha\}$ so that each h_α is a positive valued function on U_α and $h_\beta = |\dfrac{dz_\beta}{dz_\alpha}|^2 h_\alpha$. Thus $h_\alpha^{-1}\sqrt{-1}dz_\alpha \wedge d\bar{z}_\alpha$ obeys $h_\alpha^{-1}\sqrt{-1}dz_\alpha \wedge d\bar{z}_\alpha = h_\beta^{-1}\sqrt{-1}dz_\beta \wedge d\bar{z}_\beta$ and defines a volume form on M.*

For any given metric $\{h_\alpha\}$ of L we can define the $(1,1)$-form θ_L as $\theta_L = -\frac{\sqrt{-1}}{2\pi}\partial\bar{\partial}\log h_\alpha$ on U_α. Since $h_\alpha|g_{\alpha\beta}|^2 = h_\beta$ on $U_\alpha \cap U_\beta$, we have locally $\log h_\alpha + \log g_{\alpha\beta} + \log \bar{g}_{\alpha\beta} = \log h_\beta$ for some local branch of $\log g_{\alpha\beta}$. From $\bar{\partial}\log g_{\alpha\beta} = 0$ and $\partial \log \bar{g}_{\alpha\beta} = 0$, we conclude that $-\frac{\sqrt{-1}}{2\pi}\partial\bar{\partial}\log h_\alpha = -\frac{\sqrt{-1}}{2\pi}\partial\bar{\partial}\log h_\beta$ on $U_\alpha \cap U_\beta$. Thus, the form θ_L is well-defined.

Definition A2.2.6 *Let $L = \{U_\alpha, g_{\alpha\beta}\}$ be a metrized line bundle with metric $\{h_\alpha\}$. The form $\theta_L = -\frac{\sqrt{-1}}{2\pi}\partial\bar{\partial}\log h_\alpha$ on U_α is called the* **Chern form** *of L with respect to the metric $\{h_\alpha\}$. Denote θ_L by $c_1(L,h)$, or just $c_1(L)$. A holomorphic line bundle L with a metric is called* **positive** *if the Chern form θ_L for the metric of L is positive definite everywhere on M. A line bundle is said to be* **ample** *if L is positive with some metric on L. A divisor D is* **ample** *if $\mathcal{O}(D)$ is ample.*

Definition A2.2.7 *For a metrized line bundle L with metric $\{h_\alpha\}$. Given any two sections, s_i, s_j, we define the* **inner product**

$$< s_i, s_j >= s_{i\alpha}\bar{s}_{j\alpha}h_\alpha.$$

In particular, $||s||^2 = |s_\alpha|^2 h_\alpha$. By the transition properties of s_α and h_α, it is well-defined.

We notice that for a section $\{s_\alpha\}$, since $s_\alpha = g_{\alpha\beta}s_\beta$, $ds_\alpha = g_{\alpha\beta}ds_\beta + dg_{\alpha\beta}s_\beta$. So, usually the differentiation of a section does not yield a section because of the term $dg_{\alpha\beta}s_\beta$. To overcome this difficulty, one introduces a correction term and defines covariant differentiation $\mathcal{D}s = \{\mathcal{D}s_\alpha\}$ where $\mathcal{D}s_\alpha = ds_\alpha + \partial(\log h_\alpha)s_\alpha = \partial s_\alpha + \partial(\log h_\alpha)s_\alpha$. Then $\mathcal{D}s$ obeys the transition law that $\mathcal{D}s_\alpha = g_{\alpha\beta}\mathcal{D}s_\beta$. For a $(1,0)$ form A (resp. (0,1) form), we write, $|A|^2 = \frac{\sqrt{-1}}{2\pi}A \wedge \bar{A}$. So $|A|^2$ becomes a real (1.1) form. We write $||\mathcal{D}s||^2 = |\mathcal{D}s_\alpha|^2 h_\alpha$.

Lemma A2.2.8 *Let L be a metrized line bundle and s be a holomorphic section of L. Then $dd^c||s||^2 = ||\mathcal{D}s||^2 - ||s||^2 c_1(L)$.*

Proof. Write $L = \{U_\alpha, g_{\alpha\beta}\}$ and the metric $h = \{h_\alpha\}$. Then, by the definition, $||s||^2 = |s_\alpha|^2 h_\alpha = s_\alpha\bar{s}_\alpha h_\alpha$. So

$$\begin{aligned}\bar{\partial}||s||^2 &= s_\alpha\bar{\partial}\bar{s}_\alpha h_\alpha + s_\alpha\bar{s}_\alpha\bar{\partial}h_\alpha \\ &= s_\alpha(\bar{\partial}\bar{s}_\alpha + \bar{s}_\alpha\bar{\partial}\log h_\alpha)h_\alpha \\ &= s_\alpha\overline{\mathcal{D}}\bar{s}_\alpha h_\alpha =< s, \mathcal{D}s >,\end{aligned}$$

using the property that $\bar{\partial}s_\alpha = 0$ since s_α is holomorphic. Thus,

$$\begin{aligned}dd^c||s||^2 &= \frac{\sqrt{-1}}{2\pi}\partial\bar{\partial}||s||^2 \\ &= \frac{\sqrt{-1}}{2\pi}\partial(s_\alpha\overline{\mathcal{D}}\bar{s}_\alpha h_\alpha) \\ &= \frac{\sqrt{-1}}{2\pi}\partial s_\alpha \wedge \overline{\mathcal{D}}\bar{s}_\alpha h_\alpha + \frac{\sqrt{-1}}{2\pi}s_\alpha\partial(\overline{\mathcal{D}}\bar{s}_\alpha)h_\alpha + \frac{\sqrt{-1}}{2\pi}s_\alpha\partial h_\alpha \wedge \overline{\mathcal{D}}\bar{s}_\alpha \\ &= \frac{\sqrt{-1}}{2\pi}\partial s_\alpha \wedge \overline{\mathcal{D}}\bar{s}_\alpha h_\alpha + \frac{\sqrt{-1}}{2\pi}s_\alpha\partial(\overline{\mathcal{D}}\bar{s}_\alpha)h_\alpha + \frac{\sqrt{-1}}{2\pi}s_\alpha h_\alpha\partial(\log h_\alpha) \wedge \overline{\mathcal{D}}\bar{s}_\alpha \\ &= \frac{\sqrt{-1}}{2\pi}\partial s_\alpha \wedge \overline{\mathcal{D}}\bar{s}_\alpha h_\alpha + \frac{\sqrt{-1}}{2\pi}s_\alpha h_\alpha\partial(\log h_\alpha) \wedge \overline{\mathcal{D}}\bar{s}_\alpha + \frac{\sqrt{-1}}{2\pi}s_\alpha\partial(\overline{\mathcal{D}}\bar{s}_\alpha)h_\alpha \\ &= \frac{\sqrt{-1}}{2\pi}(\partial s_\alpha + s_\alpha\partial\log h_\alpha) \wedge \overline{\mathcal{D}}\bar{s}_\alpha h_\alpha + \frac{\sqrt{-1}}{2\pi}s_\alpha\partial(\overline{\mathcal{D}}\bar{s}_\alpha)h_\alpha\end{aligned}$$

$$= \quad |\mathcal{D}s_\alpha|^2 h_\alpha + \frac{\sqrt{-1}}{2\pi} s_\alpha \partial(\overline{\mathcal{D}}\bar{s}_\alpha) h_\alpha = ||\mathcal{D}s||^2 + \frac{\sqrt{-1}}{2\pi} s_\alpha \partial(\overline{\mathcal{D}}\bar{s}_\alpha) h_\alpha. \quad (2.4)$$

However,

$$\frac{\sqrt{-1}}{2\pi}\partial(\overline{\mathcal{D}}\bar{s}_\alpha) = \frac{\sqrt{-1}}{2\pi}\partial(\bar{\partial}\bar{s}_\alpha + \bar{s}_\alpha\bar{\partial}(\log h_\alpha)) = \bar{s}_\alpha \frac{\sqrt{-1}}{2\pi}\partial\bar{\partial}(\log h_\alpha) = -\bar{s}_\alpha c_1(L).$$

So

$$\frac{\sqrt{-1}}{2\pi} s_\alpha \partial(\bar{\mathcal{D}}\bar{s}_\alpha) h_\alpha = -s_\alpha \bar{s}_\alpha c_1(L) h_\alpha = -||s||^2 c_1(L). \quad (2.5)$$

Combining (2.4) and (2.5) proves the lemma. □

A2.3 Holomorphic Curves in Compact Riemann Surfaces

Let M be a compact Riemann surface. Let

$$\omega = h\frac{\sqrt{-1}}{2\pi} dz \wedge d\bar{z}$$

be a $(1,1)$ form on M. The **characteristic function** $T_{f,\omega}(r)$ of f with respect to ω is defined by

$$T_{f,\omega}(r) = \int_0^r \frac{dt}{t} \int_{|\zeta|\leq t} f^*\omega. \quad (2.6)$$

Given a positive (1,1) form ω, $T_{f,\omega}(r)$ measures the growth of f. In particular, f must be constant if $T_{f,\omega}(r)$ is bounded. If ω_1, ω_2 are two positive (1,1) forms, then there exist positive integers c, c' such that $c\omega_1 \leq \omega_2 \leq c'\omega_1$ since M is compact, so $cT_{f,\omega_1}(r) \leq T_{f,\omega_2}(r) \leq c'T_{f,\omega_1}(r)$. We now define the proximity function. A divisor $D = \sum_j n_j a_j$ is called **an effective divisor** if $n_j \geq 0$ for all j. So the canonical section $s_D = \{g_j\}$ is holomorphic, namely every g_j is holomorphic on U_j. Take a metric on $\mathcal{O}(D)$, where $\mathcal{O}(D)$ is the line bundle associated with D. The **proximity function of f with respect to D** is defined by, under the assumption that $f(\mathbf{C}) \not\subset D$,

$$m_f(r, D) = \int_0^{2\pi} \log \frac{1}{||s_D \circ f(re^{i\theta})||} \frac{d\theta}{2\pi}, \quad (2.7)$$

where s_D is a canonical meromorphic section associated with D. In particular, define $m_f(r, a)$ by taking $D = a$. We note here that, since M is compact, $m_f(r, D)$ is independent, up to a bounded term, of the choice of the section s defining D and also of the choice of the metric on $\mathcal{O}(D)$.

The **counting function of f with respect to D** is defined by, under the assumption that $f(\mathbf{C}) \not\subset D$,

$$N_f(r,D) = \int_0^r [n_f(t,D) - n_f(0,D)]\frac{dt}{t} + n_f(0,D)\log r, \tag{2.8}$$

where $n_f(t,D)$ = number of points of $f^{-1}(D)$ in the disc $|z| < t$, counting multiplicity, and $n_f(0,D) = \lim_{t\to 0} n_f(t,D)$. We note that, for any positive number $r_0 > 0$,

$$N_f(r,D) = \int_{r_0}^r \frac{n_f(t,D)}{t}dt + O(1)$$

where $O(1)$ is a constant depends on r_0. In practice, we shall use this definition for a fixed number r_0. For a divisor D, i.e. take a metric on $\mathcal{O}(D)$, where $\mathcal{O}(D)$ is the line bundle associated with D. We define the characteristic function of $T_f(r,\mathcal{O}(D))$ by

$$T_f(r,\mathcal{O}(D)) = \int_0^r \frac{dt}{t}\int_{|z|\le t} f^*c_1(\mathcal{O}(D)),$$

where $c_1(\mathcal{O}(D))$ is the Chern form of $\mathcal{O}(D)$. Again, since M is compact, $T_f(r,\mathcal{O}(D))$ does not, up to a bounded term, depend on the choice of the metric on $\mathcal{O}(D)$. If $\mathcal{O}(D)$ is ample, then $T_f(r,\mathcal{O}(D))$ is positive and it measures the growth of f.

We have the following First Main Theorem.

Theorem A2.3.1

$$T_f(r,\mathcal{O}(D)) = m_f(r,D) + N_f(r,D) + O(1).$$

Proof. By the definition,

$$N_f(r,D) = \int_{r_0}^r \frac{n_f(t,D)}{t}dt + O(1),$$

for any $r_0 > 0$, where $O(1)$ is a constant depending on r_0. We fix a $r_0 > 0$. Let s be the meromorphic section of $\mathcal{O}(D)$ associated with D. Now, $\|s\|^2 = |s_j|^2 h_j$ where s_j is holomorphic on U_j, and h_j is non-vanishing. Since

$dd^c \log h_j = -c_1(L)$, by the Poincaré-Lelong formula (Theorem A2.1.2), $[dd^c \log \|s\|^2] = -c_1(L) + [D]$. So,

$$\begin{aligned}\int_{r_0}^{r} \frac{dt}{t} \int_{|\zeta|\le t} dd^c[\log \|f^*s\|^2] &= -\int_{r_0}^{r} \frac{dt}{t} \int_{|\zeta|\le t} f^*c_1(L) + N_f(r, D) + O(1) \\ &= -T_f(r) + N_f(r, D) + O(1).\end{aligned}$$

On the other hand, by Theorem A2.1.3,

$$\begin{aligned}\int_{r_0}^{r} \frac{dt}{t} \int_{|\zeta|\le t} dd^c[\log \|f^*s\|^2] &= \int_0^{2\pi} \log \|f^*s(re^{i\theta})\|^2 \frac{d\theta}{2\pi} \\ &- \int_0^{2\pi} \log \|f^*s(r_0e^{i\theta})\|^2 \frac{d\theta}{2\pi} = -m_f(r, D) + O(1).\end{aligned}$$

So $T_f(r) = m_f(r, D) + N_f(r, D) + O(1)$. □

To state the Second Main Theorem, we write $\mathrm{Ric}(\omega) = dd^c \log h$ for $\omega = h\frac{\sqrt{-1}}{2\pi} dz \wedge d\bar{z}$. For a holomorphic map $f : \mathbf{C} \to M$, let $f^*\omega = \gamma_f \frac{\sqrt{-1}}{2\pi} d\zeta \wedge d\bar{\zeta}$ on $\mathbf{C}$. Then the zeros of γ_f define a divisor on $\mathbf{C}$. We call this divisor the **ramification divisor** and denote it by $D_{f,\mathrm{ram}}$. Then we have the following theorem.

Theorem A2.3.2 (The Second Main Theorem) *Let M be a compact Riemann surface. Let $\omega = h\frac{\sqrt{-1}}{2\pi} dz \wedge d\bar{z}$ be a positive (1,1) form on M. Let $f : \mathbf{C} \to M$ be a non-constant holomorphic map. Let $a_1, \ldots, a_q$ be distinct points on M. Then, for every $\epsilon > 0$,*

$$\sum_{j=1}^{q} m_f(r, a_j) + T_{f,\mathrm{Ric}(\omega)}(r) + N_{f,\mathrm{ram}}(r) \le \epsilon T_{f,\omega}(r) + O(\log r)$$

holds for all r outside a set $E \subset (0, +\infty)$ with finite Lebesgue measure.

By the uniformization theorem, M is either biholomorphic to the Riemann sphere $\mathbf{P}^1$, the torus or the surface of genus ≥ 2. So, before we give the proof of Theorem A2.3.2, we discuss the consequences of Theorem A2.3.2 for each case.

When $M = \mathbf{P}^1$, the **Fubini-Study** form ω on $\mathbf{P}^1$ is given in terms of an affine coordinate w by

$$\omega = \frac{1}{(1 + |w|^2)^2} \frac{\sqrt{-1}}{2\pi} dw \wedge d\bar{w} = dd^c \log(1 + |w|^2).$$

Thus $\mathrm{Ric}(\omega) = -2\omega$. So, for any meromorphic function f on $\mathbf{C}$ (also being regarded as a holomorphic map $f : \mathbf{C} \to \mathbf{P}^1$),

$$T_{f,\mathrm{Ric}(\omega)}(r) = T_{f,-2\omega}(r) = -2T_{f,\omega}(r),$$

where

$$T_{f,\omega}(r) = \int_0^r \frac{dt}{t} \int_{|\zeta| \le t} f^*\omega = \int_0^r \frac{dt}{t} \int_{|\zeta| \le t} \frac{|f'|^2}{(1+|f|^2)^2} \frac{\sqrt{-1}}{2\pi} d\zeta \wedge d\bar{\zeta}.$$

The characteristic function $T_{f,\omega}(r)$ above is called the **Ahlfors-Shimizu characteristic function**. The following Lemma says that $T_{f,\omega}(r)$ differs from the Nevanlinna's characteristic function defined in (1.8) of chapter 1 only by a constant.

Lemma A2.3.3 Let f be a meromorphic function. Let $T_{f,\omega}(r)$ be the Ahlfors-Shimizu characteristic function of f. Then

$$T_{f,\omega}(r) = m_f(r, \infty) + N_f(r, \infty) + O(1) = T_f(r) + O(1),$$

where $T_f(r)$ is the Nevanlinna's characteristic function defined in (1.8).

Proof. Since

$$\omega = \frac{1}{(1+|w|^2)^2} \frac{\sqrt{-1}}{2\pi} dw \wedge d\bar{w} = dd^c \log(1 + |w|^2),$$

$f^*\omega = dd^c \log(1 + |f|^2)$. Write $f = f_1/f_0$ where f_1, f_0 are holomorphic without common zeros. Then, by Theorem A2.1.2,

$$dd^c[\log(1 + |f|^2)] = f^*\omega - [f_0 = 0].$$

So, for a fixed positive number r_0,

$$\int_{r_0}^r \frac{dt}{t} \int_{|\zeta| \le t} dd^c[\log(1 + |f|^2)] = T_{f,\omega}(r) - N_f(r, \infty) + O(1).$$

On the other hand, by Theorem A2.1.3(Green-Jensen's formula),

$$\begin{aligned} \int_{r_0}^r \frac{dt}{t} \int_{|\zeta| \le t} dd^c[\log(1 + |f|^2)] &= \frac{1}{2} \int_0^{2\pi} \log(1 + |f(re^{i\theta})|^2) \frac{d\theta}{2\pi} + O(1) \\ &= m_f(r, \infty) + O(1). \end{aligned}$$

Hence $T_{f,\omega}(r) - N_f(r, \infty) = m_f(r, \infty) + O(1)$. That is $T_{f,\omega}(r) = m_f(r, \infty) + N_f(r, \infty) + O(1) = T_f(r) + O(1)$. □

Noting that $T_{f,\mathrm{Ric}(\omega)}(r) = T_{f,-2\omega}(r) = -2T_{f,\omega}(r) = -2T_f(r) + O(1)$, Theorem A2.3.2 implies, in this case, that

$$\sum_{j=1}^{q} m_f(r, a_j) - 2T_f(r) + N_{f,\mathrm{ram}}(r) \leq \epsilon T_f(r) + O(\log r)$$

holds for all r outside a set $E \subset (0, +\infty)$ with finite Lebesgue measure. This is the same as Theorem A1.3.1(the Second Main Theorem for meromorphic functions).

For the torus (elliptic) case, the canonical metric is a flat metric, i.e. there exists a positive (1,1) form ω such that $\mathrm{Ric}(\omega) = 0$. So in this case, Theorem A2.3.2 implies that

$$\sum_{j=1}^{q} m_f(r, a_j) + N_{f,\mathrm{ram}}(r) \leq \epsilon T_{f,\omega}(r) + O(\log r)$$

holds for all r outside a set $E \subset (0, +\infty)$ with finite Lebesgue measure.

Finally, for the surface of genus ≥ 2, there exists a positive (1,1) form ω such that $\mathrm{Ric}(\omega)$ is also a positive (1,1) form, so that $T_{f,\mathrm{Ric}(\omega)}(r) \geq 0$. Thus we have

$$T_{f,\mathrm{Ric}(\omega)}(r) \leq \epsilon T_{f,\omega}(r) + O(\log r) = \epsilon' T_{f,\mathrm{Ric}(\omega)}(r) + O(\log r)$$

holds for all r outside a set $E \subset (0, +\infty)$ with finite Lebesgue measure. This implies that $T_{f,\mathrm{Ric}(\omega)}(r)$ is bounded, hence f is constant. So there is no non-constant holomorphic map from $\mathbf{C}$ into M if its genus ≥ 2.

We define the defect

$$\delta_f(a) = \liminf_{r \to \infty} \frac{m_f(r, a)}{T_f(r)}.$$

Then, according to our discussion above, we have

Corollary A2.3.4 *(i) If $M = \mathbf{P}^1(\mathbf{C})$, then $\sum_{j=1}^{q} \delta_f(a_j) \leq 2$.*

(ii) If $M = T =$ torus then $\sum_{j=1}^{q} \delta_f(a_j) \leq 0$, in particular every non-constant holomorphic map from $\mathbf{C}$ into T is surjective.

(iii) If genus of M is greater than or equal to 2, then there is no non-constant holomorphic map from $\mathbf{C}$ *into* M.

Proof of Theorem A2.3.2.

Proof. Let

$$\omega = h\frac{\sqrt{-1}}{2\pi}dz \wedge d\bar{z}$$

be a positive (1,1) form on M. We write $f^*\omega = \gamma_f \frac{\sqrt{-1}}{2\pi} d\zeta \wedge d\bar{\zeta}$ on $\mathbf{C}$. Let $D_j = a_j, 1 \leq j \leq q$, be the divisors corresponding to the points $a_j \in M$. The line bundle $L = \mathcal{O}(D_j)$ associated to the divisor D_j is the same for $j = 1, 2, \ldots, q$. Let s_j be the meromorphic section associated with D_j. Let

$$\Gamma = \left(\prod_{j=1}^{q} \frac{1}{||f^*s_j||^2(\log ||f^*s_j||^2)^2} \right) \gamma_f.$$

By Theorem A2.1.2, $dd^c[\log \gamma_f] = dd^c \log \gamma_f + D_{f,\text{ram}} = f^*\text{Ric}(\omega) + D_{f,\text{ram}}$, so

$$\begin{aligned} dd^c[\log \Gamma] &= \sum_{j=1}^{q} -dd^c[\log ||f^*s_j||^2] + f^*\text{Ric}(\omega) + D_{f,\text{ram}} \\ &\quad - \sum_{j=1}^{q} dd^c[\log(\log ||f^*s_j||^2)^2]. \end{aligned}$$

For a fixed number $r_0 > 0$, applying the integral operator

$$\int_{r_0}^{r} \frac{dt}{t} \int_{|\zeta| \leq t}$$

to the identity above and applying Theorem A2.1.3(Green-Jensen's formula), we get

$$\begin{aligned} \frac{1}{2}\int_{|\zeta|=r} (\log \Gamma) d\theta &= \sum_{j=1}^{q} m_f(r, a_j) + \int_{r_0}^{r} \frac{dt}{t} \int_{|\zeta| \leq t} f^*\text{Ric}(\omega) + N_{f,\text{ram}}(r) \\ &\quad - \sum_{j=1}^{q} \int_{r_0}^{r} \frac{dt}{t} \int_{|\zeta| \leq t} dd^c[\log(\log ||f^*s_j||^2)^2] + O(1). \end{aligned}$$

However, it is easy to check using the definition, similar to the proof of (2.3), that

$$\int_{r_0}^r \frac{dt}{t}\int_{|\zeta|\le t} dd^c[\log(\log\|f^*s_j\|^2)^2] = \int_{r_0}^r \frac{dt}{t}\int_{|\zeta|\le t} dd^c\log(\log\|f^*s_j\|^2)^2.$$

Thus

$$\begin{aligned}\frac{1}{2}\int_{|\zeta|=r}(\log\Gamma)d\theta &= \sum_{j=1}^q m_f(r,a_j) + \int_{r_0}^r \frac{dt}{t}\int_{|\zeta|\le t} f^*\mathrm{Ric}(\omega) + N_{f,\mathrm{ram}}(r)\\ &\quad + \sum_{j=1}^q \int_{r_0}^r \frac{dt}{t}\int_{|\zeta|\le t} dd^c\log\left(\frac{1}{\log\|f^*s_j\|^2}\right)^2 + O(1). \qquad (2.9)\end{aligned}$$

By Lemma A2.1.6

$$\begin{aligned}&dd^c\log\left(\frac{1}{\log\|s_j\|^2}\right)^2\\ &= 2\left\{\frac{1}{\|s_j\|^2(\log\|s_j\|^2)^2}dd^c\|s_j\|^2 - \frac{1+\log\|s_j\|^2}{(\log\|s_j\|^2)^2}dd^c\log\|s_j\|^2\right\}.\end{aligned}$$

By definition, $dd^c\log\|s_j\|^2 = -c_1(L)$, and by Lemma A2.2.8, $dd^c\|s_j\|^2 = \|\mathcal{D}s_j\|^2 - \|s_j\|^2c_1(L)$. So

$$\begin{aligned}&dd^c\log\left(\frac{1}{\log\|f^*s_j\|^2}\right)^2\\ &= 2\left\{\frac{\|\mathcal{D}f^*s_j\|^2 - \|f^*s_j\|^2f^*c_1(L)}{\|f^*s_j\|^2(\log\|f^*s_j\|^2)^2} + \frac{1+\log\|f^*s_j\|^2}{(\log\|f^*s_j\|^2)^2}f^*c_1(L)\right\}\\ &= 2\left\{\frac{\|\mathcal{D}f^*s_j\|^2}{\|f^*s_j\|^2(\log\|f^*s_j\|^2)^2} + \frac{f^*c_1(L)}{\log\|f^*s_j\|^2}\right\}. \qquad (2.10)\end{aligned}$$

Recall on the covering $\{U_\alpha\}$ of M, the covariant differential operator is locally defined by $\mathcal{D}s = \partial s_\alpha + s_\alpha\partial(\log h_\alpha)$, for a nonzero holomorphic section $s = \{s_\alpha\}$. Hence

$$\begin{aligned}\|\mathcal{D}s\|^2 &= |\partial s_\alpha + s_\alpha\partial(\log h_\alpha)|^2 h_\alpha\\ &\ge (\frac{1}{2}|\partial s_\alpha|^2 - |s_\alpha\partial(\log h_\alpha)|^2)h_\alpha = \frac{1}{2}\|\partial s\|^2 - \|s\|^2\frac{\sqrt{-1}}{2\pi}A\wedge\bar{A}\\ &= \frac{1}{2}\|\partial s\|^2 - \|s\|^2|A|^2 \qquad (2.11)\end{aligned}$$

where A is the differential form $A = \partial(\log h_\alpha)$. Assume that the zero set of s is without multiplicity, then $||\partial s||^2 + ||s||^2 c_1(L)$ is a positive (1,1) form on M, and because M is compact,

$$\frac{1}{2}||\partial s||^2 + ||s||^2 c_1(L) > c c_1(L) \tag{2.12}$$

where c is a positive constant. Also, since L is positive and M is compact, we have

$$c' c_1(L) \leq |A|^2 \leq c'' c_1(L) \tag{2.13}$$

for some positive constant c' and c''. Combining (2.11), (2.12) and (2.13), we have $||f^*\mathcal{D}s||^2 + ||f^*s||^2 f^*c_1(L) \geq cf^*c_1(L) - c''||f^*s||^2 f^*c_1(L)$, or this simply means,

$$||f^*\mathcal{D}s||^2 \geq cf^*c_1(L) - c'''||f^*s||^2 f^*c_1(L),$$

where $c''' = 1 + c'' > 0$. The above inequality applies for holomorphic sections s_j. That is

$$||f^*\mathcal{D}s_j||^2 \geq cf^*c_1(L) - c'''||f^*s_j||^2 f^*c_1(L). \tag{2.14}$$

Combining (2.10) and (2.14) gives

$$\begin{aligned} dd^c \log\left(\frac{1}{\log||f^*s_j||^2}\right)^2 \geq\ & 2\{\frac{cf^*c_1(L)}{||f^*s_j||^2(\log||f^*s_j||^2)^2} - \frac{c'''f^*c_1(L)}{(\log||f^*s_j||^2)^2} \\ & + \frac{f^*c_1(L)}{\log||f^*s_j||^2}\}. \end{aligned} \tag{2.15}$$

When the metric of L is rescaled by a constant, so that $||s||$ becomes $\lambda||s||$, the covariant differentiation is not affected by the rescaling and $c_1(L)$ is also unchanged. Choose a rescaling such that

$$\frac{c'''}{(\log||s_j||^2)^2} - \frac{1}{\log||s_j||^2} \geq -2\epsilon.$$

Then

$$dd^c \log\left(\frac{1}{\log||f^*s_j||^2}\right)^2 \geq 2\{\frac{cf^*c_1(L)}{||f^*s_j||^2(\log||f^*s_j||^2)^2} - \epsilon f^*c_1(L)\}. \tag{2.16}$$

Since $\omega, c_1(L)$ are both positive and M is compact, there are positive constants b_1, b_2 such that $b_1\omega < c_1(L) < b_2\omega$, so (2.16) can be written as

$$dd^c \log\left(\frac{1}{\log \|f^*s_j\|^2}\right)^2 \geq 2\{\frac{cf^*\omega}{\|f^*s_j\|^2(\log\|f^*s_j\|^2)^2} - \epsilon f^*\omega\} \tag{2.17}$$

for some positive constant c. Thus, using (2.9),

$$\begin{aligned}
\frac{1}{2}\int_{|\zeta|=r}(\log\Gamma)d\theta &= \sum_{j=1}^q m_f(r,a_j) + \int_{r_0}^r \frac{dt}{t}\int_{|\zeta|\leq t} f^*\text{Ric}(\omega) + N_{f,\text{ram}}(r)\\
&\quad + \sum_{j=1}^q \int_{r_0}^r \frac{dt}{t}\int_{|\zeta|\leq t} dd^c \log\left(\frac{1}{\log\|f^*s_j\|^2}\right)^2 + O(1)\\
&\geq \sum_{j=1}^q m_f(r,a_j) + T_{f,\text{Ric}(\omega)}(r) + N_{f,\text{ram}}(r)\\
&\quad -\epsilon\int_{r_0}^r \frac{dt}{t}\int_{|\zeta|\leq t} f^*\omega + O(1).
\end{aligned}$$

Therefore

$$\sum_{j=1}^q m_f(r,a_j) - \epsilon T_{f,\omega}(r) + T_{f,\text{Ric}(\omega)}(r) + N_{f,\text{ram}}(r) \leq \frac{1}{2}\int_{|\zeta|=r}(\log\Gamma)d\theta. \tag{2.18}$$

We now estimate the term on the right-hand side of the above inequality. By the con-cavity of log

$$\int_{|\zeta|=r}(\log\Gamma)d\theta \leq \log\int_{|\zeta|=r}\Gamma d\theta. \tag{2.19}$$

Let

$$T_\Gamma(r) = \int_0^r \frac{dt}{t}\int_{|\zeta|\leq t}\Gamma(\zeta)\frac{\sqrt{-1}}{2\pi}d\zeta\wedge d\bar{\zeta}.$$

Since

$$\frac{1}{r}\frac{d}{dr}(r\frac{dT_\Gamma(r)}{dr}) = 2\int_{|\zeta|=r}\Gamma d\theta,$$

by applying Lemma A2.1.4 by $\phi(t) = t^\epsilon$,

$$\int_{|\zeta|=r}\Gamma d\theta \leq T_\Gamma(r)(T_\Gamma(r))^\epsilon[brT_\Gamma(r)(T_\Gamma(r)^\epsilon]^\epsilon, \tag{2.20}$$

for all $r \geq \gamma$ outside a set of measure $\leq 2c_0$, where $c_0 = \int_e^\infty (1/t^{1+\epsilon})dt$. Combining (2.19) and (2.20) yields that the inequality

$$\int_{|\zeta|=r} (\log \Gamma) d\theta \leq \log \int_{|\zeta|=r} \Gamma d\theta \leq (1+\epsilon) \log T_\Gamma(r) + \log r + O(1)$$

holds for all outside a set $E \subset (0, +\infty)$ with finite Lebesgue measure. $T_\Gamma(r)$ is estimated by using (2.17),

$$\begin{aligned} T_\Gamma(r) &\leq \epsilon T_{f,\omega}(r) + C \sum_{j=1}^q \int_{r_0}^r \frac{dt}{t} \int_{|\zeta|\leq t} dd^c \log[\frac{1}{\log \|f^* s_j\|^2}]^2 \\ &\leq \epsilon T_{f,\omega}(r) + C \sum_{j=1}^q \int_{|\zeta|=r} \left(\frac{1}{\log \|f^* s_j\|^2}\right)^2 \leq \epsilon T_{f,\omega}(r) + O(1), \end{aligned}$$

where, in the above inequality, $C > 0$ is a constant. Thus we have, by (2.19), (2.20) and the above inequality,

$$\int_{|\zeta|=r} (\log \Gamma) d\theta \leq O(\log r + \log T_{f,\omega}(r))$$

where the inequality holds for all outside a set $E \subset (0, +\infty)$ with finite Lebesgue measure. This, together with (2.18), implies that the inequality

$$\sum_{j=1}^q m_f(r, a_j) - \epsilon T_{f,\omega}(r) + T_{f,\mathrm{Ric}(\omega)}(r) + N_{f,\mathrm{ram}}(r) \leq O(\log r)$$

holds for all outside a set $E \subset (0, +\infty)$ with finite Lebesgue measure. The theorem is thus proven. □

Part B: Diophantine Approximation

B2.1 Integral Points on Algebraic Curves

In part A, we consider holomorphic curves in compact Riemann surfaces. According to Vojta's dictionary, a non-constant holomorphic curve corresponds to an infinite set of rational points. In this chapter, we deal with the number of rational (integral) points on M. A compact Riemann surface can be embedded into $\mathbf{P}^n$, so it is also a non-singular projective variety of dimension one. Algebraic geometry and number theory often refer to a non-singular projective variety of dimension one as **an algebraic curve**.

Therefore, the aim of this chapter is to study the number of rational (for compact curves) or integral points (for affine curves) on algebraic curves defined over some number field.

Let k be a number field and let X be an algebraic curve defined over k. Denote by $X(k)$ the set of k-rational points on X. As we discussed in chapter I part B, a number field k has a canonical set of places, denoted by M_k. The set of Archimedean places of k is denoted by M_k^∞, and the non-Archimedean places is denoted by M_k^0. We define the **height** for points on X. We first consider **heights** for points on $\mathbf{P}^n$. Let $\mathbf{x} \in \mathbf{P}^n(k), \mathbf{x} = [x_0 : \ldots : x_n]$, with $x_i \in k$ not all zero. We put

$$H_k(\mathbf{x}) = \prod_{v \in M_k} \max_{0 \leq i \leq n} \|x_i\|_v,$$

and

$$h(\mathbf{x}) = \frac{1}{[k : \mathbf{Q}]} \log H_k(\mathbf{x}). \tag{2.21}$$

By (1.44), if L/k is a finite extension,

$$H_L(\mathbf{x}) = H_k(\mathbf{x})^{[L:k]},$$

so h is independent of the ground field k, and thus extends to $\bar{k}$, the algebraic closure of k. We have thus defined a logarithmic height $h : \mathbf{P}^n(\bar{k}) \to \mathbf{R}$ with values ≥ 0.

Theorem B2.1.1 *Let $f(T) = a_0T^d + a_1T^{d-1} + \cdots + a_d = a_0(T - \alpha_1) \cdots (T - \alpha_d) \in \overline{\mathbf{Q}}(T)$ be a polynomial of degree d defined over k. Then*

$$2^{-d} \prod_{j=1}^{d} H_k(\alpha_j) \leq H_k([a_0 : \cdots : a_d]) \leq 2^{d-1} \prod_{j=1}^{d} H_k(\alpha_j).$$

Proof. First note that the inequality to be proven remains unchanged if $f(T)$ is replaced by $(1/a_0)f(T)$. It thus suffices to prove the result under the assumption that $a_0 = 1$. It is clear that it is enough to prove that, for every $v \in M_k$,

$$\epsilon(v)^{-d} \prod_{j=1}^{d} \max\{\|\alpha_j\|_v, 1\} \leq \max_{0 \leq i \leq d} \{\|a_j\|_v\} \leq \epsilon(v)^{d-1} \prod_{j=1}^{d} \max\{\|\alpha_j\|_v, 1\},$$

where $\epsilon(v) = 2$ if $v \in M_k^\infty$ and $\epsilon(v) = 1$ if $v \in M_k^0$.

The proof is by induction on $d = \deg(f)$. For $d = 1$, the inequality is clear. Assume now the result for all polynomials(with roots in k) of degree $d - 1$. Choose an index μ such that

$$\|\alpha_\mu\|_v \geq \|\alpha_j\|_v \quad \text{for all} \quad 0 \leq j \leq d,$$

and consider the polynomial

$$g(T) = f(T)/(T - \alpha_\mu) = b_0 T^{d-1} + b_1 T^{d-2} + \cdots + b_{d-1}.$$

So, comparing the coefficients yields, for $0 \leq i \leq d$

$$a_i = b_i - \alpha_\mu b_{i-1},$$

where $b_{-1} = b_d = 0$. We now prove the upper bound stated above.

$$\begin{aligned}
\max_{0 \leq i \leq d}\{\|a_j\|_v\} &\leq \max_{0 \leq i \leq d}\{\|b_i - \alpha_\mu b_{i-1}\|_v\} \\
&\leq \epsilon(v) \max_{0 \leq i \leq d}\{\|b_i\|_v, \|\alpha_\mu b_{i-1}\|_v\} \\
&\leq \epsilon(v) \max_{0 \leq i \leq d}\{\|b_i\|_v\} \max\{\|\alpha_\mu\|_v, 1\} \\
&\leq \epsilon(v)^{d-1} \prod_{j=1}^{d} \max\{\|\alpha_j\|_v, 1\}
\end{aligned}$$

where the last inequality holds because the induction hypothesis applies to g.

Next we prove the lower bound. We consider two cases. First, if $\|\alpha_\mu\|_v \leq \epsilon(v)$, then by the choice of the index μ,

$$\prod_{j=1}^{d} \max\{\|\alpha_j\|_v, 1\} \leq \max\{\|\alpha_\mu\|_v, 1\}^d \leq \epsilon(v)^d,$$

so the result is clear. Next, suppose that $\|\alpha_\mu\|_v > \epsilon(v)$, then, for $v \in M_k^0$,

$$\max_{0 \leq i \leq d}\{\|a_j\|_v\} = \max_{0 \leq i \leq d}\{\| \|b_i - \alpha_\mu b_{i-1}\|_v\} \geq \epsilon(v)^{-1} \max_{0 \leq i \leq d-1}\{\|b_i\|_v\}\{\|\alpha_\mu\|_v, 1\}.$$

For $v \in M_k^\infty$, we have, since $\|\alpha_\mu\|_v > \epsilon(v) = 2$,

$$\begin{aligned}
\max_{0 \leq i \leq d}\{\|a_j\|_v\} &= \max_{0 \leq i \leq d}\{\| \|b_i - \alpha_\mu b_{i-1}\|_v\} \\
&\geq (\|\alpha_\mu\|_v - 1) \max_{0 \leq i \leq d-1}\{\|b_i\|_v\}
\end{aligned}$$

$$> \quad \epsilon(v)^{-1}\|\alpha_\mu\|_v \max_{0\le i\le d-1}\{\|b_i\|_v\}.$$

Now applying the induction hypothesis to g gives the desired lower bound, which completes the proof. □

The following theorem is the work of D.G. Northcott [Nor].

Theorem B2.1.2 (Northcott) *Let n, d, N be integers ≥ 1. There are only finitely many points of $\mathbf{P}^n(\overline{\mathbf{Q}})$ of height $\le N$ and of degree $\le d$. Here, for $\mathbf{x} = (x_0, \ldots, x_n) \in \mathbf{P}^n(\overline{\mathbf{Q}})$, the degree of $\mathbf{x}$ is the degree of the field generated by the $x_i/x_j, 0 \le i, j \le n, x_j \ne 0$.*

Corollary B2.1.3 *Given a number field k, there are only finitely many points of $\mathbf{P}^n(k)$ of height $\le N$.*

A **morphism of degree** d between projective spaces is a map

$$F : \mathbf{P}^N \to \mathbf{P}^M, \quad F(P) = [f_0(P) : \ldots : f_M(p)],$$

where $f_0, \ldots, f_M \in \overline{\mathbf{Q}}[X_0, \ldots, X_N]$ are homogeneous polynomials of degree d without common zeros in $\overline{\mathbf{Q}}$ other than $X_0 = \cdots = X_N = 0$. If $f_i, 1 \le i \le M$, have coefficients in k, then F is said to be defined over k.

We have the following property of the height regarding the morphism.

Theorem B2.1.4 *Let*

$$F : \mathbf{P}^N \to \mathbf{P}^M$$

be a morphism of degree d. Then there are constants C_1 and C_2, depending on F, so that for all points $P \in \mathbf{P}^N(\overline{\mathbf{Q}})$,

$$dh(P) + C_1 \le h(F(P)) \le dh(P) + C_2.$$

Proof. This can be directly verified by the definition (see [Lang1], Chapter 4 Theorem 1.8 or [Sil2], Chapter VIII Theorem 5.6 for details). □

For a given algebraic curve X defined over k, let $\phi : X \to \mathbf{P}^n$ be an embedding. For $x \in \bar{k}$, we put $H_\phi(x) = H_k(\phi(x))$, and $h_\phi(x) = h(\phi(x))$. If $\psi : X \to \mathbf{P}^n$ is another embedding, then we have $c_1 h_\phi + c_2 \leq h_\phi(x) \leq c_3 h_\phi + c_4$ on $X(\bar{k})$ for some positive constants $c_1, \ldots, c_4$. Hence h_ϕ and h_ψ have roughly the same size. So we often just denote the height by h. Corollary B2.1.2 tells us that the height $h(\mathbf{x})$ measures the "growth" of $\mathbf{x}$, as $T_f(r)$ does in Nevanlinna theory.

Let D be a divisor on X. D is called **very ample** if there is an embedding $\phi : X \to \mathbf{P}^n$ with $D = \phi^* H$, where H is a hyperplane on $\mathbf{P}^n$. Note that the embedding ϕ can be obtained as follows: Let $f_1, ..., f_N$ be a basis of space of all rational functions over k with $(f) \geq -D$, the map $\phi : V \to \mathbf{P}^N$ given by $\phi(\mathbf{x}) = [f_0(\mathbf{x}) : ... : f_N(\mathbf{x})]$ is a projective embedding. For a very ample divisor, we define

$$h_D(x) = h_\phi(x).$$

Given two divisors D_1 and D_2, we say that D_1 is **rationally equivalent to** D_2 if there is a rational function f on X such that $D_1 = D_2 + (f)$ where (f) is the divisor associated with f, that is $(f) = \sum_{a \in X} ord_a(f)a$. A basic lemma of algebraic geometry tells us that any divisor D is equivalent to divisors $D_1 - D_2$, where D_1 and D_2 are very ample divisors. We then define

$$h_D(x) = h_{D_1}(x) - h_{D_2}(x).$$

$h_D(x)$ is then uniquely determined modulo a bounded function on X.

The height function can also be obtained through the Weil function. We give a short recipe for the construction of the Weil functions on projective varieties (for more details, see [Lang1], Chapter 10). Let X be a non-singular projective variety over k. Let D be a divisor on X. First construct sets of effective divisors $X_i, (i = 1, \ldots, n)$ and $Y_j, (j = 1, \ldots, m)$ such that $D + X_i$ is linearly equivalent to Y_i for every i, j and such that the $X_i, 1 \leq i \leq n$ have point in common and the $Y_j, 1 \leq j \leq m$ also have no point in common. Let $f_{ij} \in \bar{k}(X)$ be such that $(f_{ij}) = Y_j - X_i - D$ for each pair i, j. Extend the valuation to all of $\bar{k}$. For each $x \in X(\bar{k})$ with $x \notin D$, the **Weil function with respect to the divisor** D is then defined by, for each $v \in M_k$,

$$\lambda_{v,D}(x) = \max_j \min_i \log ||f_{ij}(x)||_v.$$

Of course, the Weil function defined above depends on the choice of f_{ij}, but they differ only by a bounded function.

Example *Let D be a hyperplane $D = \{[x_0 : \cdots : x_n] \mid a_0x_0 + \cdots + a_nx_n = 0\}$ in $\mathbf{P}^n(k)$. Take*

$$f_j(\mathbf{x}) = \frac{x_j}{a_0x_0 + \cdots + a_nx_n}, \qquad j = 0, ..., n.$$

Here, using the notation above, X_i is an empty set, $Y_j = [x_j = 0]$. The pole divisor of f_j is precisely D and the zero divisor is $[x_j = 0]$, so $(f_j) = Y_j - D$. For $\mathbf{x} = [x_0 : ... : x_n] \in \mathbf{P}^n(k)$, our Weil function reads

$$\lambda_{v,D}(\mathbf{x}) = \log \frac{\max_j ||x_j||_v}{||a_0x_0 + \cdots + a_nx_n||_v}. \tag{2.22}$$

By the product formula and from (2.21) and (2.22), we have,

$$h(\mathbf{x}) = \frac{1}{[k : \mathbf{Q}]} \sum_{v \in M_k} \lambda_{v,D}(\mathbf{x}).$$

So, given a divisor D on X, if we define, for $x \notin D$,

$$h_D(x) = \frac{1}{[k : \mathbf{Q}]} \sum_{v \in M_k} \lambda_{v,D}(x), \tag{2.23}$$

then we can prove (see [Lang1], chapter 10) that this definition agrees with the definition given earlier, up to a bounded term.

Fix a finite set $S \subset M_k$ containing M_k^∞, we define the proximity function $m(x, D)$ by, for $x \in X(\bar{k})$ with $x \notin D$,

$$m(x, D) = \frac{1}{[k : \mathbf{Q}]} \sum_{v \in S} \lambda_{v,D}(x). \tag{2.24}$$

The counting function $N(x, D)$ is defined by, for $x \in X(\bar{k})$ with $x \notin D$,

$$N(x, D) = \frac{1}{[k : \mathbf{Q}]} \sum_{v \notin S} \lambda_{v,D}(x). \tag{2.25}$$

Note that the sum above is still a finite sum, since the terms all vanish except for finitely many. Combining (2.23), (2.24) and (2.25) we have our First Main Theorem.

Theorem B2.1.5 (First Main Theorem)

$$h_D(x) = m(x, D) + N(x, D) + O(1).$$

Theorem B2.1.6 (Second Main Theorem) *Let k be a number field with its set of canonical places M_k. Let $S \subset M_k$ be a finite set containing all Archimedean places. Let X be a smooth algebraic curve defined over k. Let K be the canonical divisor on X. Then, for any $\epsilon > 0$,*

$$m(x, D) + h_K(x) \leq \epsilon h(x) \tag{2.26}$$

holds for all $x \in X(k)$ except for finitely many points.

In the following sections, we will discuss Theorem B2.1.6 according to the genus of X.

B2.2 Curves of Genus 0

Theorem B2.2.1 *Let X be a smooth curve defined over k. If the genus of X is zero, then X is isomorphic to a conic. If it has a k-rational point, then it is isomorphic to $\mathbf{P}^1(k)$, and thus has infinitely number of k-rational points.*

Proof. As the genus of X is zero, the canonical divisor on X has degree -2. Changing the sign, one obtains a divisor of degree 2. This divisor induces an embedding into the projective space, whose image is of degree 2, hence it is a conic. Thus, since by the assumption $X(k)$ is non-empty, $X(k)$ is isomorphic to $\mathbf{P}^1(k)$, and the canonical divisor K has degree -2. So $h_K(x) = -2h(x)$, thus (2.26) becomes, for $D = \sum_{j=1}^{q} a_j$,

$$\sum_{j=1}^{q} m(x, a_j) - 2h(x) \leq \epsilon h(x).$$

Theorem B2.1.6 in this case is equivalent to Roth's theorem. Theorem B2.1.6, is thus proved for this case. □

B2.3 Rational Points on Curves of Genus 1, Mordell-Weil Theorem

X is called an elliptic curve if X is an algebraic curve of genus 1. An elliptic curve may have infinitely many rational points. In 1901, Poincaré showed that *if a rational point on an elliptic curve X is chosen as an origin, the k-rational points of X form a group.* In 1922, Mordell proved the conjecture of Poincaré that *the group of rational points is finitely generated.* In his paper, Mordell also conjectures that *the set of integral points is finite, and that the set of rational points on a curve of genus ≥ 2 is finite.* The conjecture about integral points was proved by Siegel and the conjecture about rational points on a curve of genus ≥ 2 was recently settled by G. Faltings [Fal1]. In this section, we'll first study rational points on elliptic curves.

Definition B2.3.1 *An elliptic curve is a pair (X, O), where X is an algebraic curve of genus 1 and $O \in X$. (We often just write X for the elliptic curve, the point O being understood.) The elliptic curve X is defined over k, written X/k, if X is defined over k and $O \in X(k)$.*

The main theorem in this section is the following Mordell-Weil Theorem.

Theorem B2.3.2 (Mordell-Weil Theorem) *Let k be a number field, and X be an elliptic curve defined over k, then the group $X(k)$ is finitely generated.*

The proof of the Mordell-Weil theorem is divided into two steps. The first step is to prove the so-called "weak Mordell-Weil Theorem", and the second is the "infinite descent" method using height functions. The first part of this section is devoted to prove the weak Mordell-Weil Theorem.

Theorem B2.3.3 (Weak Mordell-Weil) *Let k be a number field, and X be an elliptic curve defined over k. Let $m \geq 2$ be an integer, then*

$$X(k)/mX(k)$$

is a finite group.

Let X be an elliptic curve defined over k. We denote by $k(X)$ the function field of X. The following theorem shows that every elliptic curve can be written as a plane cubic; and conversely, every smooth Weierstrass plane cubic curve is an elliptic curve.

Theorem B2.3.4 *Let X be an elliptic curve defined over k.*

(a) There exist functions $x, y \in k(X)$, such that the map

$$\phi : X \to \mathbf{P}^2 \qquad\qquad \phi = [x : y : 1]$$

gives an isomorphism of X/k onto a curve given by the Weierstrass equation

$$C : y^2 + a_1 xy + a_3 y = x^3 + a_2 x^2 + a_4 x + a_6 \tag{2.27}$$

with coefficients $a_1, \ldots, a_6 \in k$; and such that $\phi(O) = [0 : 1 : 0]$. (We call x, y Weierstrass coordinates on X.)

(b) Any two Weierstrass equations for X as in (a) are related by a linear change of variables of the form

$$x = u^2 x' + r, \qquad y = u^3 y' + su^2 x' + t$$

with $u, r, s, t \in k, u \neq 0$.

(c) Conversely, every smooth cubic curve C given by a Weierstrass equation as in (a) is an elliptic curve defined over k with origin $O = [0 : 1 : 0]$.

The proof of Theorem B2.3.4 uses the Riemann-Roch theorem. The details can be found in Silverman [Sil2] p.46.

Let X be an elliptic curve given by a Weierstrass equation. Remember that $X \subset \mathbf{P}^2$ consists of the points $P = (x, y)$ satisfying the Weierstrass equation together with the point $O = [0, 1, 0]$ at an infinity. Let $L \subset \mathbf{P}^2$ be a line. Since the Weierstrass equation has degree three, $L \cap X$, taken with multiplicities, consists of three points. Define an addition law for "+" on X by the following rule.

Addition Law: *Let $P, Q \in X$, L be the line connecting P and Q (tangent line to X if $P = Q$), and R be the third point of the intersection of L with X. Let L' be the line connecting R and O. Then $P + Q$ is the point such that L' intersects X at R, O, and $P + Q$.*

In terms of x, y coordinates in the Weierstrass equation, we have the following theorem.

Theorem B2.3.5 (Group Law) *Let X be an elliptic curve given by a Weierstrass equation*

$$X : y^2 + a_1xy + a_3y = x^3 + a_2x^2 + a_4x + a_6.$$

(a) Let $P_0 = (x_0, y_0) \in X$. Then

$$-P_0 = (x_0, -y_0 - a_1x_0 - a_3).$$

(b) Let $P_i = (x_i, y_i) \in X, 1 \le i \le 2$. If $x_1 = x_2$ and $y_1+y_2+a_1x_2+a_3 = 0$, then $P_1 + P_2 = O$.

(c) Otherwise, $P_3 = P_1 + P_2$ is given by

$$x_3 = \lambda^2 + a_1\lambda - a_2 - x_1 - x_2,$$

$$y_3 = -(\lambda + a_1)x_3 - \nu - a_3,$$

where

$$\lambda = \frac{y_2 - y_1}{x_2 - x_1}, \quad \nu = \frac{y_1x_2 - y_2x_1}{x_2 - x_1} \quad \text{if} \quad x_1 \neq x_2;$$

$$\text{and} \qquad \lambda = \frac{3x_1^2 + 2a_2x_1 + a_4 - a_1y_1}{2y_1 + a_1x_1 + a_3},$$

$$\nu = \frac{-x^3 + a_4x_1 + 2a_6 - a_3y_1}{2y_1 + a_1x_1 + a_3} \quad \text{if} \quad x_1 = x_2.$$

Proof. We first verify (a). Let $F(x, y) = y^2 + a_1xy + a_3y - x^3 - a_2x^2 - a_4x - a_6$. To find $-P_0$, we take the line through P_0 and O, and find its third point of intersection with X. The line L is given by:

$$L : x - x_0 = 0.$$

Substituting this into the equation for X, we see that the quadratic polynomial $F(x_0, y)$ has roots y_0 and y_0', where $-P_0 = (x_0, y_0')$. Writing out

$$F(x_0, y) = c(y - y_0)(y - y_0')$$

and comparing coefficients of y^2 gives $c = 1$, and then coefficients of y give $y_0' = -y_0 - a_1x_0 - a_3$. This yields

$$-P_0 = (x_0, -y_0 - a_1x_0 - a_3).$$

(a) is verified.

We now prove (b) and (c). Let $P_i = (x_i, y_i) \in X, 1 \le i \le 2$. If $x_1 = x_2$ and $y_1 + y_2 + a_1x_2 + a_3 = 0$, then from the above formula, $P_1 + P_2 = O$. (b) is verified. Otherwise the line L through P_1 and P_2 (tangent line to X if $P_1 = P_2$) has an equation of the form

$$L : y = \lambda x + \nu.$$

Substituting into the equation for X, we see that $F(x, \lambda x + \nu)$ has roots x_1, x_2, x_3', where $P_3' = (x_3', y_3')$ is the third point of $L \cap X$. By the addition law, $P_1 + P_2 + P_3' = O$, while writing out

$$F(x, \lambda x + \nu) = x(x - x_1)(x - x_2)(x - x_3')$$

and equating coefficients of x^3 and x^2 yields $c = -1$ and $x_1 + x_2 + x_3' = \lambda^2 + a_1\lambda - a_2$. Thus $x_3' = \lambda^2 + a_1\lambda - a_2 - x_1 - x_2$. Substituting back to the equation for X gives $y_3' = \lambda x_3 + \nu$. Note $P_3 = (-x_3', -y_3')$. (c) is proved. □

To prove the weak Mordell-Weil theorem, we start with the following reduction lemma.

Lemma B2.3.6 *Let L/k be a finite Galois extension. If $X(L)/mX(L)$ is finite, then $X(k)/mX(k)$ is also finite.*

Proof. Let Φ be the kernel of the natural map $X(k)/mX(k) \to X(L)/mX(L)$. Thus

$$\Phi = (X(k) \cap mX(L))/mX(k),$$

so for each point P (mod $mX(k)$) in Φ, there is a point $Q_P \in X(L)$ with $[m]Q_P = P$, where $[m]$ is the multiplication-by-m map, i.e. $[m]Q_P = Q_P + \ldots + Q_P$ (m terms). Denote by $X[m]$ the m-torsion subgroup of X, that is the set of points of order m in X,

$$X[m] = \{P \in X(\bar{k}) \mid [m]P = O\}. \tag{2.28}$$

Denote by $G_{L/k}$ the Galois group. We define a map of sets

$$\lambda_P : G_{L/k} \to X[m], \quad \lambda_P(\sigma) = Q_P^\sigma - Q_P,$$

where Q_P^σ is the point obtained by acting σ to the coordinates of Q_P. We note that $Q_P^\sigma - Q_P \in X[m]$, since $[m](Q_P^\sigma - Q_P) = ([m]Q_P)^\sigma - [m]Q_P = P^\sigma - P = O$.

Suppose now that $\lambda_P = \lambda_{P'}$ for two points $P, P' \in X(k) \cap mX(L)$. Then

$$(Q_P - Q_{P'})^\sigma = Q_P - Q_{P'}, \quad \text{for all } \sigma \in G_{L/k},$$

so $Q_P - Q_{P'} \in X(k)$. Therefore

$$P - P' = [m]Q_P - [m]Q_{P'} \in mX(k),$$

so $P \equiv P' \pmod{mX(k)}$. This proves that the association

$$\Phi \to \mathrm{Map}(G_{L/k}, X[m]), \qquad P \to \lambda_P,$$

is one-to-one. But $G_{L/k}$ and $X[m]$ are finite sets, so there are only a finite number of maps between them. Therefore Φ is finite.

Finally, the exact sequence

$$0 \to \Phi \to X(k)/mX(k) \to X(L)/mX(L)$$

nests $X(k)/mX(k)$ between two finite groups, so it, too, is finite. □

Lemma B2.3.6 tells us that it suffices to prove the weak Mordell-Weil theorem under the additional assumption that

$$X[m] \subset X(k). \tag{2.29}$$

For the remainder of this section we will assume, without further comment, that this inclusion is true.

Definition B2.3.7 *The Kummer paring*

$$\kappa : X(k) \times G_{\bar{k}/k} \to X[m]$$

is defined as follows. Let $P \in X(k)$, and choose any $Q \in X(\bar{k})$ satisfying $[m]Q = P$. Then

$$\kappa(P, \sigma) = Q^\sigma - Q.$$

Let C be a curve defined over a field k, using $k(C)$ to denote the function field of C.

Proposition B2.3.8 *(a) The Kummer pairing is well-defined.*

(b) The Kummer pairing is bilinear.
(c) The kernel of the Kummer pairing on the left is $mX(k)$.
(d) The kernel of the Kummer pairing on the right is $G_{\bar{k}/L}$, where

$$L = k([m]^{-1}X(k))$$

is the compositum of all fields $k(Q)$ as Q ranges over the points of $X(k)$ satisfying $[m]Q \in X(k)$.

Hence the Kummer pairing induces a perfect bilinear paring

$$X(k)/mX(k) \times G_{L/k} \to X[m],$$

where L is the field given in (d).

Proof. (a) We must show that $\kappa(P, \sigma)$ is in $X[m]$ and does not depend on the choice of Q. For the former,

$$[m]\kappa(P, \sigma) = [m]Q^{\sigma} - [m]Q = P^{\sigma} - P = O,$$

since $P \in X(k)$ and σ fixes k. For the latter, note that any other choice has the form $Q + T$ for some $T \in X[m]$. Then

$$(Q+T)^{\sigma} - (Q+Y) = Q^{\sigma} + T^{\sigma} - Q - T = Q^{\sigma} - Q,$$

because by assumption (2.29) $X[m] \subset X(k)$, so σ fixes T.

(b) The linearity in P is obvious. For the other side, let $\sigma, \tau \in G_{\bar{k}/k}$. Then

$$\kappa(P, \sigma\tau) = Q^{\sigma\tau} - Q = (Q^{\sigma} - Q)^{\tau} + Q^{\tau} - Q = \kappa(P, \sigma)^{\tau} + \kappa(P, \tau).$$

But $\kappa(P, \sigma) \in X[m]$ is contained in $X(k)$ by (2.29), so it is fixed by τ.

(c) Suppose $P \in mX(k)$, say $P = [m]Q$ with $Q \in X(k)$. Then any $\sigma \in G_{\bar{k}/k}$ fixes Q, so

$$\kappa(P, \sigma) = Q^{\sigma} - Q = O.$$

Conversely, suppose $\kappa(P, \sigma) = O$ for all $\sigma \in G_{\bar{k}/k}$. Thus choosing $Q \in X(\bar{k})$ with $[m]Q = P$, we have

$$Q^{\sigma} = Q \quad \text{for all} \quad \sigma \in G_{\bar{k}/k}.$$

Therefore, $Q \in X(k)$, so $P = [m]Q \in mX(k)$.

(d) Suppose $\sigma \in G_{\bar{k}/L}$. Then

$$\kappa(P, \sigma) = Q^{\sigma} - Q = O,$$

since $Q \in X(L)$ from the definition of L. Conversely, suppose $\sigma \in G_{\bar{k}/k}$ and $\kappa(P, \sigma) = O$ for all $P \in X(k)$. Then for every $Q \in X(\bar{k})$ satisfying $[m]Q \in mX(k)$,

$$O = \kappa([m]Q, \sigma) = Q^{\sigma} - Q.$$

But L is the composition of $k(Q)$ over all such Q, so σ fixes L. Hence $\sigma \in G_{\bar{k}/L}$.

Finally, the last statement of the Proposition is clear from what precedes it, once we note that L/k is Galois because $G_{\bar{k}/k}$ takes $[m]^{-1}X(k)$ to itself. This finishes the proof. □

Using Proposition B2.3.8, we have that $X(k)/mX(k)$ is finite if and only if $G_{L/k}$ is finite. Thus the finiteness of $X(k)/mX(k)$ is equivalent to the finiteness of the extension L/k. The next step is to analyze this extension.

To do so, we need to consider the "reduction of elliptic curve over local fields." First we recall the theory of valuations. Recall that M_k is the canonical set of places over k, M_k^{∞} is the set of Archimedean places and M_k^0 is the set of non-Archimedean places. Fix a $v \in M_k^0$. We define

$$v(x) = -\log \|x\|_v.$$

Then v is a function on k with the following properties: (a) v is a real number for any $x \neq 0$, while $v(0) = \infty$, (b) $v(x+y) \geq \min\{v(x), v(y)\}$, (c) $v(xy) = v(x) + v(y)$. Such a function is called a **valuation** on k. It is clear that the values taken on k form a subgroup Γ of the additive group of real numbers. We denote

$$\mathcal{O}_v = \{x \in k \mid v(x) \geq 0\}.$$

Then $\mathcal{O}_v$ is a ring with group of units

$$\mathcal{O}_v^* = \{x \in k \mid v(x) = 0\},$$

and the unique maximal ideal

$$\mathcal{P}_v = \{x \in k \mid v(x) > 0\}.$$

$\mathcal{O}_v$ is an integral domain with field of fractions k and has the property that for every $x \in k^*$, either $x \in \mathcal{O}_v$ or $x^{-1} \in \mathcal{O}_v$. Such ring is called a **valuation ring**. Its only maximal ideal is $\mathcal{P}_v$. In general, let k be a field, then there is a natural bijection between valuation rings on k and equivalence classes of general valuations on k. The field $\mathcal{O}_v/\mathcal{P}_v$ is called the **residue class field**. Let L be an algebraic extension of k, and we now consider the extensions of v to L. Let w be an extension of v to L. We denote by Γ the precise value group of v; by $\mathcal{O}_v$ the valuation ring in k and $\mathcal{P}_v$ its maximal ideal; and the residue class field $\mathcal{O}_v/\mathcal{P}_v$ will be written by k'. The corresponding objects for w are denoted by Δ, $\mathcal{O}_w$, $\mathcal{P}_w$ and L'. Then we have

$$\mathcal{P}_v = \mathcal{O}_v \cap \mathcal{P}_w \qquad \text{and} \qquad k' \subset L'.$$

Thus we may regard L' as an extension of k'. The degree $f = [L' : k']$ is called the **residue degree** of the extension L/k. Further, Γ is a subgroup of Δ and the index $e = (\Delta : \Gamma)$ is called the **ramification index** of the extension L/k. We have $ef \leq [L : k]$. Moreover, if k is complete with respect to v, then $ef = [L : k]$. The extension is said to be **ramified** if $e > 1$, and **unramified** otherwise. Since the problem we will consider is a local in v, we may assume that k is complete under v. In this case the extension of v to L is unique, i.e.,

$$w(\alpha) = \frac{1}{[L : k]} v(N_{L/k}(\alpha)), \qquad \text{for all} \quad \alpha \in L$$

where N denotes the norm.

We now describe a theorem of determining whether L is unramified. We first note that the residue class leads to yet another way of viewing valuations. Let k, F be two fields; by a **place** of k in F we mean a map $\phi : k \to F \cup \{\infty\}$ such that ϕ restricted to $\phi^{-1}(F)$ is a ring homomorphism and $\phi(x) = \infty$ implies that $x \neq 0$ and $\phi(x^{-1}) = 0$. There is a natural bijection between the isomorphism of places on a field k and the valuation rings in k(so sometimes we do not distinguish them). In our case, let $k' = \mathcal{O}_v/\mathcal{P}_v$ be the residue class field of k and let us write $x \mapsto \bar{x}$ for the natural homomorphism and define $\phi_0 : k \to k' \cup \{\infty\}$ by $\phi(x) = \bar{x}$ if $x \in \mathcal{O}_v$ and $\phi(x) = \infty$ otherwise. ϕ_0 is called **the canonical place**. It is clear that ϕ_0 and the valuation v are defined uniquely by each other. In fact, we have $\mathcal{O}_v = \phi_0^{-1}(k')$. So, the extension of v to L can be studied through the extension of the corresponding residue fields. Let ϕ_0 be the canonical k'-valued place on k. We assume that ϕ_0 is extended in a fixed way to the algebraic closure k^a of k, and we call this extension ϕ. Since L is finite over

k_v, L' is finite over k', and thus ϕ is k'^a-valued, where k'^a is the algebraic closure of k'. We have the following theorem.

Theorem B2.3.9 *Assume that k is complete under v. Let L be a finite extension of k with $[L:k]=n$. Then L is unramified if and only if ϕ_0 has at least n distinct extensions to places of L (in the given algebraic closure of k'), and in that case, it has exactly n.*

Proof. By uniqueness, all extensions of ϕ_0 to L are conjugate, and the number of conjugates is equal to the separable degree of L' over k'. As $[L':k'] \leq [L:k] = n$ our assertion is immediate. □

Back to our proof of the Mordell-Weil Theorem. Remember we want to prove that $L = k([m]^{-1}X(k))$ is a finite extension of k. To continue, we have the following proposition:

Proposition B2.3.10 *Let*

$$L = k([m]^{-1}X(k))$$

be the field defined in Proposition B2.3.8.

(a) L/k is an abelian extension of exponent m (i.e. $G_{L/k}$ is abelian and every element has order dividing m).

(b) There is a finite set S containing M_k^∞ such that L/k is unramified outside S.

Proof. (a) This follows immediately from Proposition B2.3.8, which implies that there is an injection

$$G_{L/k} \to \mathrm{Hom}(X(k), X[m]) \qquad \sigma \to \kappa(\cdot, \sigma).$$

(b) Suppose that $X : y^2 + a_1xy + a_3y = x^3 + a_2x^2 + a_4x + a_6$, where $a_1, \ldots, a_6$ are algebraic integers. For almost all $v \in M_k^0$, we can define an (smooth) elliptic curve $\bar{X} = X \bmod \mathcal{P}_v$, by the equation

$$y^2 + \bar{a}_1xy + \bar{a}_3y = x^3 + \bar{a}_2x^2 + \bar{a}_4x + \bar{a}_6,$$

where $\bar{a}_1, \ldots, \bar{a}_6 \in \mathbf{R}_k/\mathcal{P}_v$ are $a_j \bmod \mathcal{P}_v$, $1 \leq j \leq 6$, where $\mathbf{R}_k$ is the ring of algebraic integers of k. The algebraic formulas for addition and division of points can then be reduced mod $\mathcal{P}_v$, and give addition and division on $\bar{X}$. The map of reduction mod $\mathcal{P}_v$

$$P \to \bar{P}$$

clearly sends $X[m]$ to $\bar{X}[m]$. We claim that, for almost all $v \in M_k^0$ (all but a finite number), this gives an isomorphism $X[m] \to \bar{X}[m]$. To prove this claim, let $F_m(X)$ be the monic polynomial whose roots are x-coordinates of the points in $X[m]$. Because the group law on $\bar{X}$ is obtained by reduction mod $\mathcal{P}_v$ of the group law of X, it follows that the points $\bar{P}$ in $\bar{X}[m]$ are such that their x-coordinates are also roots of the reduced equation

$$\bar{F}_m(X) = 0.$$

If

$$F_m(X) = \sum \alpha_j X^j$$

with $\alpha_j \in k$, then by definition,

$$\bar{F}_m(X) = \sum \bar{\alpha}_j X^j.$$

For almost all(but a finite number) v, the polynomial $\bar{F}_m$ has the same degree as F_m, and has distinct roots. This gives rise to an injection

$$x(P) \to \overline{x}(P)$$

on the x-coordinates of points in $X[m]$, $P \neq 0$, whence reduction mod $\mathcal{P}_v$

$$P \to \bar{P}$$

induces an isomorphism $X[m] \to \bar{X}[m]$. This verifies the claim.

Let S be the set that, for all $v \notin S$, $X[m] \to \bar{X}[m]$ is an isomorphism. Let $v \notin S$. We will show that L/k is unramified over v. Since our statement is local in v, we may assume that k is complete under our valuation. Let $Q \in X(\bar{k})$ satisfying $[m]Q \in X(k)$, and let $k' = k(Q)$. It suffices to show that k'/k is unramified over v, since L is the composition of all such k'. Let $n = [k' : k]$ then there are n elements $a_1, \ldots, a_n$ of $X[m]$ such that the automorphisms $\sigma_i (i = 1, \ldots, n)$ satisfying

$$\sigma_i P = P + a_i$$

give all automorphisms of k' over k. Since $X[m] \to \bar{X}[m]$ is an isomorphism, $\bar{a}_i \in \bar{X}[m] (i = 1, \ldots, n)$ are distinct. If ϕ is a place of k' extending the canonical place of v, then $\phi\sigma_i$ are also places of k' and

$$\phi\sigma_i(P) = \overline{(P + a_i)} = \bar{P} + \bar{a}_i.$$

Hence $\phi\sigma_i$ are distinct. Theorem B2.3.9 implies that k' is unramified over v. □

The next Proposition is concerned with the property (a) of L, that is L/k is an abelian extension of exponent m.

Proposition B2.3.11 *Assume that the m-th roots of unity lie in k. Then there is a subgroup B of k^* containing $(k^*)^m$ such that $L = k(B^{1/m})$, where $(k^*)^m = \{a^m \mid a \in k^*\}$.*

Proof. Denote by μ_m the set of mth-roots of unity. Let $B = (L^*)^m \cap k^*$. We claim that $L = k(B^{1/m})$. In fact, if $x \in B^{1/m}$, and $x^m = \alpha^m = a \in k^*$, $\alpha \in L^*$, then $x = \xi\alpha \in L^*$, where x is an m-th root of unity which lies in k. So $k(B^{1/m}) \subset L$. On the other hand, the extension L/k is the composite of its cyclic sub-extensions of exponents m because it is the composition of its finite sub-extensions and the Galois group of a finite sub-extension is the product of cyclic groups, which may be interpreted as Galois groups of cyclic sub-extensions. Let now M/k be a cyclic sub-extension of L/k. It suffices to show that $M \subset k(B^{1/m})$. Let σ be a generator of $G_{M/k}$ where $G_{M/k}$ is the Galois group, and ζ a generator of the group μ_m. Let $[M : k] = d$, $\xi = \zeta^{n/d}$. We now need the Galois theory (**Hilbert's Theorem 90**) which says if M/k is a cyclic field extension of k, then for any element $\alpha \in M^*$ of norm $N_{M/k}(\alpha) = 1$ is of the form $\alpha = \beta^{\sigma-1}$, where $\beta \in M^*$ and σ is a generator of $G_{M/k}$. Here the norm $N_{M/k}(\alpha)$ is a map $M^* \to k^*$ defined by

$$N_{M/k}(\alpha) = \prod_{\sigma \in G_{M/k}} \alpha^\sigma, \qquad \alpha \in M^*.$$

In our case, $\xi = \zeta^{n/d}$, so $N_{M/k}(\xi) = \xi^d = \zeta^n = 1$, so Hilbert's Theorem 90 implies that $\xi = \beta^{\sigma-1}$ for some $\beta \in M^*$. Thus $k \subset k(\alpha) \subset M$. But $\beta^{\sigma^i} = \xi^i\beta$. Thus $\beta^{\sigma^i} = \beta$ is equivalent to $i \equiv 0 \bmod d$, so $k(\beta) = M$. But $(\beta^m)^{\sigma-1} = (\beta^{\sigma-1})^m = \xi^m = 1$, so that $a = \beta^m \in k^*$; then $\beta \in k(B^{1/m})$. Therefore, $M \subset k(B^{1/m})$. □

To complete the proof of the weak Mordell-Weil theorem, all that remains is to show that any field extension L/k satisfying the conditions in Proposition B2.3.10 is necessarily a finite extension. The proof of this fact relies upon the two fundamental finiteness theorems of algebraic number theory, namely the finiteness of the ideal class group and the finiteness of the group of S-units.

Proposition B2.3.12 *L/k is a finite extension.*

Proof. Let k be the given number field, and $S \subset M_k$ be a finite set of places containing M_k^∞, which appeared in Proposition B2.3.10.

Suppose that the proposition is true for some finite extension k' of k, where S' is the set of places of k' lying over S. Then L'/k', being abelian of exponent m unramified outside S', is finite; and so L/k is also finite. It thus suffices to prove that proposition under the assumption that k contains the mth-roots of unity μ_m.

Similarly, we may increase the set S, since this only has the effect of making L larger. Using the fact that the class number of k is finite, we can thus add a number of elements so that the ring of S-integers

$$R_S = \{a \in k \mid v(a) \geq 0 \text{ for all } v \notin S\}.$$

is a principal ideal domain. We may also enlarge S so that $v(m) = 0$ for all $v \notin S$.

By Proposition B2.3.11, there is a subgroup B of k^* containing $(k^*)^m$ such that,

$$L = k(a^{1/m} : a \in B),$$

and by Proposition B2.3.10, L is unramified outside S. We may assume L/k is the maximal abelian extension of k having exponent m which is unramified outside S. That is L is the largest subfield of $k(a^{1/m} : a \in k)$ which is unramified outside S.

Let $v \in M_k, v \notin S$. For every $a \in B$, looking at the equation

$$X^m - a = 0$$

over the local field k_v, and remembering that $v(m) = 0$, it can be shown that $k(a^{1/m})$ is unramified over v if and only if $ord_v(a)$ is divisible by m (see, [Lang1] Chapter 6, Proposition 1.3). Now when adjoining m^{th}-roots, it is only necessary to take one representative for each class in $k^*/(k^*)^m$. We conclude that

$$L = k(a^{1/m} : a \in T_S),$$

where

$$T_S = \{a \in k^*/(k^*)^m \mid \text{ord}_v a \equiv 0 \pmod{m} \text{ for all } v \notin S\}.$$

To finish the proof, it thus suffices to show that the set T_S is finite.

Consider the natural map

$$R_S^* \to T_S.$$

We claim that it is surjective. To see this, suppose $a \in k^*$ represents an element of T_S. Then the ideal aR_S is the mth-power of an ideal in R_S, since the prime ideals of R_S correspond to the valuations $v \notin S$. Since R_S is a principal ideal domain, there is $b \in k^*$ so that $aR_S = b^m R_S$. Hence there is $u \in R_S^*$ so that

$$a = ub^m.$$

Then a and u give the same element of T_S, so R_S^* surjects onto T_S. Now the kernel of this map certainly contains $(R_S^*)^m$, so we have a surjection

$$R_S^*/(R_S^*)^m \to T_S.$$

But Dirichlet's S-unit theorem [Lang3, V§1] says that R_S^* is finitely generated, so this proves that T_S is finite, and thereby completes the proof of the proposition. □

Proposition B2.3.8, Proposition B2.3.10, and Proposition B2.3.12 are now combined to give our proof of the weak Mordell-Weil theorem as follows.

Proof of the weak Mordell-Weil theorem:

Proof. Let $L = k([m]^{-1}X(k))$ be the field defined in Proposition B2.3.10. Since $X[m]$ is finite, the perfect pairing given in Proposition B2.3.8 shows that $X(k)/mX(k)$ is finite if and only if $G_{L/k}$ is finite. Thus $X(k)/mX(k)$ is finite if and only if L is a finite extension of k. Now Proposition B2.3.10 shows that L has certain properties, and Proposition B2.3.12 shows that any extension of k with those properties is a finite extension. This proves that L is a finite extension of k. □

Remember that our goal in this section is to prove that $X(k)$, the group of k-rational points on the elliptic curve X, is finitely generated. So far we have shown that $X(k)/mX(k)$ is finite. It is easy to see that this is not sufficient. For example, $\mathbf{R}/m\mathbf{R} = 0$ for every integer $m \geq 1$, but the set of real numbers $\mathbf{R}$ is certainly not finitely generated. The second step is the "infinite decent" method using height functions. We will first need to investigate the height properties on elliptic curves.

Let X be an elliptic curve defined over k. Any non-constant function $f \in \bar{k}(X)$ determines a surjective morphism (which we also denote by f) $f : X \to \mathbf{P}^1(\bar{k})$ with $f(P) = [1 : 0]$ if P is a pole of f, and $f(P) = [f(P) : 1]$ otherwise. Define

$$h_f(P) = h(f(P)).$$

Corollary B2.1.4 gives us the following theorem.

Theorem B2.3.13 *Let X be an elliptic curve defined over k, and $f \in \bar{k}(X)$ be a non-constant function. Then for any constant C,*

$$\{P \in X(k) \mid h_f(P) \leq C\}$$

is a finite set.

The next theorem gives a fundamental relationship between height function and the addition law on elliptic curves.

Theorem B2.3.14 *Let X be an elliptic curve defined over k, and $f \in k(X)$ be an even function (i.e. $f \circ [-1] = f$). Then for all $P, Q \in X(\bar{k})$,*

$$h_f(P+Q) + h_f(P-Q) = 2h_f(P) + 2h_f(Q) + O(1),$$

where the constants inherent in the $O(1)$ depend on X and f, but are, of course, independent of P and Q.

Proof. Choose a Weierstrass equation for X of the form

$$X : y^2 = x^3 + ax + b.$$

We start by proving the theorem for a particular function $f = x$.

Since $h_x(O) = 0$ and $h_x(-P) = h_x(P)$, the result clearly holds if $P = O$ or $Q = O$. We now assume that $P, Q \neq O$, and write

$$x(P) = [x_1 : 1], \quad x(Q) = [x_2 : 1], \quad x(P+Q) = [x_3 : 1], \quad x(P-Q) = [x_4 : 1].$$

Here, x_3 or x_4 may equal to ∞ if $P = Q$ or $P = -Q$. From the group law(Theorem B2.3.5), one gets

$$x_3 + x_4 = \frac{2(x_1 + x_2)(a + x_1x_2) + 4b}{(x_1 + x_2)^2 - 4x_1x_2},$$

$$x_3x_4 = \frac{(x_1x_2 - a)^2 - 4b(x_1 + x_2)}{(x_1 + x_2)^2 - 4x_1x_2}.$$

Define a map $g : \mathbf{P}^2 \to \mathbf{P}^2$ by

$$g([t : u : v]) = [u^2 - 4tv : 2u(at + v) + 4bt^2 : (v - at)^2 - 4btu].$$

Then the formula for x_3 and x_4 shows that there is a commutative diagram

$$\begin{array}{ccccccccccc}
 & X & \times & X & \stackrel{G}{\to} & X & \times & X & \\
 & & \downarrow & & & & \downarrow & & \\
\sigma & \mathbf{P}^1 & \times & \mathbf{P}^1 & & \mathbf{P}^1 & \times & \mathbf{P}^1 & \sigma \\
 & & \downarrow & & & & \downarrow & & \\
 & & \mathbf{P}^2 & & \stackrel{g}{\to} & & \mathbf{P}^2 & &
\end{array}$$

where $G(P, Q) = (P + Q, P - Q)$, and the vertical map σ is the composition of the two maps

$$X \times X \to \mathbf{P}^1 \times \mathbf{P}^1 \qquad (P, Q) \to (x(P), x(Q))$$

and

$$\mathbf{P}^1 \times \mathbf{P}^1 \to \mathbf{P}^2 \qquad ([\alpha_1 : \beta_1], [\alpha_2 : \beta_2]) \to [\beta_1\beta_2 : \alpha_1\beta_2 + \alpha_2\beta_1 : \alpha_1\alpha_2].$$

We note that the idea here is to treat t, u, v as $1, x_1 + x_2, x_1x_2$. Then $g([t : u : v]) = [1 : x_3 + x_4 : x_3x_4]$.

The next step is to show that g is a morphism, and this means we must show that except for $t = u = v = 0$, the three homogeneous polynomials defining g have no common zeros. Suppose now that $g([t : u : v]) = [0 : 0 : 0]$. If $t = 0$, then from

$$u^2 - 4tv = 0 \quad \text{and} \quad (v - at)^2 - 4btu = 0,$$

we see that $u = v = 0$. Thus we may assume that $t \neq 0$, and so it makes sense to define a new quantity $x = u/2t$. Notice that the equation

$u^2 - 4tv = 0$ can be written as $x^2 = v/t$. Now dividing the quantities

$$2u(at+v) + 4bt^2 = 0 \quad \text{and} \quad (v-at)^2 - 4btu = 0,$$

by t^2 and rewriting them in terms of x yields the two equations

$$\psi(x) = 4x(a+x^2) + 4b = 4x^3 + 4ax + 4b = 0,$$

and

$$\psi(x) = (x^2-a)^2 - 8bx = x^4 - 2ax^2 - 8bx + a^2 = 0.$$

To show that $\psi(X)$ and $\phi(X)$ have no common root, we use the formal identity(it is verified directly)

$$(12X^2+16a)\phi(X) - (3X^3 - 5aX - 27b)\psi(X) = 4(4a^3+27b^2) \neq 0.$$

Since the elliptic curve is non-singular, the discriminant $4a^3 + 27b^2 \neq 0$, so

$$(12X^2+16a)\phi(X) - (3X^3 - 5aX - 27b)\psi(X) = 4(4a^3+27b^2) \neq 0.$$

Thus $\psi(X)$ and $\phi(X)$ have no common root. This completes the proof that g is a morphism.

We return to our commutative diagram, and compute

$$\begin{aligned} h(\sigma(P+Q, P-Q)) &= h(\sigma \circ G(P,Q)) \\ &= h(g \circ \sigma(P,Q)) \\ &= 2h(\sigma(P,Q)) + O(1), \end{aligned} \tag{2.30}$$

using Theorem B2.1.4 that for any morphism g of degree 2, $h(g\circ\sigma(P,Q)) = 2h(\sigma(P,Q)) + O(1)$.

Now to complete the proof for $f = x$, we will show that for all $R_1, R_2 \in X(\bar{k})$ there is a relation

$$h(\sigma(R_1,R_2)) = h_x(R_1) + h_x(R_2) + O(1). \tag{2.31}$$

One immediately verifies that if either $R_1 = O$ or $R_2 = O$, then $h(\sigma(R_1,R_2)) = h_x(R_1) + h_x(R_2)$. Otherwise, we may write

$$x(R_1) = [\alpha_1 : 1], \quad x(R_2) = [\alpha_2 : 1],$$

and so

$$h(\sigma(R_1,R_2)) = h([1:\alpha_1+\alpha_2:\alpha_1\alpha_2]) \text{ and } h_x(R_1)+h_x(R_2) = h(\alpha_1)+h(\alpha_2).$$

Applying Theorem A2.1.1 to $f(T) = (T+\alpha_1)(T+\alpha_2)$ we obtain the desired estimate

$$h(\alpha_1) + h(\alpha_2) - \log 4 \le h([1 : \alpha_1 + \alpha_2 : \alpha_1\alpha_2]) \le h(\alpha_1) + h(\alpha_2) + \log 2.$$

Thus (2.31) is verified. This completes the proof for the case of $f = x$.

For the general even function f, we still denote x, y as the Weierstrass coordinates for X. Thus subfield $k(X)$ which consists of even functions is exactly $k(x)$, so we can find a rational function $\rho : \mathbf{P}^1 \to \mathbf{P}^1$ such that $f = \rho \circ x$. Hence by Theorem B2.1.4 and the fact that ρ is morphism,

$$h_f = h_x \circ \rho = (\deg\rho)h_x + O(1).$$

This proves the general case. □

Corollary B2.3.15 *Let X be an elliptic curve defined over k and $f \in k(X)$ be an even function.*

(a) Let $Q \in X(\bar{k})$. Then for all $P \in X(\bar{k})$,

$$h_f(P+Q) \le 2h_f(P) + O(1),$$

where $O(1)$ depends only on X, f and Q.

(b) Let m be an integer. Then for all $P \in X(\bar{k})$,

$$h_f([m]P) = m^2 h_f(P) + O(1),$$

where $O(1)$ depends only on X, f and m.

Proof. (a) follows Theorem B2.3.14 since $h_f(P-Q) \ge 0$.

(b) Since f is even, it suffices to consider $m \ge 0$. Further the result is trivial for $m = 0$. We finish the proof by induction. Assume it is known for $m-1$ and m. Applying Theorem B2.3.14 to $[m]P$ and P, we have

$$h_f([m+1]P) = -h_f([m-1]P) + 2h_f([m]P) + 2h_f(P) + O(1). \qquad (2.32)$$

By the induction hypothesis, $h_f([m-1]P) = (m-1)^2 h_f(P) + O(1)$ and $h_f([m]P) = m^2 h_f(P) + O(1)$. This together with (2.32) implies that

$$h_f([m+1]P) = (m+1)^2 h_f(P) + O(1).$$

□

We now complete the proof of the Mordell-Weil Theorem by going through the descent procedure.

Proof. By the weak Mordell-Weil theorem, for integer $m \geq 2$

$$X(k)/mX(k)$$

is finitely. Choose elements $Q_1, \ldots, Q_r \in X(k)$ to represent the finitely many co-sets in $X(k)/mX(k)$. Now let $P \in X(k)$. The idea is to show that by subtracting an appropriate linear combination of $Q_1, \ldots, A_r$ from P, we will be able to make the height of $Q_1, \ldots, Q_r$ less than a constant which is independent of P. Then $Q_1, \ldots, Q_r$ and the finitely many points with height less than this constant will generate $X(k)$.

Write

$$P = mP_1 + Q_{i_1}, \qquad \text{for some } 1 \leq i_1 \leq r.$$

Continuing this fashion,

$$P_1 = mP_2 + Q_{i_2},$$

$$\cdots$$

$$P_{n-1} = mP_n + Q_{i_n}.$$

Choose any even, non-constant function $f \in k(X)$, for example the x-coordinate function on a Weierstrass equation. We just write $h_f(P)$ as $h(P)$, since f will no longer be used. Now for any j, we have, by Corollary B2.3.15,

$$\begin{aligned} h(P_j) &\leq \frac{1}{m^2}[h(mP_j) + C_2] \\ &= \frac{1}{m^2}[h(P_{j-1} - Q_{i_j}) + C_2] \\ &\leq \frac{1}{m^2}[2h(P_{j-1}) + C_1' + C_2], \end{aligned}$$

where we take C_1' to be the maximum of the constants from (a) of Corollary B2.3.15 for $Q = -Q_i$, $1 \leq i \leq r$. Note that C_1' and C_2 do not depend on P.

Now we use the above inequality, starting from P_n and working back to P. This yields

$$\begin{aligned} h(P_n) &\le \left(\frac{2}{m^2}\right)^n h(P) + \left[\frac{1}{m^2} + \frac{2}{m^4} + \cdots + \frac{2^{n-1}}{m^{2n}}\right](C_1' + C_2) \\ &< \left(\frac{2}{m^2}\right)^n h(P) + \frac{C_1' + C_2}{m^2 - 2} \\ &\le 2^{-n}h(P) + (C_1' + C_2)/2. \end{aligned}$$

It follows that by taking n large enough, we will have

$$h(P_n) \le 1 + (C_1' + C_2)/2.$$

Since

$$P = m^n P_n + \sum_{j=1}^{n} m^{j-1} Q_{i_j},$$

it follows that every $P \in X(k)$ is a linear combination of the points in the set

$$\{Q_1, \ldots, Q_r\} \cup \{Q \in X(k) \ | h(Q) \le 1 + (C_1' + C_2)/2\}.$$

From Theorem B2.1.1, this is a finite set, which proves that $X(k)$ is finitely generated. □

B2.4 Integral Points on Curves of Genus 1, Siegel's Theorem

An elliptic curve may have infinitely many rational points, although the Mordell-Weil theorem at least assures us that the group of rational points is finitely generated. However, Siegel proved that there will only be a finite number of integral points on an affine elliptic curve.

Theorem B2.4.1 (Siegel) *Let k be a number field with its set of canonical places M_k. Let $S \subset M_k$ be a finite set containing all Archimedean places. Let X be a smooth algebraic curve of genus 1 defined over k. Then, for any $\epsilon > 0$,*

$$m(x, D) \le \epsilon h(x) \tag{2.33}$$

holds for all $x \in X(k)$ except for finitely many points.

When X is of genus 1, its canonical divisor is trivial. So this Theorem is equivalent to Theorem 2.1.6 in the case where the genus of the curve is equal to 1.

To prove Siegel's theorem, we define the v-distance on X for $v \in M_k$.

Definition B2.4.2 *Let X be a curve defined over k, and $P, Q \in X(k_v)$. Let t_Q be a locally defined rational function over X with values in k_v such that it vanishes at Q with order $e \geq 1$. The* **v-adic distance function from P to Q**, *denoted by $d_v(P,Q)$, is given by*

$$d_v(P,Q) = \min\{\|t_Q(P)\|_v^{1/e}, 1\}.$$

The definition above certainly depends on the choice of t_Q, so possibly a better notation would be $d_v(P, t_Q)$. However, since we will only use d_v to measure the rate at which two points approach one another, the following result shows that all of our theorems make sense.

Proposition B2.4.3 *Let $Q \in X(k_v)$, and let t_Q and t'_Q be functions vanishing at Q. Then we have the notation*

$$\lim_{P \in X(k_v), P \to Q} \frac{\log d_v(P, t'_Q)}{\log d_v(P, t_Q)} = 1,$$

here $P \to Q$ means $P \in X(k_v)$ approaches Q in the v-topology, i.e., $d_v(P, t_Q) \to 0$.

Proof. Let t_Q and t'_Q have zeros of order e and e' respectively at Q. Then the function $\phi = (t'_Q)^e/(t_Q)^{e'}$ has neither a zero nor a pole at Q. Hence $|\phi(P)|_v$ is bounded away from 0 and ∞ as $P \to Q$; so as $P \to Q$,

$$\frac{\log d_v(P, t'_Q)}{\log d_v(P, t_Q)} = 1 + \frac{\log |\phi(P)|_v^{1/ee'}}{\log d_v(P, t_Q)} \to 1.$$

□

Roth's theorem implies the following theorem.

Theorem B2.4.4 *Let X be a curve defined over k and $Q \in X(\bar{k})$. Then, for any $\epsilon > 0$, the inequality*

$$d_v(P,Q) \geq \frac{1}{H_k(P)^{2+\epsilon}}$$

holds for all $P \in X(k)$ except for finitely many points.

Proof. Let t_Q be a locally defined rational function over X values in k_v such that it vanishes at Q with order $e \geq 1$. Then by definition, we may take

$$d_v(P,Q) = \min\{\|t_Q(P)\|_v^{1/e}, 1\}.$$

Applying Theorem B2.1.6 (Roth's theorem) to $\alpha = 0$, for any $\epsilon > 0$,

$$\min\{\|t_Q(P)\|_v^{1/e}, 1\} \geq H_k(t_Q(P))^{2+\epsilon}$$

holds for all but finitely many P. The theorem is proven. □

We now prove Theorem B2.4.1.

Proof. Choose any even, non-constant function $f \in k(X)$, for example the x-coordinate function on a Weierstrass equation. We just write $h_f(x)$ as $h(x)$ for $x \in X(k)$, since f will no longer be used. Then Corollary B2.3.5 applies to the height function $h(x)$.

Let $D = \sum_{j=1}^{q}(Q_j)$. It is easy to check, by the definition, that

$$m_S(x,D) = \sum_{j=1}^{q}\sum_{v\in S} d_v(x,Q_j). \tag{2.34}$$

Let m be a positive integer such that $(m^2-1)/\epsilon \geq 3$. By the weak Mordell-Weil theorem, $X(k)/mX(k)$ is finite. For any $x \in X(k)$, there are $x', \pi \in X(k)$ such that

$$x = mx' + \pi.$$

Since $X(k)/mX(k)$ is finite, we may assume that π is independent of x. Since the morphism $\omega \to m\omega + \pi$ is etale, we have

$$d_v(x,Q_j) \leq d_v(x',Q_j') + O(1)$$

and $h(x) = m^2 h(x') + O(1)$. By Theorem B2.4.4,

$$\log d_v(x',Q_j') \leq (2+\epsilon)h(x')$$

holds for all, but finitely many, $x' \in X(k)$. So

$$\log d_v(x,Q_j) \leq (2+\epsilon)h(x') \leq \epsilon h(x)$$

holds for all, but finitely many, $x \in X(k)$. By (2.34), this is equivalent to

$$m_S(x, D) \leq \epsilon h(x)$$

which holds for all, but finitely many, $x \in X(k)$. □

Corollary B2.4.5 *Let X/k be an elliptic curve with Weierstrass coordinate functions x and y, let $S \subset M_k$ be a finite set of places containing M_k^∞. Let $\mathcal{O}_S$ be the ring of S-integers of k. Then*

$$\{P \in X(k) : x(P) \in \mathcal{O}_S\}$$

is a finite set.

Proof. Suppose Corollary B.2.4.5 is false. That is there exist distinct points $P_1, P_1, \cdots \in \{P \in X(k) : x(P) \in \mathcal{O}_S\}$. Consider the divisor consisting of a single point O. Since x has a pole of order 2 at O, we have, by the definition,

$$d_v(P, O) = \min\{\|x(P)\|_v^{-1/2}, 1\}.$$

Since for $v \notin S$ we have $\|x(P_i)\|_v \leq 1$,

$$d_v(P_i, O) = 1, \quad \text{if} \quad v \notin S.$$

So $m_S(P_i, O) = h(P_i) + O(1)$. This contradicts with (2.33) with $D = O$.□

Clearly the above proof can be applied to any rational function $f \in k(X)$. So we have the following more general corollary.

Corollary B2.4.6 *Let X/k be an elliptic curve. Let $S \subset M_k$ be a finite set of places containing M_k^∞. Let $\mathcal{O}_S$ be the ring of S-integers of k. Let $f \in k(X)$ be a non-constant function. Then*

$$\{P \in X(k) : f(P) \in \mathcal{O}_S\}$$

is a finite set.

B2.5 Curves of Genus Greater Than or Equal to Two, Theorem of Faltings

When the genus of X is greater than or equal to two, Faltings proved the following theorem.

Theorem B2.5.1 (Faltings) *Let X be an algebraic curve over $\mathbf{Q}$ whose genus of X is greater than or equal to two. Then for any number field k, the set $X(k)$ is always finite.*

This theorem is equivalent to the statement of Theorem B2.1.6 in the case that the genus of X is greater than or equal to two. To see this, we note that the canonical bundle K in this case is positive. So taking $D = \emptyset$, then Theorem B2.1.6 reads

$$h_K(x) \leq \epsilon h(x).$$

This is equivalent to, by Corollary B2.1.3, the statement that the set $X(k)$ is finite.

Faltings' [Fal1] proof of Theorem B2.5.1 used a variety of advanced techniques from modern algebraic geometry, including tools such as moduli schemes and stacks, semi-stable abelian schemes, and p-divisible groups. Vojta[Voj2] then came up with an entirely new proof of Faltings' theorem using ideas whose origins lie in the classical theory of Diophantine approximation. However, in order to obtain the precise estimates needed for the delicate arguments involved, he made use of Arakelov arithmetic intersection theory and the deep and technical Riemann-Roch theorem for arithmetic three-folds proven by Gillet and Soulé. Faltings [Fal2] then simplified Vojta's proof by eliminating the use of the Gillet and Soulé's theorem and proving a "product Lemma" especially well suited to induction. This allows Faltings to generalize Vojta's result to prove of Lang concerning rational and integral points on sub-varieties of abelian varieties(see Chapter 6). However, Faltings' proof, which uses arithmetic intersection theory and heights defined via differential geometric considerations, is far from elementary. Finally Bombieri [Bom] combined Faltings' generalization with Vojta's original proof and with other simplification of his own to give a comparatively elementary proof of Theorem B2.5.1. The tools that used in Bombieri's proof fall broadly into the following four areas: (i) *Geometric tools*: The Riemann-Roch theorem for surfaces, or more precisely, for the $C \times C$ of a curve C with itself. The theory of curves, Jacobians, and the theta divisors; (ii) *Height Functions*: Weil height functions associated to divisor classes. Canonical height functions on abelian varieties and their associated quadratic forms; (iii) *The Mordell-Weil theorem* for the Jacobian of C; (iv) *Diophantine Approximation*: The techniques used in proving Roth's theorem. In particular, the proof is divided into: (1) Construction of an auxiliary function using Siegel's lemma. (2) An elementary upper bound, essentially obtained from the triangle inequality. (3) A non-vanishing result such as Roth's lemma or Dyson's lemma. (4) A lower bound, obtained, via the product formula, from the fact that 1 is the smallest positive integer. However Bombieri's proof, although is called an elementary proof, is still very long and complicated. Thus we decide not to include his proof in this book. Fortunately, the new book (Graduate Texts series in Mathematics) [HS] written by Marc Hindry and Joseph H. Silverman contains a complete proof of Bombieri for Theorem B2.5.1.

The Correspondence Table

Nevanlinna Theory	**Diophantine Approximation**
(2.7)	(2.24)
(2.8)	(2.25)
Theorem A2.3.1	Theorem B2.1.5
Theorem A2.3.2	Theorem B2.1.6
Corollary A2.3.4 (i)	Theorem B1.2.7
Corollary A2.3.4 (ii)	Theorem B2.4.1
Corollary A2.3.4 (iii)	Theorem B2.5.1

Chapter 3

Holomorphic Curves in $\mathbf{P}^n(\mathbf{C})$ and Schmidt's Subspace Theorem

We will introduce Nevanlinna theory for holomorphic curves in part A and the Diophantine approximation results related to the Schmidt's subspace theorem in part B.

Part A: Nevanlinna Theory

There are two approaches in extending Nevanlinna theory to holomorphic curves in $\mathbf{P}^n(\mathbf{C})$. One is given by H. Cartan, another is by Ahlfors. Cartan's method uses the logarithmic derivative lemma derived in Chapter 1. In Ahlfors' approach, the negative curvature plays an important role. We note that Ahlfors' approach can be adapted to more general cases. For example, it extends Nevanlinna theory to holomorphic maps from certain parabolic manifolds to the complex projective space. In this part, we introduce both approaches. In Cartan's proof, we also carefully examine the error term that appears in the inequality.

A3.1 Cartan's Second Main Theorem

To introduce the theory, we reformulate the characteristic function $T_f(r)$, the proximity function $m_f(r, a)$ and the counting function $N_f(r, a)$ which appeared in Chapter 1 for meromorphic function f. Let $\mathbf{P}^1(\mathbf{C})$ be the complex projective space of dimension 1, that is $\mathbf{P}^1(\mathbf{C}) = \mathbf{C}^2 - \{0\}/\sim$, where $(a_1, a_2) \sim (b_1, b_2)$ if and only if $(a_1, a_2) = \lambda(b_1, b_2)$ for some $\lambda \in \mathbf{C}$. We denote by $[a_1 : a_2]$ the equivalent class of (a_1, a_2). $\mathbf{P}^1(\mathbf{C})$ is naturally identical to $\mathbf{C} \cup \{\infty\}$ by the following map

$$[a : 1] \mapsto a \text{ for } a \in \mathbf{C} \quad \text{and} \quad [1 : 0] \mapsto \{\infty\}. \tag{3.1}$$

In this way, any meromorphic function f on $\mathbf{C}$ determines a holomorphic map $f : \mathbf{C} \to \mathbf{P}^1(\mathbf{C})$ with $f(z) = [1 : 0]$ if z is a pole of f and $f(z) = [f(z) : 1]$ otherwise. We now reformulate the definitions. On $\mathbf{P}^1(\mathbf{C})$ (or $\mathbf{C} \cup \{\infty\}$) there is a natural distance called **chordal distance** defined for every $z_1, z_2 \in \mathbf{C} \cup \{\infty\}$ as

$$\|z_1, z_2\|^2 = \begin{cases} \dfrac{|z_1 - z_2|^2}{(1 + |z_1|^2)(1 + |z_2|^2)} & \text{if } z_2 \neq \infty \\ \dfrac{1}{(1 + |z_1|^2)} & \text{if } z_2 = \infty. \end{cases} \tag{3.2}$$

We note the chordal distance arises in the following way: first we project z_1, z_2 on $\mathbf{C} \cup \{\infty\}$ into the sphere of radius $1/2$ centered at the origin in $\mathbf{R}^3$ by the standard stereographic projection and then we measure the length of the chord of the sphere connecting these two points. Let $a \in \mathbf{C} \cup \{\infty\}$, and $m_f(r, a)$ be the proximity function defined by (1.7). Using the inequality $\log^+ x \leq \log(1 + x) \leq \log^+ x + \log 2$ for every $x > 0$, we can easily get

$$m_f(r, a) = \int_0^{2\pi} \log \frac{1}{\|f(re^{i\theta}), a\|} \frac{d\theta}{2\pi} + O(1). \tag{3.3}$$

Also let $N_f(r, a)$ be the counting function defined by (1.4), then

$$N_f(r, a) = \int_0^r [n_f(t, a) - n_f(0, a)] \frac{dt}{t} + n_f(0, a) \log r \tag{3.4}$$

where $n_f(t, a)$ is the number (with multiplicity counted) of zeros of $\|f, a\|$ in $|z| < r$. Let $f = f_1/f_2$, where f_1, f_2 are entire functions and without common factors. By Corollary A1.1.4,

$$N_f(r, \infty) = \int_0^{2\pi} \log |f_2(re^{i\theta})| \frac{d\theta}{2\pi} + O(1).$$

So the characteristic function is

$$\begin{aligned} T_f(r) &= m_f(r, \infty) + N_f(r, \infty) \\ &= \int_0^{2\pi} \frac{1}{2} \log(1 + |f(re^{i\theta})|^2) \frac{d\theta}{2\pi} + \int_0^{2\pi} \log |f_2(re^{i\theta})| \frac{d\theta}{2\pi} \end{aligned}$$

$$= \int_0^{2\pi} \log(|f_1(re^{i\theta})|^2 + |f_2(re^{i\theta})|^2)^{1/2} \frac{d\theta}{2\pi}. \tag{3.5}$$

We now extend the theory to holomorphic curves in $\mathbf{P}^n(\mathbf{C})$. Recall that the n-dimensional complex projective space is $\mathbf{P}^n(\mathbf{C}) = \mathbf{C}^{n+1} - \{0\}/\sim$, where $(a_0, a_1, \ldots, a_n) \sim (b_0, b_1, \ldots, b_n)$ if and only if $(a_0, \ldots, a_n) = \lambda(b_0, \ldots, b_n)$ for some $\lambda \in \mathbf{C}$. We denote by $[a_0 : \cdots : a_n]$ the equivalent class of $(a_0, \ldots, a_n)$. Let

$$f = [f_0 : \ldots : f_n] : \mathbf{C} \to \mathbf{P^n}(\mathbf{C})$$

be a holomorphic map where $f_0, \ldots, f_n$ are entire functions and without common zeros. Denote by $\mathbf{f} = (f_0, \ldots, f_n)$. $\mathbf{f}$ then is called a **reduced representation** of f. Similar to (3.5), **Cartan's characteristic function** $T_f(r)$ of f is defined by

$$T_f(r) = \int_0^{2\pi} \log \|\mathbf{f}(re^{i\theta})\| \frac{d\theta}{2\pi} - \log \|\mathbf{f}(0)\| \tag{3.6}$$

where $\mathbf{f}$ is a reduced representation of f and $\|\mathbf{f}\| = (|f_0|^2 + \cdots + |f_n|^2)^{1/2}$. Note that the characteristic does not depend on the choice of the reduced representation. For $W = [w_0 : \cdots : w_n] \in \mathbf{P}^n(\mathbf{C})$, let

$$\omega = dd^c \log \|W\|^2.$$

Then ω is a well defined $(1,1)$ form on $\mathbf{P}^n(\mathbf{C})$. Such (1,1) form is called the **Fubini-Study form**. Define

$$T_{f,\omega}(r) = \int_0^r \frac{dt}{t} \int_{|z|<t} f^*\omega.$$

The above $T_{f,\omega}(r)$ is called the **Ahlfors' characteristic function**. In fact, by Theorem A2.1.3(Green-Jensen's formula), we have

$$\begin{aligned} T_{f,\omega}(r) &= \int_0^r \frac{dt}{t} \int_{|z|<t} dd^c \log \|\mathbf{f}\|^2 \\ &= \frac{1}{2} \int_0^{2\pi} \log \|\mathbf{f}(re^{i\theta})\|^2 \frac{d\theta}{2\pi} - \frac{1}{2} \log \|\mathbf{f}(0)\|^2 = T_f(r). \end{aligned}$$

So Ahlfors' characteristic function agrees with Cartan's characteristic function.

A hyperplane in $\mathbf{P}^n(\mathbf{C})$ is given by

$$H = \left\{ [x_0 : \ldots : x_n] \in \mathbf{P}^n(\mathbf{C}) \;\middle|\; \sum_{i=0}^{n} a_i x_i = 0 \right\}$$

where $a_i \in \mathbf{C}, 0 \leq i \leq n$. Denote by $\mathbf{a} = (a_0, \ldots, a_n)$ the non-zero vector associated with H. The **Weil function $\lambda_H(f(z))$ of f with respect to the hyperplane H** is defined, if $< \mathbf{f}(z), \mathbf{a} > \not\equiv 0$, by

$$\lambda_H(f(z)) = \log \frac{\|\mathbf{f}(z)\| \|\mathbf{a}\|}{| < \mathbf{f}(z), \mathbf{a} > |}, \tag{3.7}$$

where $\mathbf{f}$ is a reduced representation of f and $< \mathbf{f}, \mathbf{a} >$ is the inner product on $\mathbf{C}^{n+1}$. The **proximity function $m_f(r, H)$ of f with respect to H** is defined as, under the assumption that $< \mathbf{f}(z), \mathbf{a} > \not\equiv 0$,

$$m_f(r, H) = \int_0^{2\pi} \lambda_H(f(re^{i\theta})) \frac{d\theta}{2\pi}. \tag{3.8}$$

Again if $< \mathbf{f}(z), \mathbf{a} > \not\equiv 0$, let $n_f(r, H)$ be the number (with multiplicity counted) of zeros of $< \mathbf{f}, \mathbf{a} >$ in $|z| < r$. Let $n_f^{(n)}(r, H)$ be the number of zeros of $< \mathbf{f}, \mathbf{a} >$ in $|z| < r$, where the multiplicity is counted only as n if the vanishing order of $< \mathbf{f}, \mathbf{a} >$ at the point is greater than or equal to n. The **counting function** $N_f(r, H)$ of f with respect to H is defined by, under the assumption that $< \mathbf{f}(z), \mathbf{a} > \not\equiv 0$,

$$N_f(r, H) = \int_0^r (n_f(t, H) - n_f(0, H)) \frac{dt}{t} + n_f(0, H) \log r, \tag{3.9}$$

and the truncated counting function is

$$N_f^{(n)}(r, H) = \int_0^r (n_f^{(n)}(t, H) - n_f^{(n)}(0, H)) \frac{dt}{t} + n_f^{(n)}(0, H) \log r, \tag{3.10}$$

where $n_f(0, H) = \lim_{t \to 0} n_f(t, H)$ and $n_f^{(n)}(0, H) = \lim_{t \to 0} n_f^{(n)}(t, H)$. Note that $\lambda_H(f(z)), m_f(r, H), N_f(r, H)$ and $N_f^{(n)}(r, H)$ are only defined under the assumption $f(\mathbf{C}) \not\subset H$, i.e., $< \mathbf{f}(z), \mathbf{a} > \not\equiv 0$. So whenever one of these functions appears, we automatically assume that $f(\mathbf{C}) \not\subset H$. Also note that the Weil function, proximity function, and the counting function introduced above depends only on f and H, and not on the choice of $\mathbf{a}$ defining H or on the choice of the reduced representation $\mathbf{f}$.

By Corollary A1.1.4,

$$N_f(r,H) = \int_0^{2\pi} \log| < \mathbf{f}(re^{i\theta}), \mathbf{a} > |\frac{d\theta}{2\pi} + O(1).$$

So, according to the definitions, we derive at the following theorem.

Theorem A3.1.1 (The First Main Theorem)

$$T_f(r) = m_f(r,H) + N_f(r,H) + O(1).$$

Theorem A3.1.2 *Let f_j, $0 \le j \le n$, be entire functions on* $\mathbf{C}$ *without common zeros. Let $f : \mathbf{C} \to \mathbf{P}^n(\mathbf{C})$ be a holomorphic curve defined by $[f_0 : \cdots : f_n]$. Then*

$$T_{f_j/f_0}(r) + O(1) \le T_f(r) \le \sum_{j=0}^{n} T_{f_j/f_0}(r) + O(1).$$

Proof. Take $H = \{[x_0 : \cdots : x_n] \,|x_0 = 0\}$. Then $N_f(r,H) = N_{f_0}(r,0) = N_{f_j/f_0}(r,\infty)$. Also

$$m_f(r,H) = \int_0^{2\pi} \log \frac{||f(re^{i\theta})||}{|f_0(re^{i\theta})|}\frac{d\theta}{2\pi}.$$

So

$$\begin{aligned} T_{f_j/f_0}(r) &= m_{f_j/f_0}(r,\infty) + N_{f_j/f_0}(r,\infty) \\ &= \int_0^{2\pi} \log^+ \frac{|f_j(re^{i\theta})|}{|f_0(re^{i\theta})|}\frac{d\theta}{2\pi} + N_{f_j/f_0}(r,\infty) \\ &\le \int_0^{2\pi} \log \frac{||f(re^{i\theta})||}{|f_0(re^{i\theta})|}\frac{d\theta}{2\pi} + N_f(r,H) \\ &= m_f(r,H) + N_f(r,H) = T_f(r) + O(1). \end{aligned}$$

This proves one direction. On the other hand,

$$\begin{aligned} T_f(r) &= m_f(r,H) + N_f(r,H) + O(1) \\ &= \int_0^{2\pi} \log \frac{||f(re^{i\theta})||}{|f_0(re^{i\theta})|}\frac{d\theta}{2\pi} + N_{f_j/f_0}(r,\infty) + O(1) \\ &\le \sum_{j=0}^{m} \int_0^{2\pi} \log^+ \frac{|f_j(re^{i\theta})|}{|f_0(re^{i\theta})|}\frac{d\theta}{2\pi} + N_{f_j/f_0}(r,\infty) + O(1) \end{aligned}$$

$$= \sum_{j=0}^{n} T_{f_j/f_0}(r) + O(1).$$

This proves the Theorem. □

The following general Second Main Theorem with a Good Error Term appeared in [Ru3].

Theorem A3.1.3 (A General SMT with a Good Error Term) *Let $f = [f_0 : \ldots : f_n] : \mathbf{C} \to \mathbf{P}^n(\mathbf{C})$ be a holomorphic curve whose image is not contained in any proper subspaces. Let $H_1, ..., H_q$ (or $\mathbf{a}_1, \ldots, \mathbf{a}_q$) be arbitrary hyperplanes in $\mathbf{P}^n(\mathbf{C})$. Denote by $W(f_0, \ldots, f_n)$ the Wronskian of $f_0, \ldots, f_n$. Then, for any $\epsilon > 0$, the inequality*

$$\int_0^{2\pi} \max_K \sum_{k\in K} \lambda_{H_k}(f(re^{i\theta}))\frac{d\theta}{2\pi} + N_W(r, 0)$$
$$\leq (n+1)T_f(r) + \frac{n(n+1)}{2}(\log T_f(r) + (1+\epsilon)\log^+\log T_f(r)) + O(1)$$

holds for all r outside a set E with finite Lebesgue measure. Here the maximum is taken over all subsets K of $\{1, \ldots, q\}$ such that $\mathbf{a}_j, j \in K$, are linearly independent.

To prove Theorem A3.1.3, we first extend Theorem A1.2.3 to higher order derivatives.

Lemma A3.1.4 (Ye) *Let g be a non-constant meromorphic function. For arbitrary α with $0 < \alpha l < 1/2$, there exist constants C, C_1, C_2 such that for any $r < \rho < R$,*

$$\frac{1}{2\pi}\int_0^{2\pi} \left|\frac{g^{(l)}(re^{i\theta})}{g(re^{i\theta})}\right|^{\alpha} d\theta \leq C\left(\frac{\rho}{r(\rho - r)}\right)^{l\alpha}\left[C_1 T_g(R) + C_2 \log\frac{R}{\rho(R-\rho)}T_g(R)\right]^{l\alpha}.$$

Proof. Note that

$$\frac{g^{(l)}}{g} = \frac{g^{(l)}}{g^{(l-1)}}\frac{g^{(l-1)}}{g^{(l-2)}}\cdots\frac{g'}{g}.$$

It turns out from the Hölder inequality and Theorem A1.2.3 that, for any $r < \rho$,

$$\begin{aligned}
\frac{1}{2\pi}\int_0^{2\pi}\left|\frac{g^{(l)}(re^{i\theta})}{g(re^{i\theta})}\right|^\alpha d\theta &= \frac{1}{2\pi}\int_0^{2\pi}\left(\left|\frac{g^{(l)}(re^{i\theta})}{g^{(l-1)}(re^{i\theta})}\right|^\alpha \cdots \left|\frac{g'(re^{i\theta})}{g(re^{i\theta})}\right|^\alpha\right) d\theta \\
&\leq \left(\int_0^{2\pi}\left|\frac{g^{(l)}(re^{i\theta})}{g^{(l-1)}(re^{i\theta})}\right|^{l\alpha}\frac{d\theta}{2\pi}\right)^{1/l} \cdots \left(\int_0^{2\pi}\left|\frac{g'(re^{i\theta})}{g(re^{i\theta})}\right|^{l\alpha}\frac{d\theta}{2\pi}\right)^{1/l} \\
&\leq C\left(\frac{\rho}{r(\rho-r)}\right)^{l\alpha} T^\alpha_{g^{(l-1)}}(\rho)\cdots T^\alpha_{g'}(\rho)T^\alpha_g(\rho). \qquad (3.11)
\end{aligned}$$

However for meromorphic function g, by Theorem A1.2.3, we have, for any $\rho \leq \rho' < \rho'' \leq R$,

$$\begin{aligned}
T_{g^{(j)}}(\rho') &= m_{g^{(j)}}(\rho',\infty) + N_{g^{(j)}}(\rho',\infty) \\
&\leq m_{g^{(j)}/g^{(j-1)}}(\rho',\infty) + m_{g^{(j-1)}}(\rho',\infty) + 2N_{g^{(j-1)}}(\rho',\infty) \\
&\leq \int_0^{2\pi}\log^+\left|\frac{g^{(j)}(\rho' e^{i\theta})}{g^{(j-1)}(\rho' e^{i\theta})}\right|\frac{d\theta}{2\pi} + 2T_{g^{(j-1)}}(\rho') \\
&\leq \frac{1}{\alpha}\log^+\int_0^{2\pi}\left|\frac{g^{(j)}(\rho' e^{i\theta})}{g^{(j-1)}(\rho' e^{i\theta})}\right|^\alpha\frac{d\theta}{2\pi} + 2T_{g^{(j-1)}}(\rho') \\
&\leq 2T_{g^{(j-1)}}(\rho') + \log\left(\frac{\rho''}{\rho'(\rho''-\rho')}T_{g^{(j-1)}}(\rho'')\right) + O(1).
\end{aligned}$$

Using the above inequality with $j = l-1$, $\rho' = \rho$ and $\rho'' = (\rho+R)/2$, we have

$$T_{g^{(l-1)}}(\rho) \leq 2T_{g^{(l-2)}}(\rho) + \log\left(\frac{2R}{\rho(R-\rho)}T_{g^{(l-2)}}((R+\rho)/2)\right) + O(1).$$

Again, with $j = l-2$, $\rho' = (R+\rho)/2$ and $\rho'' = (\rho+3R)/4$, we have

$$\begin{aligned}
T_{g^{(l-2)}}((R+\rho)/2) \leq\ & 2T_{g^{(l-3)}}(R+\rho)/2) \\
&+\log\left(\frac{8R}{\rho(R-\rho)}T_{g^{(l-3)}}((3R+\rho)/4)\right) + O(1).
\end{aligned}$$

Repeating the above process to $g^{(l-1)},\ldots,g',g$ consecutively, and combining it with (3.11) gives us the desired result. □

Proof of Theorem A3.1.3.

Proof. Let $H_1, \ldots, H_q$ be the given hyperplanes with coefficient vectors $\mathbf{a}_1, \ldots, \mathbf{a}_q$ in $\mathbf{C}^{n+1}$. Denote by $K \subset \{1, ..., q\}$ such that $\mathbf{a}_k, k \in K$, are linearly independent. Without loss of generality, we may assume that $q \geq n+1$ and that $\#K = n+1$. Let T be the set of all injective maps $\mu : \{0, 1, \ldots, n\} \to \{1, \ldots, q\}$ such that $\mathbf{a}_{\mu(0)}, \ldots, \mathbf{a}_{\mu(n)}$ are linearly independent. Then

$$
\begin{aligned}
&\int_0^{2\pi} \max_K \sum_{k \in K} \lambda_{H_k}(f(re^{i\theta})) \frac{d\theta}{2\pi} \\
= &\int_0^{2\pi} \max_{\mu \in T} \sum_{j=0}^{n} \log \left(\frac{\|\mathbf{f}(re^{i\theta})\| \|\mathbf{a}_{\mu(j)}\|}{| < \mathbf{f}(re^{i\theta}), \mathbf{a}_{\mu(j)} > |} \right) \frac{d\theta}{2\pi} \\
= &\int_0^{2\pi} \log \left\{ \max_{\mu \in T} \left(\frac{\|\mathbf{f}(re^{i\theta})\|^{n+1}}{\prod_{j=0}^{n} | < \mathbf{f}(re^{i\theta}), \mathbf{a}_{\mu(j)} > |} \right) \right\} \frac{d\theta}{2\pi} + O(1) \\
\leq &\int_0^{2\pi} \log \left\{ \sum_{\mu \in T} \frac{\|\mathbf{f}(re^{i\theta})\|^{n+1}}{\prod_{j=0}^{n} | < \mathbf{f}(re^{i\theta}), \mathbf{a}_{\mu(j)} > |} \right\} \frac{d\theta}{2\pi} + O(1) \\
= &\int_0^{2\pi} \log \left\{ \sum_{\mu \in T} \frac{|W(< \mathbf{f}, \mathbf{a}_{\mu(0)} >, \ldots, < \mathbf{f}, \mathbf{a}_{\mu(n)} >)(re^{i\theta})|}{\prod_{j=0}^{n} | < \mathbf{f}(re^{i\theta}), \mathbf{a}_{\mu(j)} > |} \right\} \frac{d\theta}{2\pi} \\
+ &\int_0^{2\pi} \log \left\{ \|\mathbf{f}(re^{i\theta})\|^{n+1} / |W(f_0, \ldots, f_n)|(re^{i\theta}) \right\} \frac{d\theta}{2\pi} + O(1), \quad (3.12)
\end{aligned}
$$

where $W(< \mathbf{f}, \mathbf{a}_{\mu(0)} >, \ldots, < \mathbf{f}, \mathbf{a}_{\mu(n)} >)$ denotes the Wronskian of functions $< \mathbf{f}, \mathbf{a}_{\mu(0)} >, \ldots, < \mathbf{f}, \mathbf{a}_{\mu(n)} >$. In the above, we use the property of Wronskian that

$$|W(f_0, \ldots, f_n)| = |W(< f, \mathbf{a}_{\mu(0)} >, \ldots, < f, \mathbf{a}_{\mu(n)} >)| \cdot C,$$

where C is a constant. We now estimate the first term on the right-hand side of (3.12). Denote by

$$g_{\mu(l)} = \frac{< \mathbf{f}, \mathbf{a}_{\mu(l)} >}{< \mathbf{f}, \mathbf{a}_{\mu(0)} >}, \ 0 \leq l \leq n.$$

Then $T_{g_{\mu(l)}}(r) \leq T_f(r) + O(1)$ for $0 \leq l \leq n$. Let $\alpha n(n+1) < 1/2$. From the con-cavity of the logarithm, the Hölder inequality, lemma A3.1.4 and the inequality $(\sum_{i,j} a_{ij})^\alpha \le C\sum_{i,j} a_{ij}^\alpha$, for $a_{ij} \ge 0$ and $\alpha > 0$, we have

$$
\begin{aligned}
&\int_0^{2\pi} \log\left\{\sum_{\mu\in T} \frac{|W(<\mathbf{f},\mathbf{a}_{\mu(0)}>,\ldots,<f,\mathbf{a}_{\mu(n)}>)(re^{i\theta})|}{\prod_{j=0}^n |<\mathbf{f}(re^{i\theta}),\mathbf{a}_\mu>|}\right\}\frac{d\theta}{2\pi}\\
\le\ & \frac{1}{\alpha}\int_0^{2\pi} \log\left\{\sum_{\mu\in T}\left(\frac{|W(<\mathbf{f},\mathbf{a}_{\mu(0)}>,\ldots,<\mathbf{f},\mathbf{a}_{\mu(n)}>)|(re^{i\theta})}{\prod_{j=0}^n |<\mathbf{f}(re^{i\theta}),\mathbf{a}_\mu>|}\right)^\alpha\right\}\frac{d\theta}{2\pi}\\
=\ & \frac{1}{\alpha}\int_0^{2\pi}\log\sum_{\mu\in T}\left(\frac{|W(1,g_{\mu(1)},\ldots,g_{\mu(n)})|}{|g_{\mu(1)}\cdots g_{\mu(n)}|}(re^{i\theta})\right)^\alpha\frac{d\theta}{2\pi}+O(1)\\
\le\ & \frac{1}{\alpha}\int_0^{2\pi}\log\left\{\sum_{\mu\in T}\sum_{i_1+\ldots+i_n\le\frac{n(n+1)}{2}}\left|\frac{g_{\mu(1)}^{(i_1)}}{g_{\mu(1)}}\cdots\frac{g_{\mu(n)}^{(i_n)}}{g_{\mu(n)}}\right|^\alpha(re^{i\theta})\right\}\frac{d\theta}{2\pi}+O(1)\\
\le\ & \frac{1}{\alpha}\log\int_0^{2\pi}\left\{\sum_{\mu\in T}\sum_{i_1+\ldots+i_n\le\frac{n(n+1)}{2}}\left|\frac{g_{\mu(1)}^{(i_1)}}{g_{\mu(1)}}\cdots\frac{g_{\mu(n)}^{(i_n)}}{g_{\mu(n)}}\right|^\alpha(re^{i\theta})\right\}\frac{d\theta}{2\pi}+O(1)\\
\le\ & \frac{1}{\alpha}\log\left\{\sum_{\mu\in T}\sum_{i_1+\ldots+i_n\le\frac{n(n+1)}{2}}\prod_{l=1}^n\left(\int_0^{2\pi}\left|\frac{g_{\mu(l)}^{(i_l)}(re^{i\theta})}{g_{\mu(l)}(re^{i\theta})}\right|^{\alpha(n+1)}\frac{d\theta}{2\pi}\right)^{1/(n+1)}\right\}\\
&+O(1)\\
\le\ & \frac{1}{\alpha}\log\Bigg\{\sum_{\mu\in T}\sum_{i_1+\ldots+i_n\le\frac{n(n+1)}{2}}\prod_{l=1}^n\Bigg(\left(\frac{\rho}{r(\rho-r)}\right)^{i_l\alpha}\Bigg[C_1T_{g_{\mu(l)}}(R)\\
&+C_2\log\Big(\frac{R}{\rho(R-\rho)}T_{g_{\mu(l)}}(R)\Big)\Bigg]^{i_l\alpha}\Bigg)\Bigg\}+O(1)\\
\le\ & \frac{n(n+1)}{2}\log\left\{\frac{\rho}{r(\rho-r)}\sum_{\mu\in T}\left[C_1T_f(R)+C_2\log\left(\frac{R}{\rho(R-\rho)}T_f(R)\right)\right]\right\}\\
&+O(1).
\end{aligned}
\tag{3.13}
$$

Taking $R = r + \frac{1}{\log^{1+\epsilon}T_f(r)}$ and $\rho = (R+r)/2 = r + \frac{1}{2\log^{1+\epsilon}T_f(r)}$, then, for all large r, $\rho/r \le 2$, $R/\rho \le 2$, $\frac{1}{\rho-r} \le 2\log^{1+\epsilon}T_f(r)$ and $\frac{1}{\rho(R-\rho)} \le 4\log^{1+\epsilon}T_f(r)$. In addition, lemma A1.2.4 implies that the inequality

$$T_f(R) \le T_f(r) + 1$$

holds for all r outside a set $E \subset (0, +\infty)$ with finite Lebesgue measure. Thus, (3.13) becomes

$$\begin{aligned}
&\int_0^{2\pi} \log\left\{\sum_{\mu\in T} \frac{|W(<\mathbf{f}, \mathbf{a}_{\mu(0)}>, \ldots, <\mathbf{f}, \mathbf{a}_{\mu(n)}>)(re^{i\theta})|}{\prod_{j=0}^n |<\mathbf{f}(re^{i\theta}), \mathbf{a}_\mu>|}\right\}\frac{d\theta}{2\pi} \\
\leq\ & \frac{n(n+1)}{2}(\log T_f(r) + (1+\epsilon)\log^+\log T_f(r)) + O(1), \quad (3.14)
\end{aligned}$$

where the inequality holds for all r outside a set $E \subset (0, +\infty)$ with finite Lebesgue measure. Now

$$\begin{aligned}
&\int_0^{2\pi} \log\ \{\|\mathbf{f}\|^{n+1}/|W(f_0, \ldots, f_n)(re^{i\theta})|\}\frac{d\theta}{2\pi} \\
=\ & \int_0^{2\pi} \log\|\mathbf{f}\|^{n+1}\frac{d\theta}{2\pi} + \int_0^{2\pi} \log\frac{1}{|W(f_0, \ldots, f_n)(re^{i\theta})|}\frac{d\theta}{2\pi} \\
=\ & (n+1)T_f(r) - N_W(0, r). \quad (3.15)
\end{aligned}$$

Combining (3.12), (3.14) and (3.15), we conclude the proof. □

Definition A3.1.5 *Given hyperplanes $H_1, \ldots, H_q$ (or $\mathbf{a}_1, \ldots, \mathbf{a}_q$). We say that $H_1, \ldots, H_q$ are* **in general position** *if for any injective map $\mu : \{0, 1, \ldots, n\} \to \{1, \ldots, q\}$, $\mathbf{a}_{\mu(0)}, \ldots, \mathbf{a}_{\mu(n)}$ are linearly independent.*

For hyperplanes $H_1, \ldots, H_q$ in general position we have the following product to the sum estimate.

Lemma A3.1.6 (Product to the sum estimate) *Let $H_1, \ldots, H_q$ be hyperplanes in $\mathbf{P}^n(\mathbf{C})$, located in general position. Denote by T the set of all injective maps $\mu : \{0, 1, \ldots, n\} \to \{1, \ldots, q\}$. Then*

$$\sum_{j=1}^q m_f(r, H_j) \leq \int_0^{2\pi} \max_{\mu\in T}\sum_{i=0}^n \lambda_{H_{\mu(i)}}(f(re^{i\theta}))\frac{d\theta}{2\pi} + O(1).$$

Proof. Let $\mathbf{a}_j$ be the coefficient vectors of $H_j, 1 \leq j \leq q$. By the definition,

$$<\mathbf{f}, \mathbf{a}_{\mu(i)}> = a_0^{\mu(i)}f_0 + \cdots + a_n^{\mu(i)}f_n, \quad 0 \leq i \leq n, \tag{3.16}$$

where $\mathbf{a}_{\mu(i)} = (a_0^{\mu(i)}, \ldots, a_n^{\mu(i)})$. By solving the system of linear equations (3.16),

$$f_i = \tilde{a}_0^{\mu(i)} < \mathbf{f}, \mathbf{a}_{\mu(0)} > + \cdots + \tilde{a}_n^{\mu(i)} < \mathbf{f}, \mathbf{a}_{\mu(n)} >, \quad 0 \leq i \leq n,$$

where $(\tilde{a}_j^{\mu(i)})$ is the inverse matrix of (a_j^i). Thus, for any $\mu \in T$,

$$\|\mathbf{f}(z)\| \leq C \max_{0 \leq i \leq n} \{| < \mathbf{f}(z), \mathbf{a}_{\mu(i)} > |\}. \tag{3.17}$$

For a given $z \in \mathbf{C}$, there is $\mu \in T$ such that

$$0 < | < \mathbf{f}(z), \mathbf{a}_{\mu(0)} > | \leq \cdots \leq | < \mathbf{f}(z), \mathbf{a}_{\mu(n)} > | \leq | < \mathbf{f}(z), \mathbf{a}_j > |,$$

for $j \neq \mu(i), i = 0, 1, \ldots, n$. Hence, by (3.17),

$$\prod_{j=1}^{q} \frac{\|\mathbf{f}(z)\|}{| < \mathbf{f}(z), \mathbf{a}_j > |} \leq C \max_{\mu \in T} \prod_{i=0}^{n} \frac{\|\mathbf{f}(z)\|}{| < \mathbf{f}(z), \mathbf{a}_{\mu(i)} > |}.$$

The lemma is thus proved. □

Combining Lemma A3.1.6 and Theorem A3.1.3 implies the following theorem known as Cartan's Second Main Theorem.

Theorem A3.1.7 (Cartan's Second Main Theorem) *Let $H_1, \ldots, H_q$ be hyperplanes in $\mathbf{P}^n(\mathbf{C})$ in general position. Let $f : \mathbf{C} \to \mathbf{P}^n(\mathbf{C})$ be a linearly non-degenerated holomorphic curve (i.e. its image is not contained in any proper subspaces). Then, for any $\epsilon > 0$, the inequality*

$$\begin{aligned} &\sum_{j=1}^{q} m_f(r, H_j) + N_W(r, 0) \\ &\leq \quad (n+1)T_f(r) + \frac{n(n+1)}{2}(\log T_f(r) + (1+\epsilon)\log^+ \log T_f(r)) + O(1) \end{aligned}$$

holds for all r outside a set E with finite Lebesgue measure.

A3.2 The Use of the Second Main Theorem with Truncated Counting Functions

We first reformulate Theorem A3.1.7. We use the following fact.

Lemma A3.2.1 *Let $H_1, \ldots, H_q$ be the hyperplanes in $\mathbf{P}^n$, located in general position. Then*

$$\sum_{j=1}^{q} N_f(r, H_j) - N_W(r, 0) \leq \sum_{j=1}^{q} N_f^{(n)}(r, H_j).$$

Proof. For each $z \in \mathbf{C}$, without loss of generality, we assume that $< \mathbf{f}, \mathbf{a}_j >$ vanishes at z for $1 \leq j \leq q_1$ and $< \mathbf{f}, \mathbf{a}_j >$ does not vanish at z for $j > q_1$. There are integers $k_j \geq 0$ and nowhere vanishing holomorphic functions g_j in a neighborhood U of z such that

$$< \mathbf{f}, \mathbf{a}_j >= (\zeta - z)^{k_j} g_j \text{ for } j = 1, \ldots, q.$$

Here $k_j = 0$ if $q_1 < j \leq q$. Also we can assume that $k_j \geq n$ if $1 \leq j \leq q_0$ and $1 \leq k_j < n$ where $0 \leq q_0 \leq q_1$. By the property of the Wronskian,

$$W = W(f_0, \ldots, f_n) = CW(< \mathbf{f}, \mathbf{a}_{\mu(1)} >, \ldots, < \mathbf{f}, \mathbf{a}_{\mu(n+1)} >),$$

and

$$W(< \mathbf{f}, \mathbf{a}_{\mu(1)} >, \ldots, < \mathbf{f}, \mathbf{a}_{\mu(n+1)} >) = \prod_{j=1}^{q_0} (\zeta - z)^{k_j - n} h(\zeta),$$

where $h(\zeta)$ is a holomorphic function defined on U. Thus W vanishes at z with order at least $\sum_{j=1}^{q_0}(k_j - n) = \sum_{j=1}^{q_0} k_j - q_0 n$. This, together with definitions of $N_f(r, H_j)$, $N_W(r, 0)$ and $N_f^{(n)}(r, H_j)$, implies the Lemma. □

We use Lemma A3.2.1 and Theorem A3.1.1(the First Main Theorem) to restate Theorem A3.1.7 as follows.

Theorem A3.2.2 (Cartan's Second Main Theorem with Truncated Counting Functions) *Let $H_1, \ldots, H_q$ be hyperplanes in $\mathbf{P}^n(\mathbf{C})$ in general position. Let $f : \mathbf{C} \to \mathbf{P}^n(\mathbf{C})$ be a holomorphic curve whose image is not contained in any proper subspaces. Then, for any $\epsilon > 0$, the inequality*

$$\begin{aligned}(q - (n+1))T_f(r) \quad \leq \quad & \sum_{j=1}^{q} N_f^{(n)}(r, H_j) \\ & + \frac{n(n+1)}{2}(\log T_f(r) + (1+\epsilon)\log^+ \log T_f(r)) + O(1)\end{aligned}$$

holds for all r outside a set E with finite Lebesgue measure.

We now use Theorem A3.2.2 to study the uniqueness problem. The following result is due to the work of W. Stoll (cf. [Sto7]).

Theorem A3.2.3 (The Uniqueness Theorem) *Let $f_1, f_2, \ldots, f_\lambda : \mathbf{C} \to \mathbf{P}^n(\mathbf{C})$ be linearly non-degenerated holomorphic curves. Let $H_1, \ldots, H_q$ be hyperplanes in $\mathbf{P}^n(\mathbf{C})$ located in general position. Assume that $f_1^{-1}(H_j) = \cdots = f_\lambda^{-1}(H_j), 1 \leq j \leq q$. Denote by $A_j = f_1^{-1}(H_j)$. Assume further that for each $i \neq j, A_i \cap A_j = \emptyset$. Let $A = \cup_{j=1}^q A_j$. Let $l, 2 \leq l \leq \lambda$, be an integer such that for any increasing sequence $1 \leq j_1 < j_2 < \ldots < j_l \leq \lambda$, $\mathbf{f}_{j_1}(z) \wedge \cdots \wedge \mathbf{f}_{j_l}(z) = 0$ for every point $z \in A$. If $q > \frac{\lambda n}{\lambda - l + 1} + n + 1$, then $\mathbf{f}_1 \wedge \cdots \wedge \mathbf{f}_\lambda \equiv 0$ over $\mathbf{C}$.*

In the case where $\lambda = 2$, Theorem A3.2.3 reads as follows.

Corollary A3.2.4 *Let $f,\ g : \mathbf{C} \to \mathbf{P}^n(\mathbf{C})$ be two linearly non-degenerated holomorphic curves. Let $H_1, \ldots, H_{3n+2}$ be hyperplanes in $\mathbf{P}^n(\mathbf{C})$ located in general position. Assume that $f^{-1}(H_j) = g^{-1}(H_j), 1 \leq j \leq 3n+2$, and for each $i \neq j$, $f^{-1}(H_i) \cap f^{-1}(H_j) = \emptyset$. Let $A = \cup_{j=1}^q f^{-1}(H_j)$. If for every point $z \in A$, $f(z) = g(z)$, then $f \equiv g$.*

When $n = 1$ and $\lambda = 2$, Theorem A3.2.3 yields the following statement of Nevanlinna.

Corollary A3.2.5 *Given two non-constant meromorphic functions f, g. Assume that there exist five distinct elements $a_1, \ldots, a_5 \in \mathbf{C} \cup \{\infty\}$ such that $f(z) = a_j$ if and only if $g(z) = a_j$, for $1 \leq j \leq 5$, then $f \equiv g$.*

Proof of Theorem A3.2.3.

Proof. We first apply Theorem A3.2.2 to $f_t, 1 \leq t \leq \lambda$, to get, for $1 \leq t \leq \lambda$,

$$(q - (n+1))T_{f_t}(r) \leq \sum_{j=1}^{q} N_{f_t}^{(n)}(r, H_j) + O(\log^+ T_{f_t}(r)), \tag{3.18}$$

where the inequality holds for all r outside a set $E \subset (0, +\infty)$ with finite Lebesgue measure.

Assume that $\mathbf{f_1} \wedge \cdots \wedge \mathbf{f}_\lambda \not\equiv 0$ on $\mathbf{C}$, where $\mathbf{f}_i, 1 \le i \le \lambda$, is the reduced representation of f_i. We denote by $\mu_{\mathbf{f_1}\wedge\cdots\wedge\mathbf{f}_\lambda}$ the divisor associated with $\mathbf{f_1} \wedge \cdots \wedge \mathbf{f}_\lambda$. Denote by $N_{\mu_{\mathbf{f_1}\wedge\cdots\wedge\mathbf{f}_\lambda}}(r)$ the counting function associated with the divisor $\mu_{\mathbf{f_1}\wedge\cdots\wedge\mathbf{f}_\lambda}$. We make the following claim.

Claim. *For every* $1 \le t \le \lambda$,

$$\sum_{j=1}^{q} N_{f_t}^{(n)}(r, H_j) \le \frac{n}{\lambda - l + 1} N_{\mu_{\mathbf{f_1}\wedge\cdots\wedge\mathbf{f}_\lambda}}(r). \tag{3.19}$$

To prove the Claim, we assume that $< \mathbf{f}_t, \mathbf{a}_{j_0} >$ vanishes at some point $z \in \mathbf{C}$ with vanishing order $m \ge 1$ for some index $1 \le j_0 \le q$. Then by the assumption of Theorem A3.2.3,

$$< \mathbf{f}_t, \mathbf{a}_j > \ne 0, \quad \text{for } j \ne j_0.$$

Also, by the assumption, since $z \in A$, for any increasing sequence $1 \le j_1 < j_2 < \ldots < j_l \le \lambda$,

$$\mathbf{f}_{j_1}(z) \wedge \cdots \wedge \mathbf{f}_{j_l}(z) = 0. \tag{3.20}$$

We verify that $\mathbf{f_1} \wedge \cdots \wedge \mathbf{f}_\lambda$ vanishes at z with the vanishing order at least $\lambda - l + 1$. In fact, by the power series expansion for each component of $\mathbf{f}_i$, we can write, for $1 \le i \le \lambda$,

$$\mathbf{f}_i(\zeta) = \mathbf{b}_i + (\zeta - z)\mathbf{h}_i(\zeta),$$

where $\mathbf{b}_i$ is a constant vector, and $\mathbf{h}_i(\zeta)$ is a holomorphic vector-valued function defined around z. Denote by $T[\alpha, \lambda]$ the set of all increasing injective maps from $\{1, 2, \ldots, \alpha\}$ to $\{1, 2, \ldots, \lambda\}$. For each $\eta \in T[\alpha, \lambda]$, there exists a unique $\hat{\eta} \in T[\lambda - \alpha, \lambda]$ such that $(Im\eta) \cap (Im\hat{\eta}) = \emptyset$. Abbreviate $\epsilon_\eta = \text{sign}\eta$. (3.20) then implies that, for any $\eta \in T[l, \lambda]$

$$\mathbf{b}_{\eta(1)} \wedge \cdots \wedge \mathbf{b}_{\eta(l)} = 0.$$

Thus,

$$\begin{aligned} \mathbf{f}_1 \wedge \cdots \wedge \mathbf{f}_\lambda &= \sum_{\alpha=1}^{l-1} (\zeta - z)^{\lambda-\alpha} \sum_{\eta \in T[\alpha,\lambda]} \epsilon_\eta \left(\bigwedge_{j=1}^{\alpha} \mathbf{b}_{\eta(j)} \right) \bigwedge \left(\bigwedge_{k=1}^{\lambda-\alpha} \mathbf{h}_{\hat{\eta}(k)} \right) \\ &\quad + (\zeta - z)^\lambda \mathbf{h}_1 \wedge \cdots \wedge \mathbf{h}_\lambda. \end{aligned}$$

The lowest exponent of $(\zeta - z)$ is $\lambda - l + 1$, so $\mathbf{f_1} \wedge \cdots \wedge \mathbf{f}_\lambda$ vanishes at z with the vanishing order at least $\lambda - l + 1$. This, together with the property of $\min\{m, n\} \leq n \leq \frac{n}{\lambda - l + 1}(\lambda - l + 1)$, concludes the claim.

We now proceed. By the First Main Theorem of the exterior product (cf. [Stol7]),

$$N_{\mu_{\mathbf{f_1} \wedge \cdots \wedge \mathbf{f}_\lambda}}(r) \leq \sum_{i=1}^{\lambda} T_{f_i}(r) + O(1). \tag{3.21}$$

Combining (3.19), and (3.21) yields

$$\sum_{j=1}^{q} N_{f_t}^{(n)}(r, H_j) \leq \frac{n}{\lambda - l + 1} \sum_{i=1}^{\lambda} T_{f_i}(r) + O(1).$$

This, together with (3.18), gives, for $1 \leq t \leq \lambda$,

$$(q - (n+1))T_{f_t}(r) \leq \frac{n}{\lambda - l + 1} \sum_{i=1}^{\lambda} T_{f_i}(r) + O(\log^+ T_{f_t}(r)),$$

where the inequality holds for all r outside a set $E \subset (0, +\infty)$ with finite Lebesgue measure. Thus, by summing them up, we have that the inequality

$$\sum_{t=1}^{\lambda} T_{f_t}(r) \leq \frac{n\lambda}{(q - (n+1))(\lambda - l + 1)} \sum_{t=1}^{\lambda} T_{f_t}(r) + O(\sum_{t=1}^{\lambda} \log^+ T_f(r)),$$

holds for all r outside a set $E \subset (0, +\infty)$ with finite Lebesgue measure. This gives a contradiction under the assumption that $q > \frac{\lambda n}{\lambda - l + 1} + n + 1$. This completes the proof of Theorem A3.2.3. □

Next we give the following result for holomorphic functions satisfying a diagonal equation.

Theorem A3.2.6 (Generalized ABC Theorem) *Let $f = [f_0 : \ldots : f_n] : \mathbf{C} \to \mathbf{P^n(C)}$ be a holomorphic map with $f_0, \ldots, f_n$ as entire functions with no common zeros. Assume that f_{n+1} is a holomorphic function and $f_0 + \ldots + f_n + f_{n+1} = 0$. If $\sum_{i \in I} f_i \neq 0$ for any proper subset I of $\{0, \ldots, n+1\}$, then the inequality*

$$T_f(r) \leq \sum_{j=0}^{n+1} N_{f_j}^{(n)}(r, 0) + O(\log^+ T_f(r))$$

holds for all r *outside a set* $E \subset (0, +\infty)$ *of finite Lebesgue measure.*

To prove Theorem A3.2.6, we recall the following lemma from [B-M].

Lemma A3.2.7 *Assume* $\sum_{i=0}^{m} f_i = 0$ *but no non-empty proper sub-sum vanishes. If some proper subset of* $\{f_0, \ldots, f_m\}$ *are linearly dependent, then we can find an integer* $l \geq 2$*, a partition*

$$\{0, 1, \ldots, m\} = I_1 \cup \ldots \cup I_l$$

into non-empty disjoint sets $I_1, \ldots, I_l$*, and non-empty sets*

$$J_1 \subseteq I_1, \; J_2 \subseteq I_1 \cup I_2, \ldots, J_{l-1} \subseteq I_1 \cup \ldots \cup I_{l-1}$$

such that

$$I_1, \; I_2 \cup J_1, \ldots, I_l \cup J_{l-1}$$

are minimal. Here, we say an index set $I \subset \{0, 1, \ldots, m\}$ *is* **minimal** *if the set* $\{f_i | \; i \in I\}$ *is linearly dependent, and for any proper subset* I' *of* I *the set* $\{f_i | \; i \in I'\}$ *is linearly independent.*

Proof. Throughout this proof, we use the term **linear forms**. Linear forms are the homogeneous polynomials of degree one in $m+1$ variables with coefficients in $\mathbf{C}$, that is $L(X) = c_0 x_0 + \cdots + c_m x_m$ where $c_0, \ldots, c_m \in \mathbf{C}$, $X = (x_0, \ldots, x_m)$. We denote by $\mathcal{L}$ the set of linear forms which vanishes on $(f_0, \ldots, f_m)$, that is $L(X) = c_0 x_0 + \cdots + c_m x_m$ is in $\mathcal{L}$ if and only if $c_0 f_0 + \ldots c_m f_m = 0$. By the assumption $f_0 + \ldots + f_m = 0$, $\mathcal{L}$ is non-empty. We make the following claim.

Claim 1 *Every linear form* L *in* $\mathcal{L}$ *can be written as*

$$L = \sum c_J L_J \text{ with } L_J \in \mathcal{L}$$

for certain minimal sets J*, where* L_J *is a linear combination of* $\{x_j \mid j \in J\}$*, and* c_J *is constant.*

We prove Claim 1 by induction on the length t of L, i.e., the number of nonzero coefficients. The case $t = 1$ is trivial. So assume that for some $t > 1$ this holds for all elements of $\mathcal{L}$ of length strictly less than t. If $L \in \mathcal{L}$ has length exactly t, we may suppose that

$$L = c_0 x_0 + \ldots + c_{t-1} x_{t-1}, \qquad c_i \neq 0, \text{ for } 0 \leq i \leq t-1.$$

If $I = \{0, 2, \ldots, t-1\}$ is minimal we are done. Otherwise, there is a linear form L' in $\mathcal{L}$ with less length. Without loss of generality we can assume that

$$L' = c_0' x_1 + \ldots + c_k' x_k,$$

lies in $\mathcal{L}$ for some k with $0 \leq k < t-1$ and $c_0' \neq 0$. Then L' and $L'' = c_0' L - c_0 L'$ are both of length strictly less than t, and so the induction hypothesis can be applied to both linear forms. Since

$$L = (c_0/c_0')L' + (1/c_0')L'',$$

L has the desired decomposition. So Claim 1 is proved.

We now prove Claim 2.

Claim 2 *Suppose that $\sum_{i=0}^m f_i = 0$ and $\sum_{i \in I} f_i \neq 0$ for some $I \subset N = \{0, 1, \ldots, m\}$. Then there is a minimal set J with $L_J \in \mathcal{L}$ such that $J \cap I \neq \emptyset$, and $J \cap I^c \neq \emptyset$ where I^c is the complement of I in N.*

In fact, the set $L = \sum_{i=0}^m x_i$ is in $\mathcal{L}$ because $\sum_{i=0}^m f_i = 0$. By Claim 1, we have

$$L = \sum c_J L_J \quad \text{with} \quad L_J \in \mathcal{L}$$

for certain minimal sets J. If Claim 2 is false, then every such J is contained either in I or in I^c. So $\sum_{i=0}^m x_i = L(x_0, \ldots, x_m) = \sum_{J \subset I} c_J L_J(x_0, \ldots, x_m) + \sum_{J \subset I^c} c_J L_J(x_0, \ldots, x_m)$. However, for those $J \subset I$, $L_J(x_0, \ldots, x_m)$ involve only $\{x_i \mid i \in I\}$ while for those $J \subset I^c$, $L_J(x_0, \ldots, x_m)$ involve only $\{x_i \mid i \in I^c\}$. Setting $x_i = 0$ for $i \in I^c$, the above equation becomes

$$\sum_{i \in I} x_i = \sum_{J \subset I} c_J L_J(x_0, \ldots, x_m).$$

Since $L_J \in \mathcal{L}$, $L_J(f_0, \ldots, f_m) = 0$. Hence $\sum_{i \in I} f_i = 0$ which leads to a contradiction that proves Claim 2.

We now pick any minimal set I_1. By hypothesis $N = \{0, 1, \ldots, m\}$ is not minimal, so $I_1 \neq N$. Hence, $\sum_{i \in I_1} f_i \neq 0$. So Claim 2 implies that there exists a minimal set I_2' with $L_{I_2'} \in \mathcal{L}$ such that $I_2' \cap I_1 \neq \emptyset$ and $I_2' \cap I_1^c \neq \emptyset$, where I_1^c is the complement of I_1 in N. Put $I_2 = I_2' \cap I_1^c$ and $J_1 = I_2' \cap I_1$. If $N = I_1 \cup I_2$ then we are done. Otherwise, let $I = I_1 \cup I_2$. Applying Claim 2 to I, there exists a minimal set I_3' with $L_{I_3'} \in \mathcal{L}$, such that $I_3' \cap I \neq \emptyset$ and $I_3' \cap I^c \neq \emptyset$. Let $I_3 = I_2' \cap (I_1 \cup I_2)^c$ and $J_2 = I_3' \cap (I_1 \cup I_2)$.

If $N = I_1 \cup I_2 \cup I_3$, then we are done. Otherwise, we repeat the same procedures until the union reaches N. $\square$

Proof of Theorem A3.2.6.

Proof. If $f_0, ..., f_n$ are linearly independent, then this is a consequence of Theorem A3.2.2. If $f_0, ..., f_n$ are linearly dependent, then by Lemma A3.2.7 we can find an integer $l \geq 2$, a partition

$$\{0, 1, ..., n+1\} = I_1 \cup ... \cup I_l$$

into non-empty disjoint sets $I_1, ..., I_l$, and non-empty sets

$$J_1 \subseteq I_1,\ J_2 \subseteq I_1 \cup I_2, ..., J_{l-1} \subseteq I_1 \cup ... \cup I_{l-1}$$

such that

$$I_1,\ I_2 \cup J_1, ..., I_l \cup J_{l-1}$$

are minimal. Let $n_i = \# I_i$. Then $\sum_{i=1}^{l} n_i = n+2$. Without loss of generality we may assume that

$$\{0, ..., n_1 - 1\} = I_1, \{n_1, \ldots, n_1 + n_2 - 1\} = I_2, ..., \{n+2-n_l, ..., n+1\} = I_l.$$

We also write

$$\hat{n}_\lambda = \sum_{\nu=1}^{\lambda} n_\nu. \tag{3.22}$$

Since I_1 is minimal, there is a linear relation among $\{f_0, \ldots, f_{n_1-1}\}$. That is

$$c_{0,1} f_0 + \cdots + c_{n_1-1,1} f_{n_1-1} = \sum_{j \in I_1} c_{j,1} f_j = 0.$$

Define $c_{j,1} = 0$ for all $j \geq n_1$. Then

$$\sum_{j=0}^{n+1} c_{j,1} f_j = 0.$$

Differentiation yields, for each positive integer ρ,

$$\sum_{j=0}^{n+1} c_{j,1} f_j^{(\rho)} = 0. \tag{3.23}$$

Take $2 \leq \lambda \leq l$. Since $I_\lambda \cup J_{\lambda-1}$ is minimal, there are non-zero complex numbers $c_{j,\lambda}$ exist such that

$$\sum_{j \in I_\lambda \cup J_{\lambda-1}} c_{j,\lambda} f_j = 0.$$

Put $c_{j,\lambda} = 0$ for all $j \notin (I_\lambda \cup J_{\lambda-1})$. Then

$$\sum_{j=0}^{n+1} c_{j,\lambda} f_j = 0.$$

Differentiation yields, for each positive integer ρ,

$$\sum_{j=0}^{n+1} c_{j,\lambda} f_j^{(\rho)} = 0. \tag{3.24}$$

We consider an $(n+1) \times (n+2)$ **master matrix** M given by

$$M = \begin{bmatrix} c_{0,1} f_0' & \cdots & c_{n+1,1} f_{n+1}' \\ \vdots & \ddots & \vdots \\ c_{0,1} f_0^{(n_1-1)} & \cdots & c_{n+1,1} f_{n+1}^{(n_1-1)} \\ c_{0,2} f_0' & \cdots & c_{n+1,2} f_{n+1}' \\ \vdots & \ddots & \vdots \\ c_{0,2} f_0^{(n_2)} & \cdots & c_{n+1,2} f_{n+1}^{(n_2)} \\ c_{0,3} f_0' & \cdots & c_{n+1,3} f_{n+1}' \\ \vdots & \ddots & \vdots \\ c_{0,3} f_0^{(n_3)} & \cdots & c_{n+1,3} f_{n+1}^{(n_3)} \\ \vdots & \ddots & \vdots \\ c_{0,l} f_0^{(n_l)} & \cdots & c_{n+1,l} f_{n+1}^{(n_l)} \end{bmatrix},$$

where, in above, we note that $n_1 + \cdots + n_l = n + 2$. We also note that, by (3.23) and (3.24), the sum of each row of M is zero. Let D_j be the determinant of the matrix obtained by deleting the j-th column of the master matrix M. Then, since the sum of each row of M is zero, we actually have

$$D_j = (-1)^j D_0. \tag{3.25}$$

We now show that

$$D_0 \not\equiv 0. \tag{3.26}$$

To show $D_0 \not\equiv 0$, we first prove that

$$D_0 = \gamma_1\gamma_2\cdots\gamma_l \tag{3.27}$$

where

$$\gamma_1 = \begin{vmatrix} c_{1,1}f_1' & \cdots & c_{n_1-1,1}f_{n_1-1}' \\ \vdots & \ddots & \vdots \\ c_{1,1}f_1^{(n_1-1)} & \cdots & c_{n_1-1,1}f_{n_1-1}^{(n_1-1)} \end{vmatrix};$$

and, for $2 \le \lambda \le l$,

$$\gamma_\lambda = \begin{vmatrix} c_{\hat{n}_{\lambda-1},\lambda}f_{\hat{n}_{\lambda-1}}' & \cdots & c_{\hat{n}_\lambda-1,\lambda}f_{\hat{n}_\lambda-1}' \\ \vdots & \ddots & \vdots \\ c_{\hat{n}_{\lambda-1},\lambda}f_{\hat{n}_{\lambda-1}}^{(n_\lambda)} & \cdots & c_{\hat{n}_\lambda-1,\lambda}f_{\hat{n}_\lambda-1}^{(n_\lambda)} \end{vmatrix},$$

where $\hat{n}_\lambda$ is defined in (3.22). (3.27) is true because the definition of D_0 and the fact that $c_{j,1} = 0$ for $j \ge n_1$ and $c_{j,\lambda} = 0$ for $j \ge \hat{n}_\lambda$ for $\lambda = 2, \cdots, l$. Now, since I_1 is minimal, $c_{i,1} \neq 0$ for $0 \le i \le n_1 - 1$ and also $\{f_1, \ldots, f_{n_1-1}\}$ is linearly independent, so $\gamma_1 \not\equiv 0$ by the property of the Wronskian. Also, since $I_\lambda \cup J_{\lambda-1}$ is minimal, $c_{i,\lambda} \neq 0$ for $\hat{n}_{\lambda-1} \le i \le \hat{n}_\lambda - 1$ and also $\{f_j, j \in I_\lambda\}$ is linearly independent. So $\gamma_\lambda \not\equiv 0$ for $2 \le \lambda \le l$. Hence $D_0 \not\equiv 0$ by (3.27). So (3.26) is verified. The rest of the proof is similar to the proof of Theorem A3.1.3(the second main theorem), replacing the Wronskian W by D_0. The following is the detail. Applying Lemma A3.1.6 (Product to the sum formula) to $f = [f_0 : \cdots : f_n]$ and to the coordinate hyperplanes $H_i = \{[x_0 : \ldots, x_n] \mid x_{i-1} = 0\}$ for $1 \le i \le n+1$ and $H_{n+2} = \{[x_0 : \ldots, x_n] \mid x_0 + \cdots + x_n = 0\}$, and noticing that these hyperplanes are in general position, we have

$$\sum_{j=1}^{n+2} m_f(r, H_j) \le \sum_{j=0}^{n+1} \int_0^{2\pi} \log \frac{\|\mathbf{f}\|^{n+1}}{|f_0 \cdots f_{j-1}f_{j+1} \cdots f_{n+1}|} \frac{d\theta}{2\pi}. \tag{3.28}$$

However, using (3.25),

$$\sum_{j=0}^{n+1} \int_0^{2\pi} \log \frac{\|\mathbf{f}\|^{n+1}}{|f_0 \cdots f_{j-1}f_{j+1} \cdots f_{n+1}|} \frac{d\theta}{2\pi}$$

$$
\begin{aligned}
&= \sum_{j=0}^{n+1} \int_0^{2\pi} \log \frac{|D_j|}{|f_0 \cdots f_{j-1} f_{j+1} \cdots f_{n+1}|} \frac{d\theta}{2\pi} \\
&\quad +(n+1) \int_0^{2\pi} \log \|\mathbf{f}(re^{i\theta})\| \frac{d\theta}{2\pi} - \int_0^{2\pi} \log |D_0(re^{i\theta})| \frac{d\theta}{2\pi} \\
&= \sum_{j=0}^{n+1} \int_0^{2\pi} \log^+ \frac{|D_j|}{|f_0 \cdots f_{j-1} f_{j+1} \cdots f_{n+1}|} \frac{d\theta}{2\pi} + (n+1)T_f(r) + O(1) \\
&\quad - \int_0^{2\pi} \log |D_0(re^{i\theta})| \frac{d\theta}{2\pi} \\
&\leq \sum_{j=0}^{n+1} \int_0^{2\pi} \log^+ \frac{|D_j|}{|f_0 \cdots f_{j-1} f_{j+1} \cdots f_{n+1}|} \frac{d\theta}{2\pi} \\
&\quad +(n+1)T_f(r) - N_{D_0}(r, 0) + O(1) \qquad (3.29)
\end{aligned}
$$

where, in the last step, we used Corollary A1.1.4. For each fixed j with $0 \leq j \leq n+1$, we now estimate

$$
\int_0^{2\pi} \log^+ \frac{|D_j|}{|f_0 \cdots f_{j-1} f_{j+1} \cdots f_{n+1}|} \frac{d\theta}{2\pi}.
$$

Note that D_j does not involve f_j, so we write

$$
D_j = D(f_0, \ldots, f_{j-1}, f_{j+1}, \ldots, f_{n+1}).
$$

Write $g_i = f_i/f_j$ for $1 \leq i \leq n+1$ and the fixed j. Similar to the property of Wronskian, it is easy to verify that

$$
\begin{aligned}
&D(f_0, \ldots, f_{j-1}, f_{j+1}, \ldots, f_{n+1}) \\
&\quad = f_j^{n+1} D(f_0/f_j, \ldots, f_{j-1}/f_j, f_{j+1}/f_j, \ldots, f_{n+1}/f_j).
\end{aligned}
$$

In fact, from (3.27) we see that D_j in fact is the product of several "small" Wronskian. So the above equation is true by the property of Wronskian. So

$$
D_j = f_j^{n+1} D(g_0, \ldots, g_{j-1}, g_{j+1}, \ldots, g_{n+1}).
$$

Hence, by Theorem A1.2.5(The Lemma of logarithmic derivative),

$$
\int_0^{2\pi} \log^+ \frac{|D_j|}{|f_0 \cdots f_{j-1} f_{j+1} \cdots f_{n+1}|} \frac{d\theta}{2\pi}
$$

$$
\begin{aligned}
&= \int_0^{2\pi} \log^+ \frac{|D(g_0, \ldots, g_{j-1}, g_{j+1}, \ldots, g_{n+1})|}{|g_0 \cdots g_{j-1} g_{j+1} \cdots g_{n+1}|} \frac{d\theta}{2\pi} \\
&\leq O\left(\sum_{i=0}^{n+1} \log T_{g_i}(r)\right)
\end{aligned}
$$

where the inequality holds for all r outside a set $E \subset (0, +\infty)$ with finite Lebesgue measure. Using Theorem A3.1.2, and the fact that $f_0 + \cdots + f_n + f_{n+1} = 0$,

$$\sum_{i=0}^{n+1} \log T_{g_i}(r) \leq O(\log^+ T_f(r)).$$

Hence

$$\int_0^{2\pi} \log^+ \frac{|D_j|}{|f_0 \cdots f_{j-1} f_{j+1} \cdots f_{n+1}|} \frac{d\theta}{2\pi} \leq O(\log^+ T_f(r)), \tag{3.30}$$

where the inequality holds for all r outside a set $E \subset (0, +\infty)$ with finite Lebesgue measure. Hence, combining (3.28), (3.29) and (3.30),

$$\sum_{j=1}^{n+2} m_f(r, H_j) + N_{D_0}(r, 0) \leq (n+1)T_f(r) + O(\log^+ T_f(r)),$$

or we can write, by the First Main Theorem, the above inequality as

$$T_f(r) \leq \sum_{j=1}^{n+2} N_f(r, H_j) - N_{D_0}(r, 0) + O(\log^+ T_f(r)),$$

here the inequality holds for all r outside a set $E \subset (0, +\infty)$ with finite Lebesgue measure. However, by the definition of H_j, we have

$$N_f(r, H_j) = N_{f_{j-1}}(r, 0).$$

So the inequality

$$T_f(r) \leq \sum_{j=0}^{n+1} N_{f_j}(r, 0) - N_{D_0}(r, 0) + O(\log^+ T_f(r))$$

holds for all r outside a set $E \subset (0, +\infty)$ with finite Lebesgue measure. Similar to the the proof of Lemma A3.2.1 and using the fact that $D_j =$

$(-1)^j D_0$, we can verify that

$$\sum_{j=0}^{n+1} N_{f_j}(r,0) - N_{D_0}(r,0) \leq \sum_{j=0}^{n+1} N_{f_j}^{(n)}(r,0).$$

Thus the theorem is proven. □

Theorem A3.2.8 *Let $f_0, \ldots, f_m$ be $m+1$ entire functions, and let $n_0, \ldots, n_m$ be positive integers such that*

$$\sum_{i=0}^{m} \frac{1}{n_i} < \frac{1}{m}.$$

If $\sum_{i=0}^{m} f_i^{n_i} = 1$, then $f_0, \ldots, f_m$ must be constants.

Proof. We prove this by induction on m. Theorem A3.2.8 is trivial if $m = 1$. Now assume that the theorem is true for $m-1$. Consider $\sum_{i=0}^{m} f_i^{n_i} - 1 = 0$. If a proper sub-sum vanishes, then the induction hypothesis applies. So we assume that $\sum_{i=0}^{m} f_i^{n_i} - 1 = 0$ and none of the proper sub-sum vanishes. Applying theorem A3.2.6 to $f = [f_0^{n_0} : \ldots : f_m^{n_m}]$, and noticing that $f_{m+1} \equiv 1$, we have

$$T_f(r) \leq \sum_{i=0}^{m} N_{f_i^{n_i}}^{(m)}(r,0) + O(\log^+ T_f(r)),$$

where the inequality holds for all r outside a set $E \subset (0, +\infty)$ with finite Lebesgue measure. However,

$$N_{f_i^{n_i}}^{(m)}(r,0) \leq m N_{f_i}^{(1)}(r,0) \leq m T_{f_i}(r) \leq \frac{m}{n_i} T_{f_i^{n_i}}(r) \leq \frac{m}{n_i} T_f(r),$$

using the property that $T_{f_i^{n_i}}(r) = n_i T_{f_i}(r)$. So the inequality

$$T_f(r) \leq \sum_{i=0}^{m} \frac{m}{n_i} T_f(r) + O(\log^+ T_f(r))$$

holds for all r outside a set $E \subset (0, +\infty)$ with finite Lebesgue measure. If f is non-constant, then

$$1 \leq \sum_{i=0}^{m} \frac{m}{n_i}.$$

Contradiction. □

The following more general case is due to Siu and Yeung ([Siu-Y3]).

Theorem A3.2.9 *Let m be a positive integer and $P_j, 1 \leq j \leq q$ be forms of degrees d_j in q variables $x_1, \ldots, x_q$, such that $\{x_j^{m-d_j} P_j(x_1, \ldots, x_q)\}_{j=1}^q$ have no common zeros except at the origin. If*

$$m \geq \sum_{j=1}^{q} d_j + q(q-2) + 1$$

and entire functions $f_1, \ldots, f_q$, satisfy

$$\sum_{j=1}^{q} f_j^{m-d_j} P_j(f_1, \ldots, f_q) = 0$$

then the set of indices $\{1, \ldots, q\}$ can be partitioned into classes, each class containing at least two elements, and such that for i and j in the same class the functions g_i and g_j are proportional.

Proof. We prove this by induction on m. The theorem is trivial if $m = 1$. Now assume that the theorem is true for $m-1$. Let $g_j = f_j^{m-d_j} P_j(f_1, \ldots, f_q)$. Consider $\sum_{j=1}^q g_j = 0$. If a proper sub-sum vanishes, then the induction hypothesis applies. So we assume that none of the sub-sum vanishes.

Elementary properties of Cartan's characteristic imply that

$$T_{[g_1:\ldots:g_{q-1}]}(r) = T_{[g_1:\ldots:g_q]}(r) = mT_{[f_1:\ldots:f_q]}(r) + O(1). \tag{3.31}$$

$$N_{P_j(f_1,\ldots,f_q)}(r, 0) \leq d_j T_{[f_1:\ldots:f_q]}(r) + O(1), 1 \leq j \leq q \tag{3.32}$$

and

$$N_{g_j}^{(q-2)}(r, 0) \leq N_{P_j(f_1,\ldots,f_q)}(r, 0) + (q-2)N_{f_j}(r, 0) \leq (d_j + q - 2)T_{[f_1:\ldots:f_q]}(r). \tag{3.33}$$

Since $g_1 + \ldots + g_q = 0$, and none of the proper sub-sum vanishes, Theorem A3.2.6 implies that the inequality

$$T_{[g_1:\ldots:g_{q-1}]}(r) \leq \sum_{j=1}^{q} N_{g_j}^{(q-2)}(r, 0) + O(\log^+ T_{[g_1:\ldots:g_{q-1}]}(r)) \tag{3.34}$$

holds for all r outside a set $E \subset (0, +\infty)$ with finite Lebesgue measure. Combining this with (3.31) and (3.33) implies

$$\sum_{j=1}^{q}(d_j + q - 2) \geq m,$$

which contradicts our assumption. □

Corollary A3.2.10 *Let $m \geq 11$. If entire functions $f_j, 0 \leq j \leq 3$, satisfy the equation*

$$x_0^m + x_1^m + x_2^m + x_3^{m-2}P(x_0, x_1, x_2, x_3) = 0$$

where P is a generic form of degree two, then $f = [f_0 : \ldots : f_3]$ must be constant.

Proof. Putting $x_j = f_j$, we conclude from Theorem A3.2.9 that one of the first three commands of the equation is proportional to the fourth one, and in addition two others of the first three commands are proportional, that is, up to renumeration

$$f_0^m = f_3^{m-2}P(f_0, f_1, f_3).$$

But for generic P this is a curve of genus greater than 1. So Picard's theorem implies that $f = [f_0 : \ldots : f_3]$ must be constant. □

A3.3 Borel's Lemma and its Applications

In this section, we derive Borel's Lemma. We will show that Borel's Lemma is very useful in the study of complex hyperbolicity problems. The counterpart of Borel's Lemma in Diophantine approximation is the so-called Unit Lemma, which is one of the fundamental results in Diophantine approximation (see B3.3 for further discussions).

Lemma A3.3.1 Let $f_1, \ldots, f_n$ be nowhere zero entire functions such that f_i/f_j are not constants for any distinct i and j. Then they are linearly independent.

Proof. We use the induction on n. Clearly the lemma is true for $n = 2$. We consider the case $n \geq 3$. If $f_1, \ldots, f_n$ are linearly dependent, then there are constants $c_i, 1 \leq i \leq n$, not all zero, such that

$$\sum_{i=1}^{n} c_i f_i = 0.$$

Without loss of generality, we assume that $c_i \neq 0$ for all i. Then

$$f_1/f_n + \cdots + f_{n-1}/f_n = -1.$$

Applying Theorem A3.1.7 to the holomorphic curve $f = [f_1 : \cdots : f_{n-1}]$ with hyperplanes $H_i = \{x_i = 0\}, 1 \leq i \leq n-1$ and $H_n = \{x_1 + \cdots + x_{n-1} = 0\}$, we conclude that the image of $f = [f_1 : \cdots : f_{n-1}] : \mathbf{C} \to \mathbf{P}^{n-2}(\mathbf{C})$ is contained in some proper subspace. That is $f_1, ..., f_{n-1}$ are linearly dependent, which contradicts the induction hypothesis. □

Theorem A3.3.2(Borel's Lemma) *Let $f_0, ..., f_{n+1}$ be nowhere zero entire functions with*

$$f_0 + \ldots + f_n + f_{n+1} = 0. \tag{3.35}$$

Consider the partition

$$\{0, 1, 2, \ldots, n+1\} = I_1 \cup I_2 \cdots \cup I_k$$

such that i and j are in the same class I_l if and only if $f_i = c_{i,j} f_j$ for some nonzero constant c_{ij}. Then

$$\sum_{i \in I_l} f_i = 0$$

for any l.

Proof. For an arbitrarily chosen $i_l \in I_l$, since $f_i = c_{i,i_l} f_{i_l}$, we can rewrite (3.35) as

$$\sum_{l=0}^{n} f_i = \sum_{l=1}^{k} \sum_{i \in I_l} c_{i,i_l} f_{i_l} = \sum_{l=1}^{k} d_l f_{i_l} = 0,$$

where $d_l = \sum_{i \in I_l} c_{i,i_l}$. By Lemma A3.3.1, $f_{i_1}, \ldots, f_{i_k}$ are linearly independent. So $d_l = 0$ for all l. Thus, for each l,

$$\sum_{i \in I_l} f_i = \sum_{i \in I_l} c_{i,i_l} f_{i_l} = d_l f_{i_l} = 0.$$

Theorem A3.3.2 is proved. □

We now give some applications of Borel's Lemma. We first consider holomorphic maps in projective space omitting hyperplanes.

Given a hyperplane $H = \{[x_0 : \cdots x_n] \in \mathbf{P}^n(\mathbf{C}) \mid a_0x_0 + \ldots + a_nx_n = 0\}$, H associates with a linear form $L \in \mathbf{C}^{*n+1}$ with $L(x_0, \ldots, x_n) = a_0x_0 + \ldots + a_nx_n$, where $\mathbf{C}^{*n+1}$ is a dual space of $\mathbf{C}^{n+1}$. L is called a **defining linear form** of H. A finite set of hyperplanes associates with a finite set of linear forms which are pairwise linearly independent. Given a finite set of hyperplanes $\mathcal{H}$, we denote by $\mathcal{L}$ the set of corresponding linear forms. Let $(\mathcal{L})$ be the vector space generated by the vectors in $\mathcal{L}$ over $\mathbf{C}$.

Definition A3.3.3 *A set of hyperplanes $\mathcal{H}$ (or linear forms $\mathcal{L}$) is called* **non-degenerate** *if*

(1) dim$(\mathcal{L}) = n + 1$

(2) For any proper non-empty subset $\mathcal{L}_1$ of $\mathcal{L}$

$$(\mathcal{L}_1) \cap (\mathcal{L} - \mathcal{L}_1) \cap \mathcal{L} \neq \emptyset. \tag{3.36}$$

In the rest of the section, we shall prove the following theorem due to Min Ru ([Ru 2]).

Theorem A3.3.4 (Ru) *Let $\mathcal{H}$ be a set of hyperplanes in $\mathbf{P}^n(\mathbf{C})$. Then $\mathcal{H}$ is non-degenerate if and only if every holomorphic map $f : \mathbf{C} \to P^n(\mathbf{C}) - \cup_{H \in \mathcal{H}} H$ is constant.*

Corollary A3.3.5 *Let $H_1, \ldots, H_q$ be hyperplanes in general position. Then every holomorphic map $f : \mathbf{C} \to \mathbf{P}^n(\mathbf{C}) - \cup_{j=1}^q H_j$ is constant if $q \geq 2n+1$.*

Proof. Let $\mathcal{L}$ be the set of linear forms corresponding to the hyperplanes in $\mathcal{H}$. The "in general position" condition for $\mathcal{H}$ means that for each subset $\mathcal{L}_0$ of $\mathcal{L}$ with $\#\mathcal{L}_0 \geq n+1$, $(\mathcal{L}_0) = \mathbf{C}^{*n+1}$. So $\dim(\mathcal{L}) = n+1$. For each proper non-empty subset $\mathcal{L}_1$ of $\mathcal{L}$, either $\#\mathcal{L}_1$ or the number of vectors in $\mathcal{L} - \mathcal{L}_1$ is greater than n, so either $(\mathcal{L}_1) = \mathbf{C}^{*n+1}$ or $(\mathcal{L} - \mathcal{L}_1) = \mathbf{C}^{*n+1}$. In any case, we have

$$(\mathcal{L}_1) \cap (\mathcal{L} - \mathcal{L}_1) \cap \mathcal{L} \neq \emptyset.$$

Thus (3.36) is satisfied, and hence $\mathcal{H}$ is non-degenerate. Theorem A3.3.4 implies that $f : \mathbf{C} \to \mathbf{P}^n(\mathbf{C}) - (H_1 \cup \cdots \cup H_q)$ is constant if $q \geq 2n+1$. This proves Corollary A3.3.5. □

We now prove Theorem A3.3.4. We first prove the following.

Proposition A3.3.6 *Let $f : \mathbf{C} \to \mathbf{P}^n(\mathbf{C})$ be a holomorphic map. If $f(\mathbf{C})$ omits at least three distinct hyperplanes in $\mathbf{P}^n(\mathbf{C})$ which are linearly dependent, then f must be degenerate (that is $f(\mathbf{C})$ is contained in some proper subspace of $\mathbf{P}^n(\mathbf{C})$).*

Proof. Let $H_1, ..., H_s, s \geq 3$ be the distinct hyperplanes which $f(\mathbf{C})$ omits. Let $L_1, ..., L_s$ be the defining linear forms of $H_i, 1 \leq i \leq s$. After rearranging indices, let $L_1, ..., L_t$ be a minimal linearly dependent subset. Then $t \geq 3$ and there exists nonzero constants c_i such that

$$\sum_{i=1}^{t} c_i L_i = 0.$$

Therefore

$$\sum_{i=1}^{t} c_i L_i(\mathbf{f}) = 0.$$

Theorem A3.3.2 implies that there exist some constants d_i, not all zero, such that

$$\sum_{i=2}^{t} d_i \frac{c_i L_i(\mathbf{f})}{L_1(\mathbf{f})} = 0$$

whence

$$\sum_{i=2}^{t} (d_i c_i) L_i(\mathbf{f}) = 0.$$

So the image of f is contained in the hyperplane of $\mathbf{P}^n(\mathbf{C})$ whose defining linear form is $\sum_{i=2}^{t}(d_i c_i)L_i$. By the minimality of t, this hyperplane(subspace) is proper. This proves proposition A3.3.6. □

Proposition A3.3.7 *A finite set of hyperplanes $\mathcal{H}$ (or linear forms $\mathcal{L}$) is non-degenerate if and only if for every $\mathcal{H}$-admissible subspace V of $\mathbf{P}^n(\mathbf{C})$ of projective dimension greater than or equal to one, $\mathcal{H} \cap V$ contains at least three distinct hyperplanes which are linearly dependent. Here $\mathcal{H} \cap V$ is the set of hyperplanes restricted on V, and V is called $\mathcal{H}$-admissible if V is not contained in any hyperplane in $\mathcal{H}$.*

Proof. Let V be a subspace of $\mathbf{P}^n(\mathbf{C})$ of projective dimension greater than or equal to one. Assume that V is $\mathcal{H}$-admissible. We first show that if $\mathcal{H}$ is non-degenerate, then $\mathcal{H}\cap V$ contains at least three distinct hyperplanes which are linearly dependent. Let $\mathbf{V}$ be the subspace of $\mathbf{C}^{n+1}$ of dimension $r \geq 2$, with $\mathbf{P}(\mathbf{V}) = V$. The inclusion map $l : \mathbf{V} \to \mathbf{C}^{n+1}$ induces a surjective map $l^* : \mathbf{C}^{*n+1} \to \mathbf{V}^*$. We denote by $l^*\mathcal{L}$ the set of all l^*L for $L \in \mathcal{L}$. Denote by $\mathcal{L}'$ a maximal set of pairwise linearly independent linear forms in $l^*\mathcal{L}$. Since $\dim(\mathcal{L}) = n+1, \dim(l^*\mathcal{L}) = r \geq 2$, $\dim(l^*\mathcal{L}) = \dim(\mathcal{L}')$, so $\#\mathcal{L}' \geq 2$. Let $\mathcal{L}_1'$ be a proper, non-empty subset of $\mathcal{L}'$. Let $\mathcal{L}_1$ be the largest subset of $\mathcal{L}$ with the property that each linear form in $l^*\mathcal{L}_1$ is linearly dependent on one of the linear forms in $\mathcal{L}_1'$ over $\mathbf{C}$. Then each vector in $l^*\mathcal{L} - l^*\mathcal{L}_1$ is linearly dependent on one of the vectors in $\mathcal{L}' - \mathcal{L}_1'$. From (3.36) we infer that

$$(\mathcal{L}_1') \cap (\mathcal{L}' - \mathcal{L}_1') \cap l^*\mathcal{L} = (l^*\mathcal{L}_1) \cap (l^*\mathcal{L} - l^*\mathcal{L}_1) \cap l^*\mathcal{L} \neq \emptyset.$$

Thus there are vectors $L_1, ..., L_p \in \mathcal{L}_1', L_{p+1}, ..., L_q \in \mathcal{L}' - \mathcal{L}_1' (q > p \geq 1)$ such that

$$\sum_{i=1}^{p} \alpha_i L_i = -\sum_{i=p+1}^{q} \alpha_i L_i \neq 0,$$

whence

$$\sum_{i=1}^{q} \alpha_i L_i = 0$$

with nonzero $\alpha_i \in \mathbf{C}$ for $i = 1, \ldots, q$. Since the vectors in $\mathcal{L}'$ are pairwise linearly independent, we have $q \geq 3$. The hyperplanes defined by $L_i, 1 \leq i \leq q$, are in $\mathcal{H} \cap V$. That is they are the restriction of hyperplanes in $\mathcal{H}$ onto V and are also distinct because of the pairwise linearly independence. So the proof of this direction is finished.

Conversely, assume that for every $\mathcal{H}$-admissible subspace V of $\mathbf{P}^n(\mathbf{C})$ of projective dimension greater than or equal to one, $\mathcal{H} \cap V$ contains at least three distinct hyperplanes whose coefficient vectors are linearly dependent, we will show that $\mathcal{H}$ is non-generate. We first prove a sub-lemma.

Sub-Lemma A3.3.8. *Let $\mathcal{L}$ be a set of linear forms in $\mathbf{C}^{r+1} (r \geq 1)$ which are pair-wisely linearly independent. Assume that dim$(\mathcal{L}) = r + 1$*

and $\mathcal{L} = \mathcal{L}_1 \cup \mathcal{L}_2$, where $\mathcal{L}_1, \mathcal{L}_2$ are two non-empty subsets and where

$$(\mathcal{L}_1) \cap (\mathcal{L}_2) = (0) \tag{3.37}$$

then there exists a $\mathcal{L}$-admissible subspace $\mathbf{V}$ of $\mathbf{C}^{r+1}$ of dimension ≥ 2 such that $\mathcal{L}$ does not contain a subset of at least three linear forms which are linearly dependent on $\mathbf{V}$, but pairwise linearly independent on $\mathbf{V}$.

Proof. Let $\mathcal{L}_j^*(j = 1, 2)$ be the maximal subsets of $\mathcal{L}_1, \mathcal{L}_2$ respectively, such that the forms in $\mathcal{L}_j^*$ are linearly independent. Then, by (3.37), the linear forms in $\mathcal{L}^* = \mathcal{L}_1^* \cup \mathcal{L}_2^*$ are linearly independent. Let $\mathcal{L}_j^* = \{L_{j,1}, ..., L_{j,r_j}\}(j = 1, 2)$. Put $\mathbf{W}_j = (0)$ if $r_j = 1$. And if $r_j \geq 2$, let $\mathbf{W}_j$ be the vector space generated by $L_{j,2} - c_{j,2}L_{j,1}, ..., L_{j,r_j} - c_{j,r_j}L_{j,1}$ over $\mathbf{C}$ for certain constants $c_{j,2}, ..., c_{j,r_j}$ which can be chosen so that

$$(\mathbf{W}_1 + \mathbf{W}_2) \cap \mathcal{L} = \emptyset. \tag{3.38}$$

Indeed, if $r_1 = r_2 = 1$, (3.38) is trivially satisfied. Suppose that $r_j \geq 2$ for some j, it is easy to see that in view of the linear independence of forms in $\mathcal{L}^*$ and the finiteness of $\mathcal{L}$, we can choose $c_{j,2}, ..., c_{j,r_j}$ to satisfy (3.38). Put $\mathbf{W} = \mathbf{W}_1 + \mathbf{W}_2$. Let $\mathbf{V}$ be the vector space defined by

$$\mathbf{V} = \{\mathbf{x} \in \mathbf{C}^{r+1} : L(\mathbf{x}) = 0 \text{ for all } L \in \mathbf{W}\}.$$

V has dimension

$$r + 1 - \dim(\mathbf{W}) = r + 1 - (\dim(\mathcal{L}^*) - 2) = r + 1 - \dim(\mathcal{L}) + 2 \geq 2.$$

It is easy to see that $\mathbf{V}$ is $\mathcal{L}$-admissible. We notice first that, all forms in $\mathcal{L}_j$ are linearly dependent on $L_{j,1}$ on $\mathbf{V}$, for $j = 1, 2$. Secondly, $L_{1,1}, L_{2,1}$ are linearly independent on $\mathbf{V}$. For suppose that $\alpha_1 L_{1,1} + \alpha_2 L_{2,1} = 0$ identically on $\mathbf{V}$, that is $\alpha_1 L_{1,1} + \alpha_2 L_{2,1} \in \mathbf{W}$. Since the forms in $\mathcal{L}^*$ are linearly independent, together with (3.37), we have $\alpha_j L_{j,1} \in W$ for $j = 1, 2$. In view of (3.38), this implies, however, that $\alpha_j = 0$ for $j = 1, 2$ (for otherwise $L_{j,1} \in \mathbf{W}_1 + \mathbf{W}_2$ and also $L_{j,1} \in \mathcal{L}$). So $\mathcal{H} \cap V$ does not contain more than three distinct hyperplanes which are linearly dependent, where $\mathcal{H}$ is the set of hyperplanes defined by the linear forms in $\mathcal{L}$. This proves the sub-lemma A3.3.8. □

We now continue proving Proposition A3.3.7. We first prove $\dim(\mathcal{L}) = n + 1$. If $\dim(\mathcal{L}) < n + 1$, then the hyperplanes have a common point P; taking V to be any line passing through P not contained in any hyperplanes in $\mathcal{H}$ will give a contradiction. So $\dim(\mathcal{L}) = n+1$. We now verify (3.36) is true. Suppose (3.36) is not true; that is there exists a proper, non-empty subset $\mathcal{L}_1$ of $\mathcal{L}$ with

$$(\mathcal{L}_1) \cap (\mathcal{L} - \mathcal{L}_1) \cap \mathcal{L} = \emptyset. \tag{3.39}$$

Let $\mathbf{V}$ be the subspace of $\mathbf{C}^{n+1}$ defined by

$$\mathbf{V} = \{\mathbf{x} \in \mathbf{C}^{n+1} : L(\mathbf{x}) = 0 \text{ for all } L \in (\mathcal{L}_1) \cap (\mathcal{L} - \mathcal{L}_1)\}.$$

By (3.39), V is $\mathcal{L}$-admissible. Denote by r the dimension of $\mathbf{V}$. Since, by (3.39), $m = \dim((\mathcal{L}_1) \cap (\mathcal{L} - \mathcal{L}_1)) < n+1$, we have $r = n+1-m \geq 1$. Let $l^* : \mathbf{C}^{n+1} \to \mathbf{V}^*$(where l is the inclusion map, l^* is the dual map). By (3.39), no form in $l^*\mathcal{L}$ is identically zero. Further, we have, by (3.39),

$$(l^*\mathcal{L}_1) \cap (l^*\mathcal{L} - l^*\mathcal{L}_1) = (0)$$

where both sets $l^*\mathcal{L}_1$ and $l^*\mathcal{L} - l^*\mathcal{L}_1$ are non-empty. But these sets consist of linear forms in r-variables, hence $r \geq 2$. Together with the sub-lemma A3.3.8, this implies that there is an $l^*\mathcal{L}$ -admissible subspace $\mathbf{W}$ of $\mathbf{V}$ of dimension at least 2 such that $l^*\mathcal{L}$ does not contain a subset of at least three linear forms which are linearly dependent on $\mathbf{W}$, but pairwise linearly independent on $\mathbf{W}$. This contradicts our assumption. So Proposition A3.3.7 is proved. □

Proof of Theorem A3.3.4.

Proof. We first prove that if $\mathcal{H}$ is non-degenerate, then every holomorphic map $f : \mathbf{C} \to \mathbf{P}^n(\mathbf{C})$ is constant. Since $\mathcal{H}$ is non-degenerate, Proposition A3.3.7 implies that $\mathcal{H}$ contains at least three distinct hyperplanes which are linearly dependent. Therefore, by Proposition A3.3.6, the image of f is contained in some proper subspace W of $\mathbf{P}^n(\mathbf{C})$. Since the image of f omits the hyperplanes in $\mathcal{H}$, W is $\mathcal{H}$-admissible. Applying Proposition A3.3.7 again, we have that $\mathcal{H} \cap W$ still contains at least three distinct hyperplanes which are linearly dependent. So we can apply Proposition A3.3.6 again to further reduce the dimension and eventually conclude that f is constant.

Conversely, if $\mathcal{H}$ is not degenerate, we are going to construct a non-constant holomorphic mapping $f : C \to \mathbf{P}^n(\mathbf{C}) - \cup_{H \in \mathcal{H}} H$. Since $\mathcal{H}$ is not degenerate, there exists an $\mathcal{H}$-admissible subspace V of $\mathbf{P}^n(\mathbf{C})$ of projective dimension greater than or equal to one such that $\mathcal{H} \cap V$ does not contain at least three distinct hyperplanes which are linearly dependent over $\mathbf{C}$. We may assume, without loss of generality, that $V = \mathbf{P}^n(\mathbf{C})$. Let $\mathcal{H} = \{H_1, ..., H_q\}$, then $q \leq n+1$ and $H_1, ..., H_q$ are linearly independent. We may assume $H_1, ..., H_q$ are first q coordinate planes, then holomorphic map f represented by $\mathbf{f} = (1, e^z, ..., e^z)$ satisfies our conditions. □

A3.4 The Linearly Degenerated Case

In section A3.1, we derive Cartan's Second Main Theorem for the holomorphic curve whose image is not contained in any proper subspaces of $\mathbf{P}^n(\mathbf{C})$. In this section, we deal with the degenerated case, that is the image of $f : \mathbf{C} \to \mathbf{P}^n(\mathbf{C})$ is contained in a proper subspace of $\mathbf{P}^n(\mathbf{C})$. We assume the image $f : \mathbf{C} \to \mathbf{P}^n(\mathbf{C})$ is contained in a subspace of dimension k, but not in any subspace of dimension lower than k. Without loss of generality, we assume that the subspace of dimension k that contains $f(\mathbf{C})$ is $\mathbf{P}^k(\mathbf{C})$. Then $f : \mathbf{C} \to \mathbf{P}^k(\mathbf{C})$ is linearly non-degenerate. However the difficulty in applying the theory in section A3.1 is that hyperplanes $H_1, \ldots, H_q$ in $\mathbf{P}^n(\mathbf{C})$ in general position may not necessarily be in general position after being restricted to $\mathbf{P}^k(\mathbf{C})$. So we have to use the techniques of Nochka to overcome this difficulty.

Let $n \geq k$ and $q \geq n+1$. We consider hyperplanes $H_j, 1 \leq j \leq q$, in $\mathbf{P}^k(\mathbf{C})$, which is given by

$$H_j = \{[x_0 : \ldots : x_k] \mid a_{j0}x_0 + \ldots a_{jk}x_k = 0\},$$

with reduced nonzero coefficient vectors $\mathbf{a}_j = (a_{j0}, \ldots, a_{jk}) \in \mathbf{C}^{k+1}$.

Definition A3.4.1 *Hyperplanes $H_1, \ldots, H_q$ (or $\mathbf{a_1}, \ldots, \mathbf{a_q}$) in $\mathbf{P}^k(\mathbf{C})$ are said to be in n-**subgeneral position** if, for every $1 \leq i_0 < \cdots < i_n \leq q$, the linear span of $\mathbf{a}_{i_0}, \ldots, \mathbf{a}_{i_n}$ is $\mathbf{C}^{k+1}$.*

The following is directly verified by the definition: *Let W be a subspace of $\mathbf{P}^n(\mathbf{C})$ of dimension k. Let $H_1, \ldots, H_q$ be hyperplanes in $\mathbf{P}^n(\mathbf{C})$ in general position. Then the hyperplanes $H_1 \cap W, \ldots, H_q \cap W$ are in n-subgeneral position in W.*

Let $H_1, \ldots, H_q$ (or $\mathbf{a}_1, \ldots, \mathbf{a}_q$) be hyperplanes in $\mathbf{P}^k(\mathbf{C})$, located in n-subgeneral position. Let $A = \{\mathbf{a}_1, \ldots, \mathbf{a}_q\}$. We introduce the **Nochka**

diagram and **Nochka Polygon** of A: For a non-empty subset B of A, we associate with B a point $P_B = (\#B, d(B))$ in $\mathbf{R}^2$, where $d(B)$ is the dimension of linear span of B. The collection of the points $\{P_B | B \subset A\}$ is called the **Nochka diagram** of A (see Figure 3.1).

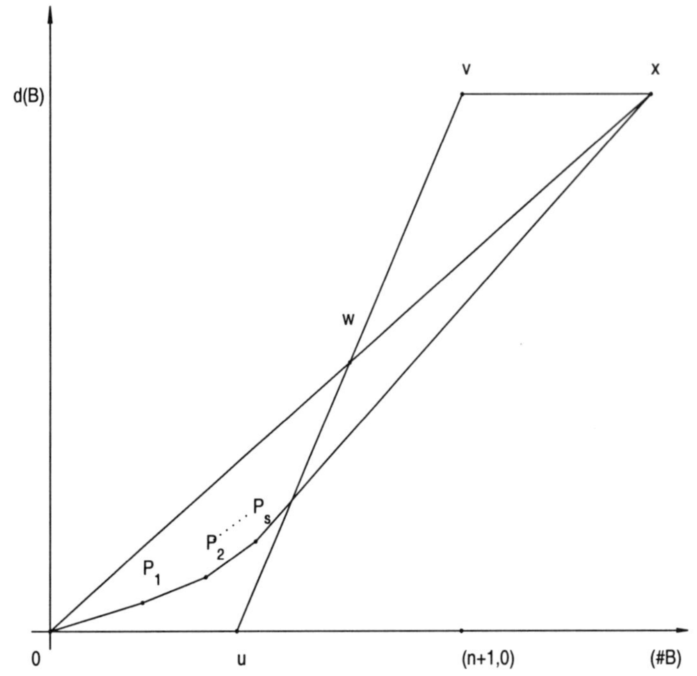

Fig. 3.1 Nochka Diagram

In Figure 3.1, the point $O = (0,0), U = (n-k, 0), V = (n+1, k+1), W = (\frac{2n-k+1}{2}, \frac{k+1}{2})$ and $X = (2n-k+1, k+1)$, so W is the midpoint of the segment UV, as well as of the segment OX. By the n-subgeneral position assumption, the points $P_B = (\#B, d(B))$ with $\#B \geq n+1$ lie on the line with $d(B) = k+1$(i.e., they lie on the horizontal line through V, and to the right of V). On the other hand, the points $P_B = (\#B, d(B))$ with $\#B \leq n+1$ lie on or above the line through U and V.

Proposition A3.4.2 *Let $A = \{\mathbf{a_1}, \ldots, \mathbf{a_q}\}$ be a set of vectors in $\mathbf{C}^{k+1}$ in n subgeneral position. Then either $\sigma(O, P_B) = \sigma(O, X)$ for all $P_B = (\#B, d(B))$ with $\#B \leq n+1$ where σ denotes the slope of the associated line segments, or there exists a uniquely determined sequence of subsets of A:*

$$A \supset B_s \supset B_{s-1} \supset \ldots \supset B_1 \supset B_0 = \emptyset$$

with the following properties:

(i) $\sigma(P_{B_{i-1}}, P_{B_i}) < \sigma(P_{B_{i-1}}, X), 1 \leq i \leq s$, *where* σ *denotes the slope of the associated line segments, and* $X = (2n - k + 1, k + 1)$.

(ii) $\sigma(O, P_{B_i}) < \sigma(O, X)$ *for* $1 \leq i \leq s$, *where* $O = (0, 0)$ *is the origin, and* X *is as above.*

(iii) $\sigma(P_{B_{i-1}}, P_{B_i}) < \sigma(P_{B_i}, P_{B_{i+1}}), 1 \leq i \leq s$, *where we set* $P_{B_{s+1}} = X$.

(iv) For $0 \leq i \leq s$, *let* $\mathcal{U}_i$ *be the collection of the set* $B \subset A$ *with* $\#B \leq n + 1$ *and where* B *strictly contains* B_i. *Then* $\sigma(P_{B_i}, P_{B_{i+1}}) \leq \sigma(P_{B_i}, P_B)$ *for any* $B \in \mathcal{U}_i$ *with strict inequality if* $i < s$ *and* $B \in \mathcal{U}_{i+1}$.

Proof. The sets $B_1, \ldots, B_s$ are constructed inductively. Suppose that $B_0, \ldots, B_j$ have been constructed. Then by induction hypothesis, (i) and (ii) are satisfied for all $0 \leq i \leq j$ (these conditions are empty if $j = 0$), and (iii), (iv) are satisfied for $0 \leq i < j$ (these conditions are empty if $j \leq 1$).

If $\sigma(P_{B_j}, P_B) \geq \sigma(P_{B_j}, X)$ for all $B \in \mathcal{U}_j$, i.e., (iv) is satisfied for $i = j$, then we set $j = s$. By (i) of the induction hypothesis, we have $\sigma(P_{B_{s-1}}, P_{B_s}) < \sigma(P_{B_{s-1}}, X)$. This implies that (consider the triangle $P_{B_{s-1}} P_{B_s} X$) $\sigma(P_{B_{s-1}}, P_{B_s}) < \sigma(P_{B_s}, X)$ which is (iii) for the case $i = j = s$.

We may now assume that there exists $B \in \mathcal{U}_j$ such that

$$\sigma(P_{B_j}, P_B) < \sigma(P_{B_j}, X). \tag{3.40}$$

Let

$$\sigma_j = \min_{B \in \mathcal{U}_j} \{\sigma(P_{B_j}, P_B)\}$$

and

$$\mathcal{M}_j = \{B \in \mathcal{U}_j \mid \sigma(P_{B_j}, P_B) = \sigma_j\}.$$

For $B \in \mathcal{M}_j$, we claim that $d(B) < (k + 1)/2$. In fact since $B \in \mathcal{M}_j$, $\sigma(P_{B_j}, P_B) < \sigma(P_{B_j}, X)$. By (ii), we also have $\sigma(O, P_{B_j}) < \sigma(O, X)$. These conditions and the remark before the proposition imply that P_B lies in the triangle OUW that appears in the Nochka diagram, but not on the segment OW. Hence $d(B) < (k + 1)/2$.

We now claim that if B and C are in $\mathcal{M}_j$, then $B \cup C$ is also in $\mathcal{M}_j$. First of all, since $d(B \cup C) \leq d(B) + d(C) < k+1$, the sub-general position condition implies that $\#(B \cup C) < n+1$. Thus, $(B \cup C) \in \mathcal{U}_j$. Now

$$\begin{aligned} d(B \cup C) - d(B_j) &= d(B) + d(C) - d(B \cap C) - d(B_j) \\ &= d(B) - d(B_j) + d(C) - d(B_j) - d(B \cap C) + d(B_j) \\ &\leq \sigma_j \{\#(B) - \#(B_j) + \#(C) - \#(B_j) - \#(B \cap C) + \#(B_j)\} \\ &= \sigma_j \{\#(B \cup C) - \#(B_j)\}, \end{aligned}$$

i.e., $\sigma(P_{B_j}, P_{B \cup C}) \leq \sigma_j$. Thus $(B \cup C) \in \mathcal{M}_j$. The claim is proved.

We now define

$$B_{j+1} = \cup_{B \in \mathcal{M}_j} B.$$

The above claim implies that $B_{j+1} \in \mathcal{M}_j$. We now check (i) to (iv). The proceeding argument shows that $B_{j+1} \in \mathcal{M}_j$. By construction, $\sigma_j = \sigma(P_{B_j}, P_{B_{j+1}}) \leq \sigma(P_{B_j}, P_B)$ for all $B \in \mathcal{M}_j$. Now that the inequality is strict if B is in $\mathcal{M}_{j+1}$. Thus (iv) is verified for $i = j$. By assumption (3.40), $\sigma(P_{B_j}, P_{B_{j+1}}) < \sigma(P_{B_j}, X)$ so (i) is verified for $i = j+1$. This together with $\sigma(O, P_{B_j}) < \sigma(O, X)$ (by induction hypothesis, (ii) holds for $i \leq j$) imply that the point $P_{B_{j+1}}$ is below the line OX. Hence $\sigma(O, P_{B_{j+1}}) < \sigma(O, X)$, which is (ii) for $i = j+1$. From (iv) with $i = j-1$ and $B = B_{j+1}$, we have $\sigma(P_{B_{j-1}}, P_{B_j}) \leq \sigma(P_{B_{j-1}}, P_{B_{j+1}})$. This implies that (by considering the triangle $P_{B_{j-1}} P_{B_j} P_{B_{j+1}}$) $\sigma(P_{B_{j-1}}, P_{B_j}) < \sigma(P_{B_j}, P_{B_{j+1}})$. Thus (iii) is verified for $i = j$.

This completes the induction step. Since the sets $B_0, B_1, \ldots$ are strictly increasing and A is a finite set, the above construction terminates after a finite number of steps, concluding the proof of the Proposition A3.4.2. □

By Proposition A3.4.2, either $\sigma(O, P(B)) = \sigma(O, X)$ for all $P_B = (\#B, d(B))$ with $\#B \leq n+1$ or there exists a sequence $B_1, \ldots, B_s$, which appears in Proposition A3.4.2. If the first case occurs, we define $\omega(\mathbf{a}) = 1$ for $\mathbf{a} \in A$. Otherwise, the sequence $B_1, \ldots, B_s$ which appears in Proposition A3.4.2 gives rise to a polygon in $\mathbf{R}^2$ that is called the **Nochka Polygon** of A (see fig. 3.1). Set $B_{s+1} = A$, then $(B_1 - B_0) \cup (B_2 - B_1) \cup \cdots \cup (B_s - B_{s-1}) \cup (A - B_s) = A$ is a partition of A. We now define the **Nochka weights** $\omega(\mathbf{a})$ of $\mathbf{a} \in A$ as follows: For any $\mathbf{a} \in A$, $\mathbf{a}$ lies in $B_{i+1} - B_i$ for some $0 \le i \le s$, define

$$\omega(\mathbf{a}) = \sigma(P_{B_i}, P_{B_{i+1}}), \tag{3.41}$$

and

$$\theta = \frac{1}{\sigma(P_{B_s}, X)} = \frac{2n - k + 1 - \#B_s}{k + 1 - d(B_s)}. \tag{3.42}$$

Example *Consider* $\mathbf{P}^2(\mathbf{C})$. *Let* $H_1 = \{[x_0 : x_1 : x_2] \mid x_0 = 0\}$, $H_2 = \{[x_0 : x_1 : x_2] \mid x_1 = 0\}$, $H_3 = \{[x_0 : x_1 : x_2] \mid x_0 + x_1 = 0\}$, $H_4 = \{[x_0 : x_1 : x_2] \mid x_2 = 0\}$, $H_5 = \{[x_0 : x_1 : x_2] \mid x_0 - x_1 + x_2 = 0\}$, $H_6 = \{[x_0 : x_1 : x_2] \mid x_0 - x_1 + 2x_2 = 0\}$. *Then these hyperplanes are in 3-subgeneral position. In this case, one can verify that* $\sigma(O, P_B) = \sigma(O, X)$ *for all* $P_B = (\#B, d(B))$ *with* $\#B \le 4$. *So* $\omega \equiv 1$.

The significance of Nochka weights is given by the following Theorem.

Theorem A3.4.3(Nochka) *Let* $H_1, \ldots, H_q$ *(or* $\mathbf{a}_1, \ldots, \mathbf{a}_q$*) be hyperplanes in* $\mathbf{P}^k(\mathbf{C})$ *in* n*-subgeneral position with* $2n - k + 1 \le q$. *Then there exists a function* $\omega : \{1, \ldots, q\} \to \mathbf{R}(0, 1]$ *called a Nochka weight and a real number* $\theta \ge 1$ *called Nochka constant satisfying the following properties:*

(i) If $j \in \{1, \ldots, q\}$, *then* $0 \le \omega(j)\theta \le 1$.

(ii) $q - 2n + k - 1 = \theta(\sum_{j=1}^{q} \omega(j) - k - 1)$.

(iii) If $\emptyset \ne B \subset \{1, \ldots, q\}$ *with* $\#B \le n + 1$, *then* $\sum_{j \in B} \omega(j) \le \dim L(B)$, *where* $L(B)$ *is the linear space generated by* $\{\mathbf{a}_j | j \in B\}$,

(iv) $1 \le (n + 1)/(k + 1) \le \theta \le (2n - k + 1)/(k + 1)$.

(v) Given real numbers $E_1, \ldots, E_q$ *with* $E_j \ge 1$ *for* $1 \le j \le q$, *and given any* $Y \subset \{1, \ldots, q\}$ *with* $0 < \#Y \le n + 1$, *there exists a subset* M *of* Y *with* $\#M = \dim L(Y)$ *such that* $\{\mathbf{a}_j\}_{j \in M}$ *is a basis for* $L(Y)$ *where* $L(Y)$ *is the linear space generated by* $\{\mathbf{a}_j | j \in Y\}$, *and*

$$\prod_{j \in Y} E_j^{\omega(j)} \le \prod_{j \in M} E_j.$$

Proof. With the $\omega(j)$ and θ defined by (3.41), and (3.42), (i) immediately follows from (iii) of Theorem A3.4.3. To verify (ii), we set $B_{s+1} = \{1, 2, \ldots, q\}$. Then, write $\sigma_i = \sigma(P_{B_i}, P_{B_{i+1}})$,

$$\begin{aligned}\sum_{j=1}^{q} \omega(j) &= \sum_{0 \le i \le s} \sigma_i(\#B_{i+1} - \#B_i) \\ &= \sum_{0 \le i \le s-1} \{d(B_{i+1}) - d(B_i)\} + \sigma_s(q - \#B_s) \\ &= d(B_s) + \sigma_s(q - \#B_s),\end{aligned}$$

where, by definition,

$$\sigma_s = \sigma(P_{B_s}, X) = \frac{k + 1 - d(B_s)}{2n - k + 1 - \#B_s} = \frac{1}{\theta}.$$

Hence

$$d(B_s) = k + 1 - \sigma_s(2n - k + 1 - \#B_s).$$

Substituting yields (ii).

For (iii), we consider the two cases (a) $\#(B \cup B_s) \ge n + 1$, (b) $\#(B \cup B_s) < n + 1$ separately.

If $\#(B \cup B_s) \ge n + 1$, then the sub-general position implies

$$k + 1 \le d(B \cup B_s). \tag{3.43}$$

By property (i), the Nochka weights satisfy $\sigma_s^{-1}\omega(\mathbf{a}) \le 1$. Thus,

$$\sum_{\mathbf{a} \in B} \omega(\mathbf{a}) \le \sigma_s \#B.$$

By the definition of subgeneral position, any set $\#B \le n + 1$, say $\#B = n + 1 - p$, satisfies $k + 1 - p \le d(B)$. Thus,

$$\#B \le d(B) + n - k.$$

Therefore,

$$\sum_{\mathbf{a} \in B} \omega(\mathbf{a}) \le \sigma_s(d(B) + n - k) = d(B)\sigma_s\left(1 + \frac{n - k}{d(B)}\right).$$

By (3.43), $k+1 \le d(B) + d(B_s)$ so that

$$\begin{aligned}\sum_{\mathbf{a}\in B}\omega(\mathbf{a}) &\le \sigma_s(d(B)+n-k) = d(B)\sigma_s\left(1+\frac{n-k}{d(B)}\right)\\ &\le d(B)\sigma_s\left(1+\frac{n-k}{k+1-d(B_s)}\right)\\ &= d(B)\sigma_s\frac{n+1-d(B_s)}{k+1-d(B_s)}\\ &\le d(B)\sigma_s\frac{2n-k+1-\#B_s}{k+1-d(B_s)}\\ &= d(B).\end{aligned}$$

So (iii) holds in the case (a).

We now assume that $\#(B\cup B_s) < n+1$. Then the set $B'_{i+1} = B_i \cup (B\cap B_{i+1})$ contains B_i and $\#B'_{i+1} \le \#(B\cup B_s) < n+1$ for all i. It follows that B'_{i+1} is in $\mathcal{U}_i$ and by part (iv) of Proposition A3.4.2,

$$\sigma_i = \sigma(P_{B_i}, P_{B_{i+1}}) \le \sigma(P_{B_i}, P_{B'_{i+1}}) = \frac{d(B_i\cup(B\cap B_{i+1})) - d(B_i)}{\#(B_i\cup(B\cap B_{i+1})) - \#B_i}.$$

Since B_{i+1} contains B_i, we have $\#(B_i\cup(B\cap B_{i+1})) - \#B_i = \#(B\cap B_{i+1}) - \#(B\cap B_i)$ and $d(B_i\cup(B\cap B_{i+1})) - d(B_i) = d(B\cap B_{i+1}) - d(B\cap B_i)$. Thus

$$\sigma_i \le \frac{d(B\cap B_{i+1}) - d(B\cap B_i)}{\#(B\cap B_{i+1}) - \#(B\cap B_i)}.$$

The sum of Nochka weights can now easily be estimated:

$$\begin{aligned}\sum_{\mathbf{a}\in B}\omega(\mathbf{a}) &\le \sum_{0\le i\le s}\sigma_i\{\#(B\cap B_{i+1}) - \#(B\cap B_i)\}\\ &\le \sum_{0\le i\le s}(d(B\cap B_{i+1}) - d(B\cap B_i))\\ &= d(B\cap B_{s+1}) = d(B).\end{aligned}$$

This completes the proof of (iii).

To verify (iv), since P_{B_s} lies below the line OX in the Nochka diagram, $\sigma_s = \sigma(P_{B_s}, X) > \sigma(O,X) = (k+1)/(2n-k+1)$. On the other hand, P_{B_s} lies below the triangle OUW, thus $\sigma(P_{B_s}, X) < \sigma(U,X) = (k+1)/(n+1)$. This proves (iv).

Finally, we prove (v). Without loss of generality, we assume that $1 \le E_q \le E_{q-1} \le \ldots \le E_1$. Define an increasing sequence of subsets of Y as follows: Let $i_1 = \min\{i \mid i \in Y\}$ and $I_1 = \{i \in Y \mid \mathbf{a}_i$ is a multiple of $\mathbf{a}_{i_1}\}$. If $Y - I_1 \neq \emptyset$, choose $i_2 \in Y - I_1$ such that $i_2 = \min\{i \mid i \in Y - I_1\}$ and $I_2 = \{i \in Y \mid \mathbf{a}_i \in$ linear span of $\mathbf{a}_{i_1}, \mathbf{a}_{i_2}\}$. Inductively, if I_{j-1} is defined and if $Y - I_{j-1} \neq \emptyset$, $i_j = \min\{i \mid i \in Y - I_{j-1}\}$ and $I_j = \{i \in Y \mid \mathbf{a}_i \in$ linear span of $\mathbf{a}_{i_1}, \ldots, \mathbf{a}_{i_{j-1}}\}$. This process stops at I_p with p =dimension of the $L(Y)$. It is clear that

$$I_p \supset I_{p-1} \supset \ldots$$

and $i_p \geq i_{p-1} \geq \ldots \geq i_1$. Let $M = \{i_1, \ldots, i_p\}$. Then, by construction, the set $\{\mathbf{a}_j\}_{j \in M}$ is a basis for $L(Y)$. Set $I_0 = \emptyset$, then $Y = \cup_{1 \leq j \leq p}(I_j - I_{j-1})$ is a disjoint union. Since $E_q \leq \ldots \leq E_1$, we have, by construction,

$$E_{i_j} = \max_{i \in I_j - I_{j-1}} E_i.$$

Thus

$$\prod_{j \in Y} E_j^{\omega(j)} \leq \prod_{1 \leq j \leq p} \prod_{i \in I_j - I_{j-1}} E_i^{\omega(i)} \leq \prod_{1 \leq j \leq p} E_{i_j}^{\alpha_j}$$

where $\alpha_j = \sum_{i \in I_j - I_{j-1}} \omega(i)$. Since the sets I_j are increasing, for any $1 \leq r \leq p$, (iii) implies that

$$\sum_{1 \leq j \leq r} \alpha_j = \sum_{1 \leq j \leq r} \sum_{i \in I_j - I_{j-1}} \omega(i) = \sum_{i \in I_r} \omega(i) \leq d(I_r) = r. \tag{3.44}$$

It remains to show that

$$\prod_{1 \leq j \leq p} E_{i_j}^{\alpha_j} \leq \prod_{1 \leq j \leq p} E_{i_j}.$$

This is easily verified by the induction on p. For $p = 1$, by (iv), $\alpha_1 \leq 1$, and since $E_1 \geq 1$, we have trivially $E_1^{\alpha_1} \leq E_1$. Assume that the inequality holds for $p = k$. Since, by (3.44) $\alpha_{k+1} \leq k + 1 - \sum_{1 \leq j \leq k} \alpha_j$, we have

$$\prod_{1 \leq j \leq k+1} E_{i_j}^{\alpha_j} \leq \left(\prod_{1 \leq j \leq k} \left(\frac{E_{i_j}}{E_{i_{k+1}}} \right)^{\alpha_j} \right) E_{i_{k+1}}^{k+1}.$$

Since $i_p \geq \ldots \geq i_1$, we have $1 \leq E_{i_j}/E_{i_{k+1}}$ for $1 \leq j \leq k$, the induction hypothesis implies that

$$\left(\prod_{1\le j\le k}\left(\frac{E_{i_j}}{E_{i_{k+1}}}\right)^{\alpha_j}\right)E_{i_{k+1}}^{k+1}\le\left(\prod_{1\le j\le k}\frac{E_{i_j}}{E_{i_{k+1}}}\right)E_{i_{k+1}}^{k+1}=\prod_{1\le j\le k+1}E_{i_j}.$$

The combination of the last two inequalities completes the proof of (v). □

We now derive the Second Main Theorem with a good error term for the holomorphic curves whose image is contained in some k-dimensional subspace of $\mathbf{P}^n(\mathbf{C})$. By using Nochka weights, we reduce the problem to Theorem A3.1.3.

Theorem A3.4.4 (Degenerated SMT) *Let* $f=[f_0:\ldots:f_n]:\mathbf{C}\to\mathbf{P}^n(\mathbf{C})$ *be a holomorphic map whose image is contained in some* k*-dimensional subspace but not in any subspace of dimension lower than* k*. Let* H_j*,* $1\le j\le q$*, be hyperplanes in general position. Assume that* $f(\mathbf{C})\not\subset H_j$ *for* $1\le j\le q$*. Then, the inequality*

$$\sum_{j=1}^{q}m_f(r,H_j)+\left(\frac{n+1}{k+1}\right)N(R_f,r)\le(2n-k+1)T_f(r)$$
$$+\frac{k(2n-k+1)}{2}(\log T_f(r)+(1+\epsilon)\log^+\log T_f(r))+O(1),$$

holds for all r *outside a set* E *with finite Lebesgue measure, here* $N(R_f,r)$ *is the ramification term defined below.*

Proof. Without loss of generality, we may assume that $f(\mathbf{C})\subset\mathbf{P}^k(\mathbf{C})$. So $f:\mathbf{C}\to\mathbf{P}^k(\mathbf{C})$ is a non-degenerate holomorphic map. We also assume that $q\ge 2n-k+1$. Denote by $\hat{H}_j=H_j\cap\mathbf{P}^k(\mathbf{C})$. Then $\hat{H}_j$ are hyperplanes in $\mathbf{P}^k(\mathbf{C})$ located in n-subgeneral position.

Since $H_1,\ldots,H_q$ (or $\mathbf{a}_1,\ldots,\mathbf{a}_q$) are hyperplanes in general position, Similar to the proof of Lemma A3.1.6 (see the proof of Lemma B3.4.4 for detail), for each $z\in\mathbf{C}$, there are indices $i(z,0),\ldots,i(z,n)\in\{1,\ldots,q\}$ such that

$$\prod_{j=1}^{q}\left(\frac{\|\mathbf{f}(z)\|\|\mathbf{a}_j\|}{|<\mathbf{f}(z),\mathbf{a}_j>|}\right)^{\omega(j)}\le C\prod_{l=0}^{n}\left(\frac{\|f(z)\|\|\mathbf{a}_{i(z,l)}\|}{|<\mathbf{f}(z),\mathbf{a}_{i(z,l)}>|}\right)^{\omega(i(z,l))}\qquad(3.45)$$

where $\omega(j)$ is the Nochka weight corresponding to $\hat{H}_j$ and $C > 0$ is a constant. Recall that the Weil function $\lambda_H(f(z))$ of f is defined by

$$\lambda_H(f(z)) = \log \frac{\|\mathbf{f}(z)\|\|\mathbf{a}\|}{|<\mathbf{f}(z), \mathbf{a}>|},$$

and $\lambda_H(f(z)) \geq 0$. Applying Theorem A3.4.3 with $E_l = e^{\lambda_{\hat{H}_{i(z,l)}}(f(z))}, 0 \leq l \leq n$, there is a subset M of $Y = \{i(z,0), \ldots, i(z,n)\}$ with $\#M = k+1$ such that $\{\hat{H}_{i(z,j)} | i(z,j) \in M\}$ is linearly independent, and

$$\prod_{l=0}^{n} e^{\omega(i(z,l))\lambda_{\hat{H}_{i(z,l)}}(f(z))} \leq \prod_{i(z,j)\in M} e^{\lambda_{\hat{H}_{i(z,j)}}(f(z))}.$$

Thus

$$\sum_{l=0}^{n} \omega(i(z,l))\lambda_{\hat{H}_{i(z,l)}}(f(z)) \leq \sum_{i(z,j)\in M} \lambda_{\hat{H}_{i(z,j)}}(f(z)) \leq \max_{\gamma\in\Gamma} \sum_{l=0}^{k} \lambda_{\hat{H}_{\gamma(l)}}(f(z)),$$

where Γ is the set of all maps $\gamma : \{0, \ldots, k\} \to \{1, \ldots, q\}$ such that $\hat{H}_{\gamma(0)}, \ldots, \hat{H}_{\gamma(k)}$ are linearly independent. Combining this with (3.45) gives us

$$\sum_{j=1}^{q} \omega(j) m_f(H_j, r) \leq \int_0^{2\pi} \max_{\gamma\in\Gamma} \sum_{l=0}^{k} \lambda_{\hat{H}_{\gamma(l)}}(f(re^{i\theta}))\frac{d\theta}{2\pi} + O(1).$$

Applying Theorem A3.1.3 yields that the inequality

$$\begin{aligned}\int_0^{2\pi} \max_{\gamma\in\Gamma} \sum_{l=0}^{k} \lambda_{\hat{H}_{\gamma(l)}}(f(re^{i\theta}))\frac{d\theta}{2\pi} &\leq (k+1)T_f(r) - N_W(r,0) \\ &+ \frac{k(k+1)}{2}(\log T_f(r) + (1+\epsilon)\log^+ \log T_f(r)) + O(1)\end{aligned}$$

holds for all r outside a set $E \subset (0, +\infty)$ with finite Lebesgue measure. Denote by $N(R_f, r) = N_W(r, 0)$, then the inequality

$$\begin{aligned}\sum_{j=1}^{q} \omega(j) m_f(r, H_j) &\leq (k+1)T_f(r) - N(R_f, r) \\ &\quad + \frac{k(k+1)}{2}(\log T_f(r) + (1+\epsilon)\log^+ \log T_f(r)) + O(1)\end{aligned}$$

holds for all r outside a set $E \subset (0, +\infty)$ with finite Lebesgue measure. Combining this with Theorem A3.4.3, and recalling that $m_f(r, H_j) \leq T_f(r) + O(1)$, we have

$$\begin{aligned}
\sum_{j=1}^{q} m_f(r, H_j) &= \sum_{j=1}^{q}(1 - \theta\omega(j))m_f(r, H_j) + \sum_{j=1}^{q}\theta\omega(j)m_f(r, H_j) \\
&\leq \sum_{j=1}^{q}(1 - \theta\omega(j))m_f(r, H_j) + \theta(k+1)T_f(r) - \theta N_W(r, 0) \\
&\quad + \theta\frac{k(k+1)}{2}(\log T_f(r) + (1+\epsilon)\log^+\log T_f(r)) + O(1) \\
&\leq \sum_{j=1}^{q}(1 - \theta\omega(j))T_f(r) + \theta(k+1)T_f(r) - \binom{n+1}{k+1}N(R_f, r) \\
&\quad + \frac{(2n-k+1)k}{2}(\log T_f(r) + (1+\epsilon)\log^+\log T_f(r)) + O(1) \\
&= \left\{q - \theta\left(\sum_{1\leq j\leq q}\omega(j) - k - 1\right)\right\}T_f(r) - \binom{n+1}{k+1}N(R_f, r) \\
&\quad + \frac{(2n-k+1)k}{2}(\log T_f(r) + (1+\epsilon)\log^+\log T_f(r)) + O(1) \\
&= (2n-k+1)T_f(r) - \binom{n+1}{k+1}N(R_f, r) \\
&\quad + \frac{(2n-k+1)k}{2}(\log T_f(r) + (1+\epsilon)\log^+\log T_f(r)) + O(1),
\end{aligned}$$

where the inequality holds for all r outside a set $E \subset (0, +\infty)$ with finite Lebesgue measure. □

Corollary A3.4.5 *Let $H_1, \ldots, H_q$ be hyperplanes in $\mathbf{P}^n(\mathbf{C})$, located in general position. Let $f : \mathbf{C} \to \mathbf{P}^n(\mathbf{C}) - \cup_{j=1}^{q} H_j$ be a holomorphic map. If $q \geq 2n+1$, then f is constant.*

A3.5 Ahlfors' Approach

In this section, we give another proof of the Second Main Theorem using Ahlfors' method. The method also extends to holomorphic maps from certain parabolic manifolds to the projective space. For details, see [Sto4]. Let f be a holomorphic map from $\mathbf{C}$ to $\mathbf{P}^n$, and let $\mathbf{f} : \mathbf{C} \to \mathbf{C}^{n+1} - \{0\}$ be a reduced representation of f. Consider the holomorphic map $\mathbf{F}_k$ defined by

$$\mathbf{F}_k = \mathbf{f} \wedge \mathbf{f}' \wedge \cdots \wedge \mathbf{f}^{(k)} : \mathbf{C} \to \bigwedge^{k+1} \mathbf{C}^{n+1}.$$

Evidently $\mathbf{F}_{n+1} \equiv 0$. Throughout this section, we assume that f is linearly non-degenerated, so $\mathbf{F}_k \not\equiv 0$ for $0 \leq k \leq n$. The map $F_k = \mathbf{P}(\mathbf{F}_k) : \mathbf{C} \to \mathbf{P}(\bigwedge^{k+1} \mathbf{C}^{n+1}) = \mathbf{P}^{N_k}$, where $N_k = \frac{(n+1)!}{(k+1)!(n-k)!} - 1$ and $\mathbf{P}$ is the natural projection, is called **the k-th associated map**. Let $\omega_k = dd^c \log \|Z\|^2$ be the Fubini-Study form on $\mathbf{P}^{N_k}(\mathbf{C})$, where $Z = [x_0 : \ldots : x_{N_k}] \in \mathbf{P}^{N_k}(\mathbf{C})$. Let

$$\Omega_k = F_k^* \omega_k = \frac{\sqrt{-1}}{2\pi} h_k dz \wedge d\bar{z}, \quad 0 \leq k \leq n \tag{3.46}$$

be the pull-back via the k-th associated curve.

Lemma A3.5.1 *In terms of homogeneous coordinates,*

$$\Omega_k = F_k^* \omega_k = dd^c \log \|\mathbf{F}_k\|^2 = \frac{\sqrt{-1}}{2\pi} \frac{\|\mathbf{F}_{k-1}\|^2 \|\mathbf{F}_{k+1}\|^2}{\|\mathbf{F}_k\|^4} dz \wedge d\bar{z}$$

for $0 \leq k \leq n$, and by convention $\|\mathbf{F}_{-1}\| \equiv 1$.

Proof. Recall that, in chapter 2, we introduce the differential operators $d = \partial + \bar{\partial}$ and $d^c = \frac{\sqrt{-1}}{4\pi}(\bar{\partial} - \partial)$ so that

$$dd^c = \frac{\sqrt{-1}}{2\pi} \partial \bar{\partial}.$$

The Fubini-Study form ω_k on $\mathbf{P}^{N_k}(\mathbf{C})$ is $\omega_k = dd^c \log \|Z\|^2$. So

$$\begin{aligned} \Omega_k &= F_k^* \omega_k = dd^c \log \|\mathbf{F}_k\|^2 = \frac{\sqrt{-1}}{2\pi} \partial \bar{\partial} \log(\mathbf{F}_k, \mathbf{F}_k) \\ &= \frac{\sqrt{-1}}{2\pi} \frac{(\mathbf{F}'_k, \mathbf{F}'_k)(\mathbf{F}_k, \mathbf{F}_k) - (\mathbf{F}'_k, \mathbf{F}_k)(\mathbf{F}_k, \mathbf{F}'_k)}{\|\mathbf{F}_k\|^4} dz \wedge d\bar{z}, \end{aligned}$$

where $(\cdot, \cdot)$ is the inner product on $\mathbf{C}^{N_k+1}$. Since

$$\mathbf{F}_k = \mathbf{f} \wedge \mathbf{f}' \wedge \cdots \wedge \mathbf{f}^{(k)},$$

$$\mathbf{F}'_k = \mathbf{f} \wedge \mathbf{f}' \wedge \cdots \wedge \mathbf{f}^{(k-1)} \wedge \mathbf{f}^{(k+1)}.$$

For any $z \in \mathbf{C}$ with $\mathbf{F}_{k+1}(z) \neq 0$, by Gram Schmidt there exist numbers A_{ji} for $0 \leq i \leq j \leq n+1$ with $A_{ii} > 0$ and an orthonormal basis $\mathbf{e}_0, \ldots, \mathbf{e}_{k+1}$ such that

$$\mathbf{f}^{(j)}(z) = \sum_{i=0}^{j} A_{ji}\mathbf{e}_i \qquad \text{for } j = 0, \ldots, k+1.$$

So

$$\mathbf{F}_j = A_{00} \cdots A_{jj}\mathbf{e}_0 \wedge \ldots \wedge \mathbf{e}_j \qquad \text{for } j = 0, \ldots, k+1.$$

Thus

$$||\mathbf{F}_j|| = A_{00} \ldots A_{jj} \qquad \text{for } j = 0, \ldots, k+1.$$

Also

$$\mathbf{F}'_k(z) = \left(\prod_{q=0}^{k-1} A_{qq}\right) \mathbf{e}_0 \wedge \ldots \wedge \mathbf{e}_{k-1} \wedge (A_{k+1,k}\mathbf{e}_k + A_{k+1,k+1}\mathbf{e}_{k+1}).$$

Since $\mathbf{e}_0, \ldots, \mathbf{e}_{k+1}$ is an orthonormal basis,

$$(\mathbf{F}'_k, \mathbf{F}'_k) = \left(\prod_{q=0}^{k-1} A_{qq}^2\right) (A_{k+1,k}^2 + A_{k+1,k+1}^2);$$

$$(\mathbf{F}_k, \mathbf{F}_k) = A_{00}^2 \cdots A_{kk}^2;$$

and

$$(\mathbf{F}'_k, \mathbf{F}_k)(\mathbf{F}_k, \mathbf{F}'_k) = \left(\prod_{q=0}^{k-1} A_{qq}^4\right) A_{k,k}^2 A_{k+1,k}^2.$$

Therefore,

$$\begin{aligned}&(\mathbf{F}'_k, \mathbf{F}'_k)(\mathbf{F}_k, \mathbf{F}_k) - (\mathbf{F}'_k, \mathbf{F}_k)(\mathbf{F}_k, \mathbf{F}'_k)\\ &= A_{00}^4 \ldots A_{k-1,k-1}^4 A_{k,k}^2 A_{k+1,k+1}^2 = ||\mathbf{F}_{k-1}||^2 ||\mathbf{F}_{k+1}||^2.\end{aligned}$$

This proves Lemma A3.5.1. □

So, by Lemma A3.5.1 and (3.46),

$$h_k = \frac{||\mathbf{F}_{k-1}||^2 ||\mathbf{F}_{k+1}||^2}{||\mathbf{F}_k||^4}. \tag{3.47}$$

Note that $\Omega_n \equiv 0$. It follows that

$$dd^c \log h_k = \Omega_{k-1} + \Omega_{k+1} - 2\Omega_k.$$

Define the kth characteristic function

$$T_{F_k}(r,s) = \int_s^r \frac{dt}{t} \int_{|z|\le t} F_k^* \omega_k.$$

Denote by

$$N_{d_k}(r,s) = \int_s^r n_{d_k}(t) \frac{dt}{t}$$

where $n_{d_k}(t)$ is the number of zeros of the k-th associated map $\mathbf{F}_k$ in $|z| < t$, counting multiplicities. Note that $N_{d_k}(r,s)$ does not depend on the choice of the reduced representation. Define

$$S_k(r) = \frac{1}{2} \int_0^{2\pi} \log h_k(re^{i\theta}) \frac{d\theta}{2\pi}. \tag{3.48}$$

Then Lemma A3.5.1 and Theorem A2.1.3 (Green-Jensen's formula) imply the following.

Theorem A3.5.2 (Plücker's Formula) *For $0 < s < r$ and integers k with $0 \le k \le n$,*

$$N_{d_k}(r,s) + T_{F_{k-1}}(r,s) - 2T_{F_k}(r,s) + T_{F_{k+1}}(r,s) = S_k(r) - S_k(s),$$

where $T_{F_{-1}}(r,s) \equiv 0$ and $T_{F_0}(r,s) = T_f(r,s)$.

The Plücker formula implies the following theorem.

Theorem A3.5.3 *For $0 \le k \le n-1$,*

$$T_{F_k}(r,s) \le (n+2)^2 T_f(r,s) + O(\log r)$$

holds for all r outside a set E with finite Lebesgue measure.

Proof. Fix a number s and write $T(r) = \sum_{k=0}^{n-1} T_{F_k}(r,s)$. Recall that

$$T_{F_k}(r,s) = \int_s^r \frac{dt}{t} \int_{|z|\le t} F_k^* \omega_k = \int_s^r \frac{dt}{t} \int_{|z|\le t} \frac{\sqrt{-1}}{2\pi} h_k dz \wedge d\bar{z}.$$

Observe that

$$r\frac{dT_{F_k}(r,s)}{dr} = \int_{|z|\le r} \frac{\sqrt{-1}}{2\pi} h_k dz \wedge d\bar{z}$$

and using spherical coordinates we see further that

$$\frac{d}{dr}\left(\int_{|z|\le r} \frac{\sqrt{-1}}{2\pi} h_k dz \wedge d\bar{z}\right) = 2r \int_0^{2\pi} h_k(re^{i\theta})\frac{d\theta}{2\pi}.$$

Hence

$$\frac{1}{r}\frac{d}{dr}\left(r\frac{dT_{F_k}(r,s)}{dr}\right) = 2\int_0^{2\pi} h_k(re^{i\theta})\frac{d\theta}{2\pi}.$$

Applying Lemma A2.1.5 (Calculus Lemma) with $\phi(t) = t^\epsilon$, we get

$$\int_{|z|=r} h_k(re^{i\theta})\frac{d\theta}{2\pi} \le T_{F_k}(r,s)(T^\epsilon_{F_k}(r,s))[brT_{F_k}(r,s)(T_{F_k}(r,s))^\epsilon]^\epsilon,$$

for all r outside a set E with finite Lebesgue measure. This implies

$$\begin{aligned} S_k(r) &= \frac{1}{2}\int_0^{2\pi} \log h_k(re^{i\theta})\frac{d\theta}{2\pi} \\ &\le \frac{1}{2}\log\int_0^{2\pi} h_k(re^{i\theta})\frac{d\theta}{2\pi} \\ &\le (1+\epsilon)\log T_{F_k}(r,s) + \log r + O(1) \\ &\le (1+\epsilon)\log T(r) + \log r + O(1) \end{aligned}$$

where the inequality holds for r outside a set E with finite Lebesgue measure. From Theorem A3.5.2, we claim that, for $0 \le q \le p$,

$$T_{F_p}(r,s)+(p-q)T_{F_{q-1}}(r,s) \le (p-q+1)T_{F_q}(r,s)+\sum_{j=q}^{p-1}(p-j)S_j(r)-\sum_{j=q}^{p-1}(p-j)S_j(s).$$

In fact, the claim is true for $p = q$. Assume that the claim is true for $q, q+1, \ldots, p$. If $p = n$, the proof is done. If $p < n$, we proceed, by Theorem A3.5.2, for any k with $0 \le k \le n$,

$$T_{F_{k+1}}(r,s) = T_{F_k}(r,s) - T_{F_0}(r,s) + \sum_{j=0}^{k} S_j(r) - \sum_{j=0}^{k} S_j(s) - \sum_{j=q}^{p} N_{d_j}(r,s).$$

So

$$T_{F_{p+1}}(r,s)+T_{F_{q-1}}(r,s) = T_{F_p}(r,s)+T_{F_q}(r,s)+\sum_{j=q}^{p} S_j(r)-\sum_{j=q}^{p} S_j(s)-\sum_{j=q}^{p} N_{d_j}(r,s).$$

Thus

$$\begin{aligned}
T_{F_{p+1}}(r,s) + (p+1-q)T_{F_{q-1}}(r,s) &= T_{F_{p+1}}(r,s) + T_{F_{q-1}}(r,s) + (p-q)T_{F_{q-1}}(r,s) \\
&= T_{F_p}(r,s) + T_{F_q}(r,s) + \sum_{j=q}^{p} S_j(r) - \sum_{j=q}^{p} S_j(s) - \sum_{j=q}^{p} N_{d_j}(r,s) \\
&\quad +(p-q)T_{F_{q-1}}(r,s) \\
&\leq T_{F_p}(r,s) + (p-q)T_{F_{q-1}}(r,s) + T_{F_q}(r,s) + \sum_{j=q}^{p} S_j(r) - \sum_{j=q}^{p} S_j(s) \\
&\leq (p-q+1)T_{F_q}(r,s) + \sum_{j=q}^{p-1}(p-j)S_j(r) + T_{F_q}(r,s) + \sum_{j=q}^{p} S_j(r) - \sum_{j=q}^{p} S_j(s) \\
&= (p-q+2)T_{F_q}(r,s) + \sum_{j=q}^{p}(p+1-j)S_j(r) - \sum_{j=q}^{p}(p+1-j)S_j(s).
\end{aligned}$$

So this proves our claim. Now take $q = 0$ and $p = k$ and notice that $T_{F_{-1}}(r,s) \equiv 0$, then

$$T_{F_k}(r,s) \leq (k+1)T_f(r,s) + \sum_{j=0}^{k-1}(k-j)S_j(r) - \sum_{j=0}^{k-1}(k-j)S_j(s).$$

Thus, for $0 \leq k \leq n$, the inequality

$$T_{F_k}(r,s) \leq (k+1)T_f(r) + \frac{1}{2}k(k+1)(1+\epsilon)\left(\log T(r) + \log r + O(1)\right)$$

holds for all r outside a set E with finite Lebesgue measure. Therefore,

$$T(r) \leq (n+1)^2 T_f(r) + \frac{1}{2}n(n+1)^2(1+\epsilon)\left(\log T(r) + O(\log r)\right)$$

holds for all r outside a set E with finite Lebesgue measure. Because $\frac{1}{2}n(n+1)^2(1+\epsilon)\log T(r) \leq \frac{1}{2}T_f(r)$ where r is big enough, we have

$$T(r) \leq (n+2)^2 T_f(r) + O(\log r),$$

which holds for all r outside a set E with finite Lebesgue measure. □

We now define the projective distance. For integer $1 \leq q \leq p \leq n+1$, the **interior product** $\xi \lfloor \alpha \in \bigwedge^{p-q} \mathbf{C}^{n+1}$ of vectors $\xi \in \bigwedge^{p+1} \mathbf{C}^{n+1}$ and $\alpha \in \bigwedge^{q+1} \mathbf{C}^{*n+1}$ is defined by

$$\beta(\xi \lfloor \alpha) = (\alpha \wedge \beta)(\xi)$$

for any $\beta \in \bigwedge^{p-q} \mathbf{C}^{*n+1}$. Let

$$H = \{[x_0 : \cdots : x_n] \mid a_0 x_0 + \cdots + a_n x_n = 0\}$$

be a hyperplane in $\mathbf{P}^n(\mathbf{C})$with unit normal vector $\mathbf{a} = (a_0, \cdots, a_n)$. In the rest of this section, we will regard $\mathbf{a}$ as a vector in $\mathbf{C}^{*n+1}$ which is defined by $\mathbf{a}(\mathbf{x}) = a_0 x_0 + \cdots + a_n x_n$ for each $\mathbf{x} = (x_0, \cdots, x_n) \in \mathbf{C}^{n+1}$, where $\mathbf{C}^{*n+1}$ is the dual space of $\mathbf{C}^{n+1}$. Let $x \in \mathbf{P}(\bigwedge^{k+1} \mathbf{C}^{n+1})$, the **projective distance** is defined by

$$||x; H|| = \frac{||\xi \lfloor \mathbf{a}||}{||\xi|| ||\mathbf{a}||}. \tag{3.49}$$

where $\xi \in \mathbf{C}^{n+1}$ with $\mathbf{P}(\xi) = x$. Define

$$m_{F_k}(r, H) = \int_0^{2\pi} \log \frac{1}{||F_k(re^{i\theta}); H||} \frac{d\theta}{2\pi}. \tag{3.50}$$

Note that when $k = 0$, this definition is the same as the proximity function defined in (3.8).

We have, by the definition, the following First Main Theorem.

Theorem A3.5.4 (First Main Theorem)

$$m_{F_k}(r, H) + N_{F_k}(r, H) = T_{F_k}(r) + O(1).$$

We shall need the following product to sum estimate. It is an extension of the estimate of the geometric mean by the arithmetic mean. First, we prove the following lemma.

Lemma A3.5.5 *Assume that $H_1, \ldots, H_q$ (or $\mathbf{a}_1, \ldots, \mathbf{a}_q$) are hyperplanes in $\mathbf{P}^n(\mathbf{C})$ in general position. Take $p \in \mathbf{Z}[0, n]$ and $x \in \mathbf{P}(\bigwedge^p \mathbf{C}^{n+1})$. Define*

$$I_x = \{j \in \mathbf{N}[1, q] \mid ||x; H_j|| = 0\}.$$

Then $\# I_x \leq n - p$.

Proof. Let $k = \#I_x$ and let $E = \cap_{j \in I_x} H_j$. If $k \geq n+1$, then, by the assumption of general position, $E = \emptyset$. If $k \leq n+1$, then $\dim E = n - k$. There are $\xi_j \in \mathbf{C}^{n+1}$ such that $0 \neq \xi_0 \wedge \ldots \wedge \xi_p$ and $x = \mathbf{P}(\xi)$. Let

$$E(x) = E(\xi) = \{\zeta \in \mathbf{C}^{n+1} \mid \zeta \wedge \xi = 0\}.$$

Then $E(x)$ is a linear subspace of dimension $p+1$ of $\mathbf{C}^{n+1}$ with base $\xi_0, \ldots, \xi_p$. Since $\|x, H_j\| = 0$ implies $\xi \lfloor \mathbf{a}_j = 0$ or $< \xi_\mu, \mathbf{a}_j >= 0$ for $\mu = 0, \ldots, p$. Hence $\mathbf{P}(E(x)) \subset E$ which implies

$$p = \dim \mathbf{P}(E(x)) \leq \dim E = n - k,$$

or $k \leq n - p$. □

Lemma A3.5.6 *Assume that $H_1, \ldots, H_q$ (or $\mathbf{a}_1, \ldots, \mathbf{a}_q$) are hyperplanes in $\mathbf{P}^n(\mathbf{C})$ in general position. For $x \in \mathbf{P}(\bigwedge^p \mathbf{C}^{n+1})$ and a real number $d > 0$ define*

$$I_x(d) = \{j \in \mathbf{N}[1,q] \mid \|x; H_j\|^2 \leq d\}.$$

Then there is a number $c > 0$ such that $\#I_x(c) \leq n - p$ for all $x \in \mathbf{P}(\bigwedge^p \mathbf{C}^{n+1})$.

Proof. Take $x \in \mathbf{P}(\bigwedge^p \mathbf{C}^{n+1})$. Let I_x be the set defined in Lemma A3.5.5. A number $c_x > 0$ exists such that $\|x, H_j\|^2 > c_x$ for all $j \in \mathbf{N}[1,q] - I_x$. An open neighborhood U_x of x in $\mathbf{P}(\bigwedge^p \mathbf{C}^{n+1})$ exists such that $\|y, H_j\|^2 > c_x$ for all $y \in U_x$ and $j \in \mathbf{N}[1,q] - I_x$. Since $\mathbf{P}(\bigwedge^p \mathbf{C}^{n+1})$ is compact, there are only finitely many $U_{x_1}, \ldots, U_{x_s}$ whose union covers $\mathbf{P}(\bigwedge^p \mathbf{C}^{n+1})$. Define

$$c = \min\{c_{x_1}, \ldots, c_{x_s}\} > 0.$$

Take $y \in \mathbf{P}(\bigwedge^p \mathbf{C}^{n+1})$. Then $\lambda \in \mathbf{N}[1,s]$ exists such that $y \in U_{x_\lambda}$. Take $j \in I_y(c)$. Assume that $j \notin I_{x_\lambda}$. Then

$$\|y, H_j\|^2 > c_{x_\lambda} \geq c \geq \|y, H_j\|^2$$

which is impossible. Hence $I_y(c) \subset I_{x_\lambda}$ and $\#I_y(c) \leq \#I_{x_\lambda} \leq n - p$. □

Theorem A3.5.7(Product to the sum estimate) *Let $H_1, \ldots, H_q$ (or $\mathbf{a}_1, \ldots, \mathbf{a}_q$) be hyperplanes in $\mathbf{P}^n(\mathbf{C})$ in general position. Take $k \in \mathbf{Z}[0, n-1]$ with $n - k \leq q$. Then there exists a constant $c_k > 0$ such that for every $0 <$*

$\lambda < 1$ and $x \in \mathbf{P}(\bigwedge^k \mathbf{C}^{n+1})$ with $x \not\subset H_j, 1 \le j \le q$ and $y \in \mathbf{P}(\bigwedge^{k+1} \mathbf{C}^{n+1})$ we have

$$\prod_{j=1}^{q} \frac{||y; H_j||^2}{||x; H_j||^{2-2\lambda}} \le c_k \left(\sum_{j=1}^{q} \frac{||y; H_j||^2}{||x; H_j||^{2-2\lambda}} \right)^{n-k}.$$

Proof. A constant $c > 0$ exists such that $\# I_x(c) \le n - k$ for all $x \in \mathbf{P}(\bigwedge^k \mathbf{C}^{n+1})$. Define $u = (q - n + k)/(n - k)$ and $c_k = \frac{1}{(n-k)c^u} > 0$. Take x, y and λ as indicated. Take a set I with $I_x(c) \subset I \subset \mathbf{N}[1, q]$ such that $\# I = n - k$. For $j \in \mathbf{N}[1, q] - I$ we have

$$||x; H_j||^{2-2\lambda} \ge ||x; H_j||^2 > c,$$

$$\frac{||y; H_j||^2}{||x; H_j||^{2-2\lambda}} < \frac{1}{c}.$$

Therefore

$$\begin{aligned} \frac{1}{c_k} \left(\prod_{j=1}^{q} \frac{||y; H_j||^2}{||x; H_j||^{2-2\lambda}} \right)^{\frac{1}{n-k}} &\le \left(\frac{1}{c_k c^u} \right) \left(\prod_{j \in I} \frac{||y; H_j||^2}{||x; H_j||^{2-2\lambda}} \right)^{\frac{1}{n-k}} \\ &= (n-k) \left(\prod_{j \in I} \frac{||y; H_j||^2}{||x; H_j||^{2-2\lambda}} \right)^{\frac{1}{n-k}} \le \sum_{j \in I} \frac{||y; H_j||^2}{||x; H_j||^{2-2\lambda}} \\ &\le \sum_{j=1}^{q} \frac{||y; H_j||^2}{||x; H_j||^{2-2\lambda}}. \end{aligned}$$

□

Let $\phi_k(H) = ||F_k; H||^2$. Define

$$h_k(H) = \frac{\phi_{k-1}(H)\phi_{k+1}(H)}{\phi_k^2(H)} \Omega_k.$$

We need the following lemma.

Lemma A3.5.8 *For a hyperplane H in $\mathbf{P}^n$ and constant λ with $0 < \lambda < 1$, we have, for $0 \le k \le n$,*

(i) $dd^c \log \phi_k(H) = h_k(H) - \Omega_k$

(ii) $0 \leq \phi_{k-1}(H) \leq \phi_k(H) \leq \phi_n(H) = 1$.

(iii) $d\phi_k(H) \wedge d^c\phi_k(H) = (\phi_{k+1}(H) - \phi_k(H))(\phi_k(H) - \phi_{k-1}(H))\Omega_k$.

(iv) $dd^c\phi_k(H) = (\phi_{k+1}(H) - 2\phi_k(H) + \phi_{k-1}(H))\Omega_k$.

Proof. Since

$$\phi_k(H) = \frac{||\mathbf{F}_k \lfloor \mathbf{a}||^2}{||\mathbf{F}_k||^2||\mathbf{a}||^2},$$

$$dd^c \log \phi_k(H) = dd^c \log ||\mathbf{F}_k \lfloor \mathbf{a}||^2 - dd^c \log ||\mathbf{F}_k||^2 = dd^c \log ||\mathbf{F}_k \lfloor \mathbf{a}||^2 - \Omega_k.$$

Now, applying Lemma A3.5.1,

$$\begin{aligned} dd^c \log ||\mathbf{F}_k \lfloor \mathbf{a}||^2 &= dd^c \log ||(\mathbf{F}_1 \lfloor \mathbf{a})_{k-1}||^2 \\ &= \frac{||(\mathbf{F}_1 \lfloor \mathbf{a})_{k-2}||^2||(\mathbf{F}_1 \lfloor \mathbf{a})_k||^2}{||(\mathbf{F}_1 \lfloor \mathbf{a})_{k-1}||^4} \frac{\sqrt{-1}}{2\pi} dz \wedge d\bar{z} \\ &= \frac{||\mathbf{F}_{k-1} \lfloor \mathbf{a}||^2||\mathbf{F}_{k+1} \lfloor \mathbf{a}||^2}{||\mathbf{F}_k \lfloor \mathbf{a}||^4} \frac{\sqrt{-1}}{2\pi} dz \wedge d\bar{z} \\ &= \frac{\phi_{k-1}(H)\phi_{k+1}(H)}{\phi_k^2(H)} \frac{||\mathbf{F}_{k-1}||^2||\mathbf{F}_{k+1}||^2}{||\mathbf{F}_k||^4} \frac{\sqrt{-1}}{2\pi} dz \wedge d\bar{z} \\ &= \frac{\phi_{k-1}(H)\phi_{k+1}(H)}{\phi_k^2(H)} \Omega_k. \end{aligned}$$

So (i) is proved by combining the above two inequalities.

We now prove (ii). For any $z \in \mathbf{C}$ with $\mathbf{F}_n(z) \neq 0$. By Gram Schmidt there exists numbers A_{ji} for $0 \leq i \leq j \leq n+1$ with $A_{ii} > 0$ and an orthonormal basis $\mathbf{e}_0, \ldots, \mathbf{e}_n$ such that

$$\mathbf{f}^{(j)}(z) = \sum_{i=0}^{j} A_{ji}\mathbf{e}_i \qquad \text{for } j = 0, \ldots, n.$$

So

$$\mathbf{F}_j = A_{00} \ldots A_{jj}\mathbf{e}_0 \wedge \ldots \wedge \mathbf{e}_j \qquad \text{for } j = 0, \ldots, n, \tag{3.51}$$

and

$$||\mathbf{F}_j|| = A_{00} \ldots A_{jj} \qquad \text{for } j = 0, \ldots, n.$$

Let $\mathbf{a} \in \mathbf{C}^{*^{n+1}}$ be the unit normal vector associated with the hyperplane H. Define $B_j = \mathbf{a}(\mathbf{e}_j)$, and $\mathbf{e}_q^j = (-1)^j \mathbf{e}_0 \wedge \ldots \wedge \mathbf{e}_{j-1} \wedge \mathbf{e}_{j+1} \wedge \ldots \wedge \mathbf{e}_q$ for $0 \leq j \leq q \leq n$. We note that

$$(\mathbf{e}_0 \wedge \ldots \wedge \mathbf{e}_q) \lfloor \mathbf{a} = \sum_{j=0}^{q} \mathbf{a}(\mathbf{e}_j)\mathbf{e}_q^j = \sum_{j=0}^{q} B_j \mathbf{e}_q^j. \tag{3.52}$$

So we have

$$\mathbf{F}_q \lfloor \mathbf{a} = ||\mathbf{F}_q|| \sum_{j=0}^{q} B_j \mathbf{e}_q^j. \tag{3.53}$$

Thus

$$||\mathbf{F}_q \lfloor \mathbf{a}||^2 = ||\mathbf{F}_q||^2 || \sum_{j=0}^{q} |B_j|^2. \tag{3.54}$$

Hence

$$\phi_q(H) = \sum_{j=0}^{q} |B_j|^2, \quad \phi_q(H) - \phi_{q-1}(H) = |B_q|^2 \geq 0. \tag{3.55}$$

This gives $0 \leq \phi_{q-1}(H) \leq \phi_q(H)$ for $q = 1, \ldots, n$. Also

$$\phi_n(H) = \sum_{j=0}^{n} |B_j|^2 = \sum_{j=0}^{n} |\mathbf{a}(\mathbf{e}_j)|^2 = ||\mathbf{a}||^2 = 1.$$

Thus (ii) is proved. We proceed to verify (iii). First of all

$$\partial \phi_k(H) = \left(\frac{(\mathbf{F}'_k \lfloor \mathbf{a}, \mathbf{F}_k \lfloor \mathbf{a})}{||\mathbf{F}_k||^2} - \frac{(\mathbf{F}'_k, \mathbf{F}_k)||\mathbf{F}_k \lfloor \mathbf{a}||^2}{||\mathbf{F}_k||^4} \right) dz.$$

We now calculate each individual term that appears in the above inequality.

$$\begin{aligned} \mathbf{F}'_k &= \mathbf{f} \wedge \mathbf{f}' \wedge \ldots \wedge \mathbf{f}^{(k-1)} \wedge \mathbf{f}^{(k+1)} \\ &= \left(\prod_{q=0}^{k-1} A_{qq} \right) \mathbf{e}_0 \wedge \ldots \wedge \mathbf{e}_{k-1} \wedge (A_{k+1,k}\mathbf{e}_k + A_{k+1,k+1}\mathbf{e}_{k+1}). \end{aligned} \tag{3.56}$$

Using (3.52) and (3.56)

$$\mathbf{F}'_k \lfloor \mathbf{a} = \left(\prod_{q=0}^{k-1} A_{qq} \right) \sum_{j=0}^{k-1} B_j \mathbf{e}_{k-1}^j \wedge (A_{k+1,k}\mathbf{e}_k + A_{k+1,k+1}\mathbf{e}_{k+1})$$

$$+ \left(\prod_{q=0}^{k-1} A_{qq}\right)(B_k A_{k+1,k} + B_{k+1}A_{k+1,k+1})\mathbf{e}_k^k.$$

Combining the above identity with (3.53),

$$(\mathbf{F}'_k \lfloor \mathbf{a}, \mathbf{F}_k \lfloor \mathbf{a}) = \left(\prod_{q=0}^{k} A_{qq}^2\right) G \quad \text{with}$$

$$G = A_{k,k}\left(\sum_{j=0}^{k} |B_j|^2 A_{k+1,k} + B_{k+1}A_{k+1,k+1}\bar{B}_k\right).$$

Using (3.56) and (3.51)

$$(\mathbf{F}'_k, \mathbf{F}_k) = \left(\prod_{q=0}^{k-1} A_{qq}^2\right) A_{k,k}A_{k+1,k}.$$

This, together with (3.54), gives

$$(\mathbf{F}'_k, \mathbf{F}_k)\frac{||\mathbf{F}_k \lfloor \mathbf{a}||^2}{||\mathbf{F}_k||^2} = \left(\sum_{j=0}^{k} |B_j|^2\right)\left(\prod_{q=0}^{k-1} A_{qq}^2\right) A_{k,k}A_{k+1,k}.$$

Therefore

$$\partial \phi_k(H) = \frac{||\mathbf{F}_{k-1}||||\mathbf{F}_{k+1}||}{||\mathbf{F}_k||^2} B_{k+1}\overline{B}_k dz.$$

We obtain, by the above identity, (3.55) and Lemma A3.5.1

$$\begin{aligned} d\phi_k(H) \wedge d^c\phi_k(H) &= \frac{\sqrt{-1}}{2\pi}\partial\phi_k(H) \wedge \bar{\partial}\phi_k(H) \\ &= \frac{\sqrt{-1}}{2\pi}\frac{||\mathbf{F}_{k-1}||^2||\mathbf{F}_{k+1}||^2}{||\mathbf{F}_k||^4}|B_{k+1}|^2|B_k|^2 dz \wedge d\bar{z} \\ &= (\phi_{k+1}(H) - \phi_k(H))(\phi_k(H) - \phi_{k-1}(H))\Omega_k. \end{aligned}$$

Hence (iii) follows. Also we have, using (i) and (iii)

$$\begin{aligned} dd^c\phi_k(H) &= \phi_k(H) dd^c \log \phi_k(H) + \frac{1}{\phi_k(H)} d\phi_k(H) \wedge d^c\phi_k(H) \\ &= \phi_k(H)(h_k(H) - \Omega_k) \\ &\quad + \frac{1}{\phi_k(H)}(\phi_{k+1}(H) - \phi_k(H))(\phi_k(H) - \phi_{k-1}(H))\Omega_k \end{aligned}$$

$$
\begin{aligned}
&= \frac{1}{\phi_k(H)}(\phi_{k-1}(H)\phi_{k+1}(H) - \phi_k^2(H))\Omega_k \\
&+ \frac{1}{\phi_k(H)}(\phi_{k+1}(H) - \phi_k(H))(\phi_k(H) - \phi_{k-1}(H))\Omega_k \\
&= (\phi_{k+1}(H) - 2\phi_k(H) + \phi_{k-1}(H))\Omega_k.
\end{aligned}
$$

Thus (iv) is verified. □

Theorem A3.5.9 (Ahlfors Estimate) *Let H be a hyperplane in $\mathbf{P}^n$. Then for any $0 < \lambda < 1$, $0 < s < r$, we have*

$$\int_s^r \int_{|z|<t} \frac{\phi_{k+1}(H)}{\phi_k(H)^{1-\lambda}}\Omega_k \frac{dt}{t} \le \frac{1}{\lambda^2}(8T_{F_k}(r,s) + 2\log 2).$$

To prove Ahlfors' estimate, the following lemma plays a crucial role.

Lemma A3.5.10 *For a hyperplane H in $\mathbf{P}^n$ constant λ with $0 < \lambda < 1$, then, for $0 \le k \le n$, the following inequality*

$$\frac{\lambda^2}{4}\frac{\phi_{k+1}(H)}{\phi_k^{1-\lambda}(H)}\Omega_k - \lambda(1+\lambda)\Omega_k \le dd^c \log(1 + \phi_k(H)^\lambda)$$

holds on $\mathbf{C} - \{z \mid \phi_k(H)(z) = 0\}$.

Proof. We have

$$
\begin{aligned}
dd^c\phi_k(H)^\lambda &= \lambda(\lambda-1)\phi_k(H)^{\lambda-2}d\phi_k(H)\wedge d^c\phi_k(H) + \lambda\phi_k(H)^{\lambda-1}dd^c\phi_k(H) \\
&= \lambda(\lambda-1)\phi_k(H)^{\lambda-2}(\phi_{k+1}(H) - \phi_k(H))(\phi_k(H) - \phi_{k-1}(H))\Omega_k \\
&\quad +\lambda\phi_k(H)^{\lambda-1}(\phi_{k+1}(H) - 2\phi_k(H) + \phi_{k-1}(H))\Omega_k \\
&= \lambda(\lambda-1)\phi_k(H)^{\lambda-1}(\phi_{k+1}(H) - \phi_k(H) + \phi_{k-1}(H))\Omega_k \\
&\quad +\lambda(1-\lambda)\phi_{k+1}(H)\phi_{k-1}(H)\phi_k(H)^{\lambda-2}\Omega_k \\
&\quad +\lambda\phi_k(H)^{\lambda-1}(\phi_{k+1}(H) - 2\phi_k(H) + \phi_{k-1}(H))\Omega_k \\
&= \lambda^2\phi_k(H)^{\lambda-1}(\phi_{k+1}(H) + \phi_{k-1}(H))\Omega_k - \lambda(1+\lambda)\phi_k^\lambda(H)\Omega_k \\
&\quad +\lambda(1-\lambda)\phi_{k+1}(H)\phi_{k-1}(H)\phi_k(H)^{\lambda-2}\Omega_k \\
&\ge (\lambda^2\phi_k(H)^{\lambda-1}\phi_{k+1}(H) - \lambda(1+\lambda)\phi_k^\lambda(H))\Omega_k.
\end{aligned}
$$

Hence

$(1 + \phi_k(H)^\lambda)^2 dd^c \log(1 + \phi_k(H)^\lambda)$

$$
\begin{aligned}
&= (1+\phi_k(H)^\lambda)dd^c\phi_k(H)^\lambda - d\phi_k(H)^\lambda \wedge d^c\phi_k(H)^\lambda \\
&= dd^c\phi_k(H)^\lambda + \phi_k(H)^{2\lambda}dd^c \log \phi_k^\lambda(H) \\
&\geq (\lambda^2\phi_k(H)^{\lambda-1}\phi_{k+1}(H) - \lambda(1+\lambda)\phi_k^\lambda(H))\Omega_k + \lambda\phi_k(H)^{2\lambda}(h_k(H)-\Omega_k) \\
&\geq (\lambda^2\phi_k(H)^{\lambda-1}\phi_{k+1}(H) - \lambda(1+\lambda)(1+\phi_k^\lambda(H))^2)\Omega_k \\
&\geq (1+\phi_k^\lambda(H))^2\left(\frac{\lambda^2}{4}\phi_{k+1}(H)\phi_k(H)^{\lambda-1} - \lambda(1+\lambda)\right)\Omega_k,
\end{aligned}
$$

using $0 \leq \phi_k(H) \leq 1$. □

We now prove Theorem A3.5.9(Ahlfors' Estimate).

Proof. By Lemma 3.5.10,

$$dd^c \log(1+\phi_k(H)^\lambda) \geq \frac{\lambda^2}{4}\frac{\phi_{k+1}(H)}{\phi_k^{1-\lambda}(H)}\Omega_k - \lambda(1+\lambda)\Omega_k.$$

Thus

$$\frac{\lambda^2}{4}\frac{\phi_{k+1}(H)}{\phi_k^{1-\lambda}(H)}\Omega_k \leq dd^c \log(1+\phi_k(H)^\lambda) + \lambda(1+\lambda)\Omega_k. \tag{3.57}$$

By Theorem A2.1.2(Green-Jensen's formula),

$$
\begin{aligned}
&\int_s^r \frac{dt}{t}\int_{|z|\leq t} dd^c \log(1+\phi_k(H)^\lambda) \\
&\leq \frac{1}{2}\int_{|z|=r} \log(1+\phi_k(H)^\lambda)\frac{d\theta}{2\pi} - \frac{1}{2}\int_{|z|=s} \log(1+\phi_k(H)^\lambda)\frac{d\theta}{2\pi}.
\end{aligned}
$$

This, together with (3.57) implies that

$$
\begin{aligned}
&\frac{\lambda^2}{4}\int_s^r \frac{dt}{t}\int_{|z|\leq t} \frac{\phi_{k+1}(H)}{\phi_k^{1-\lambda}(H)}\Omega_k \\
&\leq \int_s^r \frac{dt}{t}\int_{|z|\leq t} dd^c \log(1+\phi_k(H)^\lambda) + \lambda(1+\lambda)T_{F_k}(r) \\
&\leq \frac{1}{2}\int_{|z|=r} \log(1+\phi_k(H)^\lambda)\frac{d\theta}{2\pi} - \frac{1}{2}\int_{|z|=s} \log(1+\phi_k(H)^\lambda)\frac{d\theta}{2\pi} \\
&+\lambda(1+\lambda)T_{F_k}(r) \\
&\leq \lambda(1+\lambda)T_{F_k}(r) + \frac{1}{2}\log 2 \leq T_{F_k}(r) + \frac{1}{2}\log 2,
\end{aligned}
$$

using $0 \leq \phi_k(H) \leq 1$. □

To prove the Second Main Theorem, we need the following consequence of Ahlfors' estimate.

Theorem A3.5.11 *For $0 < s < r$ and any $0 < \lambda < 1$,*

$$\int_0^{2\pi} \log^+ \left(\frac{||F_{k+1}(re^{i\theta}); H||^2}{||F_k(re^{i\theta}); H||^{2-2\lambda}} h_k \right) \frac{d\theta}{2\pi} \leq O(\log T_{F_k}(r,s) + \log r) - 2\log\lambda,$$

where the inequality holds for all r outside of a set E with finite Lebesgue measure.

Proof. First of all, by Ahlfors' estimate(Theorem A3.5.9),

$$\int_s^r \left(\int_{|z|<t} \frac{||F_{k+1}; H||^2}{||F_k; H||^{2-2\lambda}} h_k \frac{\sqrt{-1}}{2\pi} dz \wedge d\bar{z} \right) \frac{dt}{t} \leq \frac{1}{\lambda^2}(8T_{F_k}(r,s) + 2\log 2). \tag{3.58}$$

Observe that, using the concavity of log,

$$\begin{aligned} &\int_0^{2\pi} \log^+ \left(\frac{||F_{k+1}(re^{i\theta}); H||^2}{||F_k(re^{i\theta}); H||^{2-2\lambda}} h_k(re^{i\theta}) \right) \frac{d\theta}{2\pi} \\ \leq\; & \int_0^{2\pi} \log^+ \left(\frac{||F_{k+1}(re^{i\theta}); H||^2}{||F_k(re^{i\theta}); H\;||^{2-2\lambda}} h_k(re^{i\theta}) + 1 \right) \frac{d\theta}{2\pi} \\ =\; & \int_0^{2\pi} \log \left(\frac{||F_{k+1}(re^{i\theta}); H||^2}{||F_k(re^{i\theta}); H\;||^{2-2\lambda}} h_k(re^{i\theta}) + 1 \right) \frac{d\theta}{2\pi} \\ \leq\; & \log \int_0^{2\pi} \left(\frac{||F_{k+1}(re^{i\theta}); H||^2}{||F_k(re^{i\theta}); H\;||^{2-2\lambda}} h_k(re^{i\theta}) + 1 \right) \frac{d\theta}{2\pi}. \end{aligned} \tag{3.59}$$

Let

$$T(r) = \int_s^r \left(\int_{|z|<t} \frac{||F_{k+1}; H||^2}{||F_k; H||^{2-2\lambda}} h_k \frac{\sqrt{-1}}{2\pi} dz \wedge d\bar{z} \right) \frac{dt}{t}. \tag{3.60}$$

Then, by Lemma A2.1.4 with $\phi(t) = \log^{1+\epsilon} t$,

$$r^{-1}\frac{d}{dr}(r\frac{dT}{dr}) \leq T(r)\log^{1+\epsilon}(T(r))\log^{1+\epsilon}[c_1 rT(r)\log^{1+\epsilon}(T(r))], \tag{3.61}$$

where the inequality holds for all r outside of a set E with finite Lebesgue measure. Since

$$\int_0^{2\pi} \frac{||F_{k+1}(re^{i\theta}); H||^2}{||F_k(re^{i\theta}); H\;||^{2-2\lambda}} h_k(re^{i\theta}) \frac{d\theta}{2\pi} = r^{-1}\frac{d}{dr}(r\frac{dT}{dr}), \tag{3.62}$$

Theorem A3.5.7 is derived by combining (3.58), (3.59), (3.60), (3.61) and (3.62). □

Theorem A3.5.12 (A General Form of the SMT) *Let $f : \mathbf{C} \to \mathbf{P}^n(\mathbf{C})$ be a linearly non-degenerated holomorphic map. Let $H_1, ..., H_q$ (or $\mathbf{a}_1, \ldots, \mathbf{a}_q$) be arbitrary hyperplanes in $\mathbf{P}^n(\mathbf{C})$. Fix a positive number s_0. Then*

$$\int_0^{2\pi} \max_K \sum_{k\in K} \lambda_{H_k}(f(re^{i\theta}))\frac{d\theta}{2\pi} \leq (n+1)T_f(r) - N_{d_n}(r, s_0) + c_n(\log T_f(r) + \log r),$$

where the inequality holds for all $r \geq s_0$ outside of a set E with finite Lebesgue measure, the max is taken over all subsets K of $\{1, \ldots, q\}$ such that $\mathbf{a}_i, i \in K$, are linearly independent, and c_n is a positive constant dependent only on n.

Proof. Denote by $K \subset \{1, ..., q\}$ such that vectors $\{\mathbf{a}_k, k \in K\}$, are linearly independent. Without loss of generality, we may assume $q \geq n+1$ and that $\#K = n+1$. Let T be the set of all the injective maps $\mu : \{0, 1, \ldots, n\} \to \{1, \ldots, q\}$ such that $\mathbf{a}_{\mu(0)}, \ldots, \mathbf{a}_{\mu(n)}$ are linearly independent.

Denote by

$$\Gamma = \max\{\sum_{k=0}^{n-1} m_{F_k}(s_0, H_j), 1 \leq j \leq q\}, \tag{3.63}$$

$$\beta(r) = \frac{1}{2(q+1)!(\sum_{k=0}^{n-1}(n+1)T_{F_k}(r, s_0) + 1 + \Gamma)}. \tag{3.64}$$

Observe that $0 < \beta(r) < 1$ for $r > s_0$.

For any $\mu \in T, z \notin I_f$, the Product to Sum Estimate (Theorem A3.5.7), with $\lambda = \beta(r)$, reads

$$\prod_{j=0}^{n} \frac{||F_{k+1}; H_{\mu(j)}||^2}{||F_k; H_{\mu(j)}||^{2-2\beta(r)}} \leq c_k \sum_{j=0}^{n} \left(\frac{||F_{k+1}; H_{\mu(j)}||^2}{||F_k; H_{\mu(j)}||^{2-2\beta(r)}}\right)^{n-k},$$

where $\beta(r)$ is defined in (3.64). Since $||F_n; a_{\mu(j)}||$ is a constant for any

$0 \leq j \leq n$ and $F_0 = f$, we have

$$\prod_{j=0}^{n} \frac{1}{||f; H_{\mu(j)}||^2}$$
$$\leq \quad c \prod_{k=0}^{n-1} \left(\sum_{j=0}^{n} \left(\frac{||F_{k+1}; H_{\mu(j)}||^2}{||F_k; H_{\mu(j)}||^{2-2\beta(r)}} \right)^{n-k} \right) \cdot \prod_{k=0}^{n-1} \prod_{j=0}^{n} \frac{1}{||F_k; H_{\mu(j)}||^{2\beta(r)}}.$$

Therefore, for $r > s_0$, we have, noticing that $F_0 = f$,

$$\int_0^{2\pi} \max_K \sum_{k \in K} \log \frac{1}{||f(re^{i\theta}); H_k||^2} \frac{d\theta}{2\pi}$$
$$= \quad \int_0^{2\pi} \max_{\mu \in T} \sum_{j=0}^{n} \log \frac{1}{||f(re^{i\theta}); H_{\mu(j)}||^2} \frac{d\theta}{2\pi}$$
$$= \quad \int_0^{2\pi} \max_{\mu \in T} \log \left(\prod_{j=0}^{n} \frac{1}{||f(re^{i\theta}); H_{\mu(j)}||^2} \right) \frac{d\theta}{2\pi}$$
$$\leq \quad \sum_{k=0}^{n-1} \int_0^{2\pi} \max_{\mu \in T} \log \left(\sum_{j=0}^{n} \left(\frac{||F_{k+1}(re^{i\theta}); H_{\mu(j)}||^2}{||F_k(re^{i\theta}); H_{\mu(j)}||^{2-2\beta(r)}} \right)^{n-k} \right) \frac{d\theta}{2\pi}$$
$$+ \sum_{k=0}^{n-1} \sum_{j=0}^{n} \int_0^{2\pi} \max_{\mu \in T} \log \frac{1}{||F_k(re^{i\theta}); H_{\mu(j)}||^{2\beta(r)}} \frac{d\theta}{2\pi} + O(1)$$
$$= \quad \sum_{k=0}^{n-1} \int_0^{2\pi} \max_{\mu \in T} \log \left(\sum_{j=0}^{n} \left(\frac{||F_{k+1}(re^{i\theta}); H_{\mu(j)}||^2}{||F_k(re^{i\theta}); H_{\mu(j)}||^{2-2\beta(r)}} h_k \right)^{n-k} \right) \frac{d\theta}{2\pi} + O(1)$$
$$-2 \sum_{k=0}^{n-1} (n-k) S_k(r) + \sum_{k=0}^{n-1} \sum_{j=0}^{n} \int_0^{2\pi} \max_{\mu \in T} \log \frac{1}{||F_k(re^{i\theta}); H_{\mu(j)}||^{2\beta(r)}} \frac{d\theta}{2\pi}$$
$$\leq \quad \sum_{k=0}^{n-1} \int_0^{2\pi} \sum_{\mu \in T} \log^+ \left(\sum_{j=0}^{n} \left(\frac{||F_{k+1}(re^{i\theta}); H_{\mu(j)}||^2}{||F_k(re^{i\theta}); H_{\mu(j)}||^{2-2\beta(r)}} h_k \right)^{n-k} \right) \frac{d\theta}{2\pi} + O(1)$$
$$-2 \sum_{k=0}^{n-1} (n-k) S_k(r) + \sum_{k=0}^{n-1} \sum_{j=0}^{n} \int_0^{2\pi} \max_{\mu \in T} \log \frac{1}{||F_k(re^{i\theta}); H_{\mu(j)}||^{2\beta(r)}} \frac{d\theta}{2\pi}$$
$$\leq \quad \sum_{k=0}^{n-1} (n-k) \sum_{\mu \in T} \sum_{j=0}^{n} \int_0^{2\pi} \log^+ \left(\frac{||F_{k+1}(re^{i\theta}); H_{\mu(j)}||^2}{||F_k(re^{i\theta}); H_{\mu(j)}||^{2-2\beta(r)}} h_k \right) \frac{d\theta}{2\pi}$$

$$-2\sum_{k=0}^{n-1}(n-k)S_k(r)$$
$$+\sum_{k=0}^{n-1}\sum_{j=0}^{n}\int_0^{2\pi}\max_{\mu\in T}\log\frac{1}{||F_k(re^{i\theta});H_{\mu(j)}||^{2\beta(r)}}\frac{d\theta}{2\pi}+O(1), \tag{3.65}$$

where h_k is defined by (3.46) and $S_k(r)$ is defined by (3.48).

We now estimate each term above. By theorem A3.5.11, the inequality

$$\int_0^{2\pi}\log^+\left(\frac{||F_{k+1}(re^{i\theta});H_{\mu(j)}||^2}{||F_k(re^{i\theta});H_{\mu(j)}||^{2-2\beta(r)}}h_k\right)\frac{d\theta}{2\pi}\le O(\log T_{F_k}(r,s)+\log r) \tag{3.66}$$

holds for all r outside of a set E with finite Lebesgue measure. Using the Plücker formula (Theorem A3.5.2), we have

$$N_{d_k}(r,s_0)+T_{F_{k-1}}(r,s_0)-2T_{F_k}(r,s_0)+T_{F_{k+1}}(r,s_0)=S_k(r)-S_k(s_0).$$

Noticing that $T_{F_n}(r,s_0)=0$ and $T_{F_0}(r,s_0)=T_f(r,s_0)$,

$$\sum_{k=0}^{n-1}(n-k)S_k(r)=N_{d_n}(r,s_0)-(n+1)T_f(r,s_0)+O(1). \tag{3.67}$$

By the First Main Theorem (Theorem A3.5.4)

$$\begin{aligned}
&\sum_{k=0}^{n-1}\sum_{j=0}^{n}\int_0^{2\pi}\max_{\mu\in T}\log\frac{1}{||F_k(re^{i\theta});H_{\mu(j)}||^{2\beta(r)}}\frac{d\theta}{2\pi}\\
&\le \sum_{\mu\in T}\sum_{k=0}^{n-1}\sum_{j=0}^{n}\int_0^{2\pi}2\beta(r)\log\frac{1}{||F_k(re^{i\theta});H_{\mu(j)}||}\frac{d\theta}{2\pi}+O(1)\\
&= \sum_{\mu\in T}\sum_{k=0}^{n-1}\sum_{j=0}^{n}2\beta(r)m_{F_k}(r,H_{\mu(j)})+O(1)\\
&\le \sum_{k=0}^{n-1}\sum_{j=0}^{n}2q!\beta(r)(T_{F_k}(r,s_0)+m_{F_k}(s_0,H_{\mu(j)}))+O(1)\le O(1).
\end{aligned}$$

Combining (3.65), (3.66), (3.67) and (3.68), and by Theorem A3.5.3, we have that the inequality

$$\int_0^{2\pi}\max_K\sum_{k\in K}\log\frac{1}{||f(re^{i\theta});H_k||}\frac{d\theta}{2\pi} \le (n+1)T_f(r)-N_{d_n}(r,s_0)$$

$$+c_n(\log T_f(r) + \log r),$$

holds for all r outside a set E with finite Lebesgue measure, where c_n is a positive constant dependent only on n. Since, by the definition,

$$\lambda_{H_k}(f(re^{i\theta})) = \log \frac{1}{||f(re^{i\theta}); H_k||},$$

the theorem is proven. □

Part B: Diophantine Approximation

B3.1 Schmidt's Subspace Theorem

Let k be a number field. In section B1.2, we introduce the concept of absolute values on k. Let k be a number field and M_k be the canonical set of non-equivalent almost-absolute values on k. Almost-absolute values satisfy the product formula

$$\prod_{v \in M_k} ||x||_v = 1 \quad \text{for } x \in k^*. \tag{3.68}$$

Let $\mathbf{x} = (x_0, \cdots, x_n) \in k^{n+1}$. For $v \in M_k$ put

$$||\mathbf{x}||_v := \max(||x_0||_v, \ldots, ||x_n||_v). \tag{3.69}$$

Let $S \subset M_k$ be a finite set of absolute values on k containing the infinite places, we define

$$\mathcal{O}_S := \{x \in k : ||x||_v \leq 1 \quad \text{for } v \notin S\} \tag{3.70}$$

the ring of S-integers and

$$\mathcal{O}_S^* = \{x \in K : ||x||_v = 1 \quad \text{for } v \notin S\}$$

the multiplicative group of S-units. Schlickewei [Sch1] extended Schmidt's subspace theorem to the following version.

Theorem B3.1.1 (Schmidt-Schlickewei) *Let k be a number field. Let $S \subset M_k$ be a finite set containing all Archimedean absolute values. Suppose that for each $v \in S$ we are given $n+1$ linearly independent linear forms $L_1^{(v)}, \ldots, L_{n+1}^{(v)}$ in $n+1$ variables with algebraic coefficients. Then, for any*

$\epsilon > 0$, *there exists proper subspaces* $W_1, \ldots, W_l$ *of* $\mathbf{P}^n(k)$ *such that, the inequality*

$$\prod_{v \in S} \prod_{j=1}^{n+1} \frac{\|\mathbf{x}\|_v}{\|L_j^{(v)}(\mathbf{x})\|_v} \leq \{H_k(\mathbf{x})\}^{n+1+\epsilon} \tag{3.71}$$

holds for all $\mathbf{x} \in \mathcal{O}_S^{n+1} - (\cup_{i=1}^l W_i)$.

We recall that, for $\mathbf{x} = [x_0 : \cdots : x_n]$, the **height** of $\mathbf{x}$ is defined by

$$H_k(\mathbf{x}) = \prod_{v \in M_k} \max_{0 \leq i \leq n} \{\|x_i\|_v\}. \tag{3.72}$$

By (3.68) this does not depend on the choice of the coordinates of $\mathbf{x}$. The **logarithmic height** of $\mathbf{x}$ is defined by

$$h(\mathbf{x}) = \frac{1}{[k : \mathbf{Q}]} \log H_k(\mathbf{x}). \tag{3.73}$$

Note that the logarithmic height is independent of the number field k. Given a hyperplane

$$H = \{[x_0 : \ldots : x_n] \mid a_0 x_0 + \cdots + a_n x_n = 0\}$$

with coefficients $a_i \in k, 0 \leq i \leq n$. Denote by $\mathbf{a} = (a_0, \ldots, a_n)$ the non-zero vector associated with H. Let $\mathbf{x} = [x_0 : \ldots : x_n] \in \mathbf{P}^n(k)$ with $\mathbf{x} \notin H$, i.e., $a_0 x_0 + \cdots + a_n x_n \neq 0$. Write $\|\mathbf{a}\|_v = \max_{0 \leq i \leq n} \|a_i\|_v$ and $\|\mathbf{x}\|_v = \max_{0 \leq i \leq n} \|x_i\|_v$, then the **Weil Function** $\lambda_{v,H}$ is defined by

$$\lambda_{v,H}(\mathbf{x}) = \frac{1}{[k : \mathbf{Q}]} \log \frac{(n+1)\|\mathbf{a}\|_v \|\mathbf{x}\|_v}{\| < \mathbf{x}, \mathbf{a} > \|_v}, \tag{3.74}$$

where $< \cdot, \cdot >$ is the standard inner product on k^{n+1}. The extra term $\max_{0 \leq i \leq n} \|a_i\|_v$ ensures that $\lambda_{v,H}(\mathbf{x})$ depends only on H, $\mathbf{x}$, and not on $a_0, \cdots, a_n$ or on the choice of homogeneous coordinates $[x_0 : \cdots : x_n]$. We note that we always have $\lambda_{v,L}(\mathbf{x}) \geq 0$. Fix a finite set $S \subset M_k$, we define the **proximity function**

$$m(\mathbf{x}, H) = \sum_{v \in S} \lambda_{v,H}(\mathbf{x}) \tag{3.75}$$

and **counting function**

$$N(\mathbf{x}, H) = \sum_{v \notin S} \lambda_{v,H}(\mathbf{x}). \tag{3.76}$$

We note that the Weil function, the proximity function and the counting function are only defined for points $\mathbf{x}$ with $\mathbf{x} \notin H$. So, in the rest of the book, whenever one of these functions appears, we automatically assume that $\mathbf{x} \notin H$. By (3.73), (3.74), (3.75), (3.76) and the product formula (3.68), we have

$$m(\mathbf{x}, H) + N(\mathbf{x}, H) = \frac{1}{[k : \mathbf{Q}]} \sum_{v \in M_k} \log \|\mathbf{x}\|_v + O(1) = h(\mathbf{x}) + O(1).$$

Thus we have the following analogue of the First Main Theorem (compare to Theorem A3.1.1).

Theorem B3.1.2 *If H is a hyperplane in $\mathbf{P}^n(k)$ and $\mathbf{x} \in \mathbf{P}^n(k)$ with $\mathbf{x} \notin H$, then*

$$m(\mathbf{x}, H) + N(\mathbf{x}, H) = h(\mathbf{x}) + O(1). \tag{3.77}$$

Theorem B3.1.1(Schmidt's subspace theorem) is formulated as follows.

Theorem B3.1.3 *Let k be a number field, and $S \subset M_k$ be a finite set containing all Archimedean absolute values. Given hyperplanes $H_1, \ldots, H_q$ (or $\mathbf{a}_1, \ldots, \mathbf{a}_q$) in $\mathbf{P}^n(k)$. Then for every $\epsilon > 0$, there exists a finite set of proper subspaces $W_1, \ldots, W_l$ of $\mathbf{P}^n(k)$ such that the inequality*

$$\sum_{v \in S} \max_K \sum_{j \in K} \lambda_{v, H_j}(\mathbf{x}) \leq (n + 1 + \epsilon) h(\mathbf{x}) \tag{3.78}$$

holds for all points $\mathbf{x} \in \mathbf{P}^n(k) - \cup_{j=1}^l W_j$. Here the maximum is taken over all subsets K of $\{1, \ldots, q\}$ such that $\mathbf{a}_j, j \in K$, are linearly independent.

Proof. Let $H_1, \ldots, H_q$ be the given hyperplanes with coefficient vectors $\mathbf{a}_1, \ldots, \mathbf{a}_q$ in k^{n+1}. Denote by $K \subset \{1, ..., q\}$ such that $\mathbf{a}_j, j \in K$, are linearly independent. Without loss of generality, we may assume that $q \geq n + 1$ and that $\#K = n + 1$. Let T be the set of all injective maps $\mu : \{0, 1, \ldots, n\} \to \{1, \ldots, q\}$ such that $\mathbf{a}_{\mu(0)}, \ldots, \mathbf{a}_{\mu(n)}$ are linearly independent. Then Theorem B3.1.1 implies that for any $\epsilon > 0$, there exists proper subspaces $W_1, \ldots, W_l$ of $\mathbf{P}^n(k)$ such that, the inequality

$$\prod_{v \in S} \prod_{i=0}^{n} \max_{\mu \in T} \frac{\|\mathbf{x}\|_v}{\| < \mathbf{x}, \mathbf{a}_{\mu(i)} > \|_v} \leq H_k(\mathbf{x})^{n+1+\epsilon}$$

holds for all $\mathbf{x} \in \mathbf{P}^n(k) - (\cup_{i=1}^l W_i)$. Hence

$$\begin{aligned}\sum_{v\in S} \max_K \sum_{j\in K} \lambda_{v,H_j}(\mathbf{x}) &= \sum_{v\in S} \max_{\mu\in T} \sum_{i=0}^{n} \lambda_{v,H_{\mu(i)}}(\mathbf{x}) \\ &= \frac{1}{[k:\mathbf{Q}]} \sum_{v\in S} \sum_{i=0}^{n} \max_{\mu\in T} \log \frac{\|\mathbf{x}\|_v}{\|<\mathbf{x}, \mathbf{a}_{\mu(i)}>\|_v} + O(1) \\ &\le (n+1+\epsilon)h(\mathbf{x}) + O(1). \end{aligned} \tag{3.79}$$

This finishes the proof. □

Lemma B3.1.4 (Product to the Sum Estimate) *Let $H_1, \ldots, H_q$ (or $\mathbf{a}_1, \ldots, \mathbf{a}_q$) be hyperplanes in $\mathbf{P}^n(k)$, located in general position. Denote by T the set of all injective maps $\mu : \{0, 1, \ldots, n\} \to \{1, \ldots, q\}$ such that $\mathbf{a}_{\mu(0)}, \ldots, \mathbf{a}_{\mu(n)}$ are linearly independent. Then*

$$\sum_{j=1}^{q} m(\mathbf{x}, H_j) \le \sum_{v\in S} \max_{\mu\in T} \sum_{i=0}^{n} \lambda_{v,H_{\mu(i)}}(\mathbf{x}).$$

Proof. Let $\mathbf{a}_j$ be the coefficient vector of $H_j, 1 \le j \le q$. By the definition,

$$<\mathbf{x}, \mathbf{a}_{\mu(i)}> = a_0^{\mu(i)} x_0 + \cdots + a_n^{\mu(i)} x_n, \quad 0 \le i \le n. \tag{3.80}$$

By solving the system of linear equations (3.80),

$$x_i = \tilde{a}_0^{\mu(i)} <\mathbf{x}, \mathbf{a}_{\mu(0)}> + \cdots + \tilde{a}_n^{\mu(i)} <\mathbf{x}, \mathbf{a}_{\mu(n)}>, \quad 0 \le i \le n, \tag{3.81}$$

where $(\tilde{a}_j^{\mu(i)})$ is the inverse matrix of (a_j^i). Thus, for any $\mu \in T$,

$$\|\mathbf{x}\|_v \le C \max_{0\le i\le n} \|<\mathbf{x}, \mathbf{a}_{\mu(i)}>\|_v. \tag{3.82}$$

For a given $v \in S$, there is $\mu \in T$ such

$$0 < \|<\mathbf{x}, \mathbf{a}_{\mu(0)}>\|_v \le \cdots \le \|<\mathbf{x}, \mathbf{a}_{\mu(n)}>\|_v \le \|<\mathbf{x}, \mathbf{a}_{\mu(j)}>\|_v$$

for $j \ne \mu(0), \ldots, \mu(n)$. This, together with (3.82), implies that

$$\frac{1}{[k:\mathbf{Q}]} \sum_{j=1}^{q} \log \frac{(n+1)\|\mathbf{x}\|_v \|\mathbf{a}_j\|_v}{\|<\mathbf{x}, \mathbf{a}_j>\|_v}$$

$$= \frac{1}{[k:\mathbf{Q}]}\sum_{i=0}^{n}\log\frac{(n+1)||\mathbf{x}||_v||\mathbf{a}_{\mu(i)}||_v}{||<\mathbf{x},\mathbf{a}_{\mu(i)}>||_v} + O(1)$$

$$\leq \frac{1}{[k:\mathbf{Q}]}\max_{\mu\in T}\sum_{i=0}^{n}\log\frac{(n+1)||\mathbf{x}||_v||\mathbf{a}_{\mu(i)}||_v}{||<\mathbf{x},\mathbf{a}_{\mu(i)}>||_v} + O(1).$$

The lemma holds by taking the sum for $v \in S$. □

Combining Lemma B3.1.4 with Theorem B3.1.3, we have the following theorem.

Theorem B3.1.5 *Let k be a number field, and $S \subset M_k$ be a finite set containing all Archimedean absolute values. Given hyperplanes $H_1, \ldots, H_q$ in $\mathbf{P}^n(k)$, located in general position. Then for every $\epsilon > 0$, there exists a finite set of proper subspaces $W_1, \ldots, W_l$ of $\mathbf{P}^n(k)$ such that the inequality*

$$\sum_{j=1}^{q} m(\mathbf{x}, H_j) \leq (n+1+\epsilon)h(\mathbf{x}) \tag{3.83}$$

holds for all points $\mathbf{x} \in \mathbf{P}^n(k) - \cup_{j=1}^{l} W_j$.

B3.2 The abc-Conjecture

Consider $a + b = c, \quad a, b, c \in \mathbf{Z}$ and consider the following table.

$a+b=c$	$\max\{\|a\|, \|b\|, \|c\|\}$	$\prod_{p\|abc} p$
2+3=5	5	30
9+16=25	25	30
3+125=128	128	30
$19 \cdot 1307 + 7 \cdot 29^2 \cdot 31^8 = 2^8 \cdot 3^{22} \cdot 5^4$	$e^{36.15}$	$e^{22.26}$

We see that in the last two lines, the height $h(a,b,c) = \max\{|a|,|b|,|c|\}$ is larger than the radical $r(a,b,c) = \prod_{p|abc} p$, where the product is taken over primes p. The abc conjecture says that the height cannot be much larger than the radical.

Conjecture B3.2.1 (abc Conjecture) *For every* $\epsilon > 0$ *there exists a constant* $K(\epsilon)$ *such that*

$$h(a,b,c) \leq r(a,b,c) + \epsilon h(a,b,c) + K(\epsilon), \tag{3.84}$$

for every sum $a + b = c$ *of coprime nonzero integers.*

Define, for $[a:b:c] \in \mathbf{P}^2(\mathbf{Q})$ with integer coefficients,

$$h([a:b:c]) = \log \max\{|a|,|b|,|c|\}$$

$$N^{(1)}(x,0) = \sum_{p|x} \log p, \qquad x \in \mathbf{Q}^*.$$

Then we can restate abc Conjecture as

$$(1-\epsilon)h([a:b:c]) \leq N^{(1)}(abc,0) + K(\epsilon).$$

The abc conjecture was formulated in 1983 by Masser and Oesterlé as a possible approach to Fermat's conjecture (now Wiles' theorem): For $n \geq 3$, the equation $x^n + y^n = z^n$ has no solution in positive integers x, y, z. Indeed, this is a simple consequence of the abc conjecture. Let $x^n + y^n = z^n$ be a solution. Without loss of generality, we assume that $\max\{|x|,|y|,|z|\} = |z|$. So

$$h([x^n : y^n : z^n]) = \log |z|^n.$$

The radical is composed of the prime factors of $x^n y^n z^n$, hence of the prime factors of xyz. Thus $r = \prod_{p|xyz} p \leq \log |xyz| \leq \log |z|^3$. Take $\epsilon = 1/2$, then, (3.84) implies that $n \leq 6 + 2K(1/2)\log|z|$. Since it is known that there are no solutions for $n = 3, 4, 5$ or 6, this leaves only finitely many values of x, y, z and n to check.

Corresponding to (3.10), we define, for x,

$$N^{(n)}(x,0)=\sum_p(\min\{\mathrm{ord}_p^+ x, n\})\log p, \tag{3.85}$$

where the sum is taken over all primes p, and $ord_p x$ is defined as follows: $ord_p x = k$ if and only if $x = p^k q$, where p, q are coprime.

Conjecture B3.2.2 (Generalized abc Conjecture) *For every $\epsilon > 0$ there exists a constant $K(\epsilon)$ such that the following is true: If $a_0 + \cdots + a_n + a_{n+1} = 0$ with $a_i \in \mathbf{Z}$, $0 \le i \le n+1$, and no common factors. Assume that $\sum_{i\in I} a_i \ne 0$ for any proper subset I of $\{0, ..., n+1\}$. Then there exists a constant $K(\epsilon)$ such that*

$$(1-\epsilon)h([a_0:\ldots:a_n]) \le \sum_{j=0}^{n+1} N^{(n)}(a_j,0)+K(\epsilon).$$

Corresponding to Theorem A3.2.6, according to Vojta's dictionary, we raise the following conjecture.

Conjecture B3.2.3 *Let $n_0, \ldots, n_m$ be positive integers such that*

$$\sum_{i=0}^{m}\frac{1}{n_i}<\frac{1}{m}.$$

Then the equation

$$\sum_{i=0}^{m} x_i^{n_i}=1, \quad \textit{with} \quad x_i\in\mathbf{Z}$$

has only finitely many solutions.

Corresponding to Corollary A3.2.8, according to Vojta's dictionary, we raise the following conjecture.

Corollary B3.2.4 *If $m \ge 11$ the surface defined by*

$$x_0^m+x_1^m+x_2^m+x_3^{m-2}P(x_0,x_1,x_2,x_3)=0$$

has only finitely many rational points, where P is a generic form of degree two.

B3.3 The S-Unit Lemma

Let k be a number field. Let S be a finite subset of $M(k)$ containing $M_\infty(k)$. Recall that an element $x \in k$ is said to be **S-integer** if $||x||_v \leq 1$ for each $v \in M(K) - S$. Denote by $\mathcal{O}_S$ the set of S-integers. The units of $\mathcal{O}_S$ are called **S-units**. They form a multiplicative group which is denoted by $\mathcal{O}_S^*$. For example, if $k = \mathbf{Q}$ the field of rational numbers and $S = \{| \ |, p_1, \ldots, p_s\}$ where $p_1, \ldots, p_s$ are prime numbers. Then the set $\mathcal{O}_S$ of S-integers consists of the numbers $p_1^{k_1} \cdots p_s^{k_s} x$ where $k_1, \ldots, k_s$ and x are integers, and the set $\mathcal{O}_S^*$ of S-units consists of the numbers $\pm p_1^{k_1} \cdots p_s^{k_s}$ where $k_1, \ldots, k_s$ are integers.

Theorem B3.3.1 (S-unit Lemma) *Let k be a number field. Let S be a finite subset of $\subset M(k)$ containing $M_\infty(k)$. Then the equation*

$$x_1 + \ldots + x_n = 1$$

has finitely many solutions in S-units $x_1, \ldots, x_n$ such that no proper sub-sum $x_{i1} + \ldots + x_{im}$ vanishes.

The proof of Theorem B3.3.1 is the same as the proof of Theorem A3.3.2.

As an application of the S-unit lemma, we study integer solutions for the decomposable homogeneous polynomial equation.

Definition B3.3.2 *Let k be a number field, and S be a finite subset of M_k containing M_k^∞. Let $F(\mathbf{X}) = F(x_0, \ldots, x_n)$ be a homogeneous polynomial in $n+1$ variables with coefficient in $\mathcal{O}_S$. F is called* **decomposable** *if F factors into linear forms $L_1, \ldots, L_q$ over some finite extension of k.*

For $n = 1$, every homogeneous polynomial is decomposable, but for $n > 1$ this is not always the case. Important classes of decomposable homogeneous polynomial equations are Thue's equations, when $n = 1$, and norm form equations, discriminant form equations, and index form equations for $n > 1$.

Thue's equation,

$$F(x_0, x_1) = a, \quad a \in \mathcal{O}_S$$

is well-studied: If F contains three distinct linear factors over some finite extension of k, then the number of integer solutions of the equation is finite and the explicit bound (even an effective bound) is obtained (see [Eve1]).

Here we study the general case of $n > 1$. Given a decomposable homogeneous polynomial $F(x_0, ..., x_n)$. F then factors into linear forms $L_1, \ldots, L_q$ over some finite extension of k. Since by enlarging k, the finiteness does not change, we assume that F factors into linear forms $L_1, \ldots, L_q$ over k. We denote by $\mathcal{L}$ the set of all linear factors of F. Then $\mathcal{L}$ defines a set of hyperplanes in $\mathbf{P}^n(k)$.

Definition B3.3.3 *A decomposable homogeneous polynomial F is called* **non-degenerate** *if $\mathcal{L}$ is non-degenerate, where $\mathcal{L}$ is the set of linear factors of F. Here, similar to Definition A3.3.3, $\mathcal{L}$ is non-degenerate if and only if $\dim(\mathcal{L}) = n+1$, and $(\mathcal{L}_1) \cap (\mathcal{L} - \mathcal{L}_1) \cap \mathcal{L} \neq \emptyset$ for every subset $\mathcal{L}_1$ of $\mathcal{L}$ where $(\mathcal{L})$ is the linear span over k of the linear forms in $\mathcal{L}$.*

Theorem B3.3.4 (Evertse and Győry [Ev-G2]) *If a decomposable homogeneous polynomial $F(x_0, ..., x_n)$ is non-degenerate, then for every $b \in k^*$ and every finite subset S of M_k containing M_k^∞, the equation*

$$F(\mathbf{x}) = b \quad in \quad \mathbf{x} = (x_0, ..., x_n) \in \mathcal{O}_S^{n+1} \tag{3.86}$$

has only finitely many solutions. In particular, if $F \in \mathbf{Z}[X]$ factors into linear forms over $\overline{\mathbf{Q}}$ which is in general position and the degree of $q \geq 2n+1$, then the equation

$$F(\mathbf{x}) = b \quad in \quad \mathbf{x} = (x_0, ..., x_n) \in \mathbf{Z}^n \tag{3.87}$$

has only finitely many solutions.

Proof. By extending k, if necessary, we may assume that

$$F(x_0, ..., x_n) = L_1(x_0, ..., x_n) \cdots L_q(x_0, ..., x_n),$$

where $L_1, ..., L_q$ are linear forms coefficients in k. Since F is non-degenerate, by Proposition A3.3.7 (we note that Proposition A3.3.7 can be easily extended to k), there are at least three linear forms among $L_1, \ldots, L_q$ that are linearly dependent over k, but not pair-wisely dependent. Without loss of generality, we may assume that L_1, L_2, L_3 are linearly dependent. So there are non-zero constants $c_1, c_2, c_3 \in k$ such that

$$c_1L_1 + c_2L_2 + c_3L_3 = 0.$$

We assume that coefficients of L_j, c_1, c_2, c_3, $1/c_1$, $1/c_2$, $1/c_3$ b, and $1/b$ are S-integers by enlarging S, if necessary. Denote by the solutions of (3.86) by I. For $\mathbf{x} \in I$,

$$L_1(\mathbf{x})...L_q(\mathbf{x})/b = 1.$$

It follows that $1/L_j(\mathbf{x})$ is an S-integer because $L_j(\mathbf{x})$ is an S-integer for $\mathbf{x} \in I$ and b^{-1} is also an S-integer. Thus $L_j(\mathbf{x}), 1 \leq j \leq q, \mathbf{x} \in I$, are S-units. So $c_jL_j(\mathbf{x}), 1 \leq j \leq 3, \mathbf{x} \in I$, are S-units. The S-unit lemma implies that I is obtained in a proper subspace of k^{n+1}. Proposition A3.3.7 (we note that Proposition A3.3.7 can be easily extended to k) thus allows us to use induction to reduce I to a finite set. □

B3.4 The Degenerated Schmidt's Subspace Theorem

Definition B3.4.1 *Let $1 \leq l \leq n$ be an integer. Hyperplanes $H_1, \ldots, H_q$ (or $\mathbf{a}_1, \ldots, \mathbf{a}_q$) in $\mathbf{P}^l(k)$ are in* n**-subgeneral position** *if for every $1 \leq i_0 < \cdots < i_n \leq q$ the linear span over k of $\mathbf{a}_{i_0}, \ldots, \mathbf{a}_{i_n}$ is k^{l+1}.*

The following is directly verified by the definition.

Lemma B3.4.2 Let W be a subspace of $\mathbf{P}^n(k)$ of dimension l. Let $H_1, \ldots, H_q$ be hyperplanes in $\mathbf{P}^n(k)$ in general position. Then the hyperplanes $H_1 \cap W, \ldots, H_q \cap W$ are in n-subgeneral position in W.

Nochka's Theorem A3.4.3 easily extends to any field of characteristic 0, in particular to the number field k.

Theorem B3.4.3(Nochka) Let $H_1, \ldots, H_q$ (or $\mathbf{a}_1, \ldots, \mathbf{a}_q$) be hyperplanes in $\mathbf{P}^l(k)$, located in n-subgeneral position. Assume that $q \geq 2n - l + 1$. For each $\emptyset \neq J \subset \{1, \ldots, q\}$ let $L(J)$ be the linear space generated by $\{\mathbf{a}_j | j \in J\}$. Then there exists a function $\omega : \{1, \ldots, q\} \to \mathbf{R}(0, 1]$ called the Nochka weight and a real number $\theta \geq 1$ called the Nochka constant satisfying the following properties:

(i) If $j \in \{1, \ldots, q\}$, then $0 < \omega(j)\theta \leq 1$.

(ii) $q - 2n + l - 1 = \theta(\sum_{j=1}^{q} \omega(j) - l - 1)$.

(iii) If $\emptyset \neq J \subset \{1, \ldots, q\}$ with $\#J \leq l+1$, then $\sum_{j \in J} \omega(j) \leq \dim L(J)$.

(iv) $1 \leq (l+1)/(n+1) \leq \theta \leq (2n - l + 1)/(n+1)$.

(v) Given $E_1, \ldots, E_q$, real numbers ≥ 1, and given any $Y \subset \{1, \ldots, q\}$ with $0 < \#Y \leq n+1$, there exists a subset M of Y with $\#M = \dim L(Y)$ such that $\{\mathbf{a}_j\}_{j \in M}$ is a basis for $L(Y)$ and

$$\prod_{j \in Y} E_j^{\omega(j)} \leq \prod_{j \in M} E_j.$$

Lemma B3.4.4 (Product to the Sum Estimate) *Let $H_1, \ldots, H_q$ (or $\mathbf{a}_1, \ldots, \mathbf{a}_q$) be hyperplanes in $\mathbf{P}^l(k)$, located in n-subgeneral position. Let $\omega : \{1, \ldots, q\} \to \mathbf{R}(0,1]$ be the Nochka weight associated with H_j. Denote by T the set of all injective maps $\mu : \{0, 1, \ldots, l\} \to \{1, \ldots, q\}$ such that $\mathbf{a}_{\mu(0)}, \ldots, \mathbf{a}_{\mu(l)}$ are linearly independent. Then*

$$\sum_{j=1}^{q} \omega(j) m(\mathbf{x}, H_j) \leq \sum_{v \in S} \max_{\mu \in T} \sum_{i=0}^{l} \lambda_{v, H_{\mu(i)}}(\mathbf{x}). \tag{3.88}$$

Proof. Let $\mathbf{a}_j$ be the coefficient vectors of $H_j, 1 \leq j \leq q$. For every $\mathbf{x} \in \mathbf{P}^l(k)$ and $v \in S$, we rearrange $\{1, 2, \ldots, q\}$ as $i_1, \ldots, i_q$ such that

$$0 < \| < \mathbf{x}, \mathbf{a}_{i_1} > \|_v \leq \| < \mathbf{x}, \mathbf{a}_{i_2} > \|_v \leq \cdots \leq \| < \mathbf{x}, \mathbf{a}_{i_q} > \|_v. \tag{3.89}$$

We note that the indices $i_1, \ldots, i_q$ depend on $\mathbf{x}$ and v. The n-subgeneral position implies that $\mathbf{a}_{i_1}, \ldots, \mathbf{a}_{i_{n+1}}$ span k^{l+1}. Thus there is an injective map $\alpha : \{0, 1, \ldots l\} \to \{i_1, \ldots, i_{n+1}\}$ such that $\{\mathbf{a}_{\alpha(0)}, \ldots, \mathbf{a}_{\alpha(l)}\}$ is a basis of k^{l+1}. Consider

$$< \mathbf{x}, \mathbf{a}_{\alpha(i)} >= a_{0,\alpha(i)} x_0 + \cdots + a_{l,\alpha(i)} x_l, \quad 0 \leq i \leq l, \tag{3.90}$$

with $\mathbf{a}_{\alpha(i)} = (a_{0,\alpha(i)}, \ldots, a_{l,\alpha(i)})$. By solving the system of linear equations (3.90),

$$x_i = \tilde{a}_{0,\alpha(i)} < \mathbf{x}, \mathbf{a}_{\alpha(0)} > + \cdots + \tilde{a}_{l,\alpha(i)} < \mathbf{x}, \mathbf{a}_{\alpha(l)} >, \quad 0 \leq i \leq l,$$

where $(\tilde{a}_{j,\alpha(i)})$ is the inverse matrix of $(a_{\alpha(j),i})$.Thus,

$$\|\mathbf{x}\|_v \leq C_0 \max_{0\leq i\leq l} \| < \mathbf{x}, \mathbf{a}_{\alpha(i)} > \|_v \leq C_0 \| < \mathbf{x}, \mathbf{a}_j > \|_v \tag{3.91}$$

for $j \geq i_{n+1}$, where C_0 is a positive constant, independent of $\mathbf{x}$ and v. Combining (3.89) and (3.91) we have

$$\prod_{j=1}^{q} \left(\frac{(l+1)\|\mathbf{x}\|_v \|\mathbf{a}_j\|_v}{\| < \mathbf{x}, \mathbf{a}_j > \|_v} \right)^{\omega(j)} \leq C_1 \prod_{t=1}^{n+1} \left(\frac{(l+1)\|\mathbf{x}\|_v \|\mathbf{a}_{i_t}\|_v}{\| < \mathbf{x}, \mathbf{a}_{i_t} > \|_v} \right)^{\omega(i_t)}, \tag{3.92}$$

where c_1 is a positive constant, independent of $\mathbf{x}$ and v. Let

$$E_t = \frac{(l+1)\|\mathbf{x}\|_v \|\mathbf{a}_{i_t}\|_v}{\||< \mathbf{x}, \mathbf{a}_{i_t} > \|_v}.$$

Note that $E_t \geq 1$. Lemma B3.4.3 (v) applied to $E_t, 1 \leq t \leq n+1$, implies that there is an injective map $\mu : \{0, \ldots, l\} \rightarrow \{i_1, \ldots, i_{n+1}\}$ such that $\{\mathbf{a}_{\mu(0)}, \ldots, \mathbf{a}_{\mu(l)}\}$ is linearly independent and

$$\prod_{t=1}^{n+1} E_t^{\omega(i_t)} \leq \prod_{i=0}^{l} E_{\mu(i)}.$$

That is, by the definition of E_t,

$$\prod_{t=1}^{n+1} \left(\frac{(l+1)\|\mathbf{x}\|_v \|\mathbf{a}_{i_t}\|_v}{\| < \mathbf{x}, \mathbf{a}_{i_t} > \|_v} \right)^{\omega(i_t)} \leq \prod_{i=0}^{l} \frac{(l+1)\|\mathbf{x}\|_v \|\mathbf{a}_{\mu(i)}\|_v}{\| < \mathbf{x}, \mathbf{a}_{\mu(i)} > \|_v}. \tag{3.93}$$

Thus the lemma follows by combining (3.92), (3.93), the definition of $m(\mathbf{x}, H_j)$ and the definition of $\lambda_{v,H_j}(\mathbf{x})$. □

Theorem B3.4.5 (Ru-Wong) *Let k be a number field, let S be a finite set of places of k containing M_k^∞. Let $H_1, \ldots, H_q$ be hyperplanes in $\mathbf{P}^l(k)$, located in n-subgeneral position ($1 \leq l \leq n$). Then for any $\epsilon > 0$, there exists a finite set of proper subspaces $W_1, \ldots, W_t$ of $\mathbf{P}^l(k)$, such that*

$$\sum_{j=1}^{q} m(\mathbf{x}, H_j) \leq (2n - l + 1 + \epsilon)h(\mathbf{x})$$

holds for all $\mathbf{x} \in \mathbf{P}^l(k) - \cup_{\alpha=1}^t W_\alpha$.

Proof. Let $\mathbf{a}_j$ be the coefficient vectors of $H_j, 1 \leq j \leq q$. If $q \leq 2n-l+1$ then Theorem B3.4.5 is a consequence of the First Main Theorem and $h(\mathbf{x})$ is unbounded. We may now assume that $q > 2n-l+1$. By Lemma B3.4.4,

$$\sum_{j=1}^{q} \omega(j) m(\mathbf{x}, H_j) \leq \sum_{v \in S} \max_{\gamma \in T} \sum_{i=0}^{l} \lambda_{v, H_{\mu(i)}}(\mathbf{x}) \tag{3.94}$$

where T is the set of all injective maps $\mu : \{0, 1, \ldots, l\} \to \{1, \ldots, q\}$ such that $\mathbf{a}_{\mu(0)}, \ldots, \mathbf{a}_{\mu(l)}$ are linearly independent. However, Theorem B3.1.4 implies that for any $\epsilon > 0$, there exists a finite set of proper subspaces $W_1, \ldots, W_t$ of $\mathbf{P}^l(k)$, such that

$$\sum_{v \in S} \max_{\gamma \in T} \sum_{i=0}^{l} \lambda_{v, H_{\mu(i)}}(\mathbf{x}) \leq (l + 1 + \epsilon) h(\mathbf{x})$$

holds for all $\mathbf{x} \in \mathbf{P}^l(k) - \cup_{\alpha=1}^{t} W_\alpha$. This, together with (3.94), yields that for all $\mathbf{x} \in \mathbf{P}^l(k) - \cup_{\alpha=1}^{t} W_\alpha$,

$$\sum_{j=1}^{q} \omega(j) m(\mathbf{x}, H_j) \leq (l + 1 + \epsilon) h(\mathbf{x}). \tag{3.95}$$

Combining this with Lemma B3.4.3, and recalling that $m(\mathbf{x}, H_j) \leq h(\mathbf{x}) + O(1)$, we have

$$\begin{aligned}
\sum_{j=1}^{q} m(\mathbf{x}, H_j) &= \sum_{j=1}^{q} (1 - \theta\omega(j)) m(\mathbf{x}, H_j) + \sum_{j=1}^{q} \theta\omega(j) m(\mathbf{x}, H_j) \\
&\leq \sum_{j=1}^{q} (1 - \theta\omega(j)) m(\mathbf{x}, H_j) + \theta(l + 1 + \epsilon) h(\mathbf{x}) \\
&\leq \sum_{j=1}^{q} (1 - \theta\omega(j)) h(\mathbf{x}) + \theta(l + 1 + \epsilon) h(\mathbf{x}) \\
&= \left\{ q - \theta \left(\sum_{1 \leq j \leq q} \omega(j) - l - 1 - \epsilon \right) \right\} h(\mathbf{x}) \\
&= (2n - l + 1 + \epsilon) h(\mathbf{x}),
\end{aligned}$$

where the inequality holds for all $\mathbf{x} \in \mathbf{P}^l(k) - \cup_{\alpha=1}^{t} W_\alpha$. □

Theorem B3.4.6 (Ru-Wong) *Let k be a number field, let S be a finite set of places of k, let $q \in \mathbf{Z}_{>0}$. Let $H_1, \ldots, H_q$ (or $\mathbf{a}_1, \ldots, \mathbf{a}_q$) be hyperplanes in $\mathbf{P}^n(k)$, located in general position. Then for any $\epsilon > 0$, there exists a finite set of proper subspaces $W_1, \ldots, W_t$ of $\mathbf{P}^n(k)$ of dimension $l-1$, such that*

$$\sum_{j=1}^{q} m(\mathbf{x}, H_j) \leq (2n - l + 1 + \epsilon)h(\mathbf{x})$$

holds for all $\mathbf{x} \in \mathbf{P}^{l+1}(k) - \cup_{\alpha=1}^{t} W_\alpha$ with $L_j(\mathbf{x}) \neq 0$ for $1 \leq j \leq q$. In particular, the set of points $\mathbf{x} \in \mathbf{P}^n(k)$ with $< \mathbf{x}, \mathbf{a}_j > \neq 0$ and satisfying

$$\sum_{j=1}^{q} m(\mathbf{x}, H_j) > (2n + \epsilon)h(\mathbf{x})$$

is a finite set.

Proof. Consider the points $\mathbf{x} \in \mathbf{P}^n(k)$ with $< \mathbf{x}, \mathbf{a}_j > \neq 0$. We use induction for l. The $l = n$ case is just Theorem B3.1.6, so the theorem is true for this case. Assume that the theorem holds for $l+1$, that is there exists a finite set of proper subspaces $W_1, \ldots, W_t$ of $\mathbf{P}^n(k)$ of dimension l, such that

$$\sum_{j=1}^{q} m(\mathbf{x}, H_j) \leq (2n - l + \epsilon)h(\mathbf{x}) \tag{3.96}$$

holds for all $\mathbf{x} \in \mathbf{P}^n(k) - \cup_{\alpha=1}^{t} W_\alpha$. Now consider each subspace W_α of $\mathbf{P}^n(k)$ of dimension l. Lemma B3.4.2 tells us that the hyperplanes $H_1, \ldots, H_q$ restricted to W_α are in n-subgeneral position in $W_\alpha \cong \mathbf{P}^l(k)$. So Theorem B3.4.5 applies to this case. Therefore the theorem holds for l. Thus Theorem B3.4.6 is verified by induction. □

Theorem B3.4.6 implies Wirsing's Theorem.

Theorem B3.4.7 (Wirsing's Theorem) *Let k be a number field and let S be a finite set of places of k containing M_k^∞. Let r be a positive integer and let $a_v(\alpha) \in k$ for each $v \in S$. Then the inequality*

$$\prod_{v \in S} \prod_{w \in M_{k(x)},\, w|v} \frac{\|x - a_v\|_w}{\max(1, \|x\|_w) \cdot \max(1, \|a_v\|_w)} \geq \frac{1}{H_{k(x)}(x)^{2r+\epsilon}} \tag{3.97}$$

holds for all, except for a finite number, $x \in \bar{k}$ *with* $[k(x):k] \le r$.

Proof. It will suffice to prove the statement under the additional assumption that $[k(x):k] = r$. For those points let

$$f(X) = A_r X^r + \ldots + A_0$$

be the minimal polynomial of x over k. This defines a collection of points $P_x := [A_0 : \ldots : A_r]$ in $\mathbf{P}^r(k)$. It is easy to verify that

$$\begin{aligned} C_{1,v} \cdot \frac{\max(\|A_0\|_v, \ldots, \|A_r\|_v)}{\|A_r\|_v} &\le \prod_{w \in M_{k(x)},\, w|v} \max(1, \|x\|_w) \\ &\le C_{2,v} \cdot \frac{\max(\|A_0\|_v, \ldots, \|A_r\|_v)}{\|A_r\|_v} \end{aligned}$$

for all $v \in M_k$, where $C_{1,v}$ and $C_{2,v}$ are positive constants. Moreover, $C_{1,v} = C_{2,v} = 1$ if v is non-Archimedean. Hence the heights of x and P_x are related by

$$C_1 H_k(P_x) \le H_{k(x)}(x) \le C_2 H_k(P_x) \tag{3.98}$$

where C_1 and C_2 are positive constants.

For each $v \in S$, a_v determines a hyperplane H_v defined by the equation

$$H_v = \{[x_0 : \cdots : x_n] \mid x_0 + a_v x_1 + \ldots + a_v^r x_r = 0\}. \tag{3.99}$$

The hyperplanes H_v are defined over k, and the set $\{H_v\}_{v \in S}$ lies in general position by the non-vanishing of the van der Monde determinant. Moreover, it is easy to check that the left-hand side of (3.97) is related to the Weil functions of P_x with respect to H_v, as follows. First of all,

$$\begin{aligned} \prod_{w \in M_{k(x)},\, w|v} \max(1, \|a_v\|_w) &= \max(1, \|a_v\|_v)^r \\ &= \max(1, \|a_v\|_v, \|a_v\|_v^2, \ldots, \|a_v\|_v^r). \end{aligned} \tag{3.100}$$

Also, we have

$$\prod_{w \in M_{k(x)},\, w|v} \|x - a_v\|_w = \|N_k^{k(x)}(x - a_v)\|_v = \|f(a_v)\|_v. \tag{3.101}$$

Combining (3.98), (3.99), (3.100) and (3.101) then gives

$$-\log \prod_{w\in M_{k(x)},\ w|v} \frac{\|x-a_v\|_w}{\max(1,\|x\|_w)\cdot\max(1,\|a_v\|_w)} = \lambda_{v,H_v}(P_x)+O(1).$$

So, (3.97) is equivalent to

$$\frac{1}{[k:\mathbf{Q}]}\sum_v \lambda_{v,H_v}(P_x) \le (2r+\epsilon)h(P_x)+O(1).$$

This follows immediately from the last assertion of Theorem B3.4.6. □

Theorem B3.4.6 also gives the finiteness of number of integer solutions of decomposable form equations.

Theorem B3.4.8 (Györy-Ru) *Given positive integers q, m with $q > 2m$, and a polynomial $G(\mathbf{X}) \in \mathcal{O}_S[\mathbf{X}]$ in $\mathbf{X} = (X_0, \ldots, X_m)$ with total degree less than $q - 2m$. Let $F(\mathbf{X}) \in \mathcal{O}_S[\mathbf{X}]$ be a decomposable form of degree q whose linear factors are in general position. Then the equation*

$$F(\mathbf{X}) = G(\mathbf{X}) \tag{3.102}$$

has only finitely many solutions $\mathbf{x} = (x_0, \ldots, x_m) \in \mathcal{O}_S^{m+1}$ with $G(\mathbf{x}) \neq 0$.

Proof. By extending k, if necessary, we may assume that

$$F(X) = L_1(X)\cdots F_q(X)$$

over k, where $L_1, \ldots, L_q$ are linear forms, located in general position. Assume that there is an infinite sequence $\mathbf{x}_n = (x_{0,n}, \ldots, x_{m,n}) \in \mathcal{O}_S^{m+1}$ which satisfies (3.102).

First consider the case when the values $h(\mathbf{x}_n)$ are bounded. We may assume without loss of generality that $x_{0,n} \neq 0$ for each n. Then the $h(\mathbf{x}_n/x_{0,n})$ are bounded and this implies that $\mathbf{x}_n/x_{0,n}$ may assume only finitely many values in k^{m+1}. Hence there are infinitely many n such that $\mathbf{x}_n = x_{0,n}\mathbf{x}_0$ for some $\mathbf{x}_0 \in k^{m+1}$. For these n we deduce from (3.102) that $h(x_{0,n})$ are bounded. This is a contradiction.

Next consider the case when $h(\mathbf{x}_n)$ are not bounded. We may assume that $h(\mathbf{x}_n) \to \infty$ as $n \to \infty$. Let $\epsilon > 0$ with $0 < \epsilon < q - 2m - \nu$, where $\nu = \deg G$. Then by Theorem B3.4.6, there is an infinite subsequence

$\mathbf{x}_{n_k} \in \mathcal{O}_S^{m+1}, k = 1, 2, \ldots$, of $\{\mathbf{x}_n\}$, without loss of generality, we assume $\{\mathbf{x}_n\}$ itself, such that

$$\frac{1}{[k:\mathbf{Q}]} \sum_{v\in S} \sum_{j=1}^{q} \log \frac{||\mathbf{x}_n||_v \cdot ||L_j||_v}{||L_j(\mathbf{x}_n)||_v} \leq (2m+\epsilon)h(\mathbf{x}_n).$$

However, $F(\mathbf{x}_n) = \prod_{j=1}^{q} L_j(\mathbf{x}_n)$. Furthermore, in view of $\mathbf{x}_n \in \mathcal{O}_S^{m+1}$, we have

$$h(\mathbf{x}_n) \leq \frac{1}{[k:\mathbf{Q}]} \log H_S(\mathbf{x}_n),$$

where $H_S(\mathbf{x}_n) = \prod_{v\in S} ||x_n||_v$. Hence it follows that

$$\prod_{v\in S} \frac{||\mathbf{x}_n||_v^q \cdot \prod_{j=1}^{q} ||L_j||_v}{||F(\mathbf{x}_n)||_v} \leq H_S(\mathbf{x}_n)^{2m+\epsilon},$$

whence

$$\frac{H_S^q(\mathbf{x}_n) \cdot \prod_{v\in S} \prod_{j=1}^{q} ||L_j||_v}{\prod_{v\in S} ||F(\mathbf{x}_n)||_v} \leq H_S(\mathbf{x}_n)^{2m+\epsilon}. \tag{3.103}$$

Since the coefficients of L_j are S-integers,

$$\prod_{v\in S} \prod_{j=1}^{q} ||L_j||_v \geq 1, \text{ for } n = 1, 2 \ldots . \tag{3.104}$$

Furthermore, since $F(\mathbf{x}_n) = G(\mathbf{x}_n)$ and $\deg G = \nu$.

$$\prod_{v\in S} ||F(\mathbf{x}_n)||_v \leq c_2 H_S(\mathbf{x}_n)^{\nu} \text{ for } n = 1, 2 \ldots . \tag{3.105}$$

Combining (3.103), (3.104) and (3.105) gives

$$H_S(\mathbf{x}_n)^q \leq c_2 H_S(\mathbf{x}_n)^{\nu+2m+\epsilon}.$$

Since $H_S(\mathbf{x}_n) \to \infty$ as $n \to \infty$, and $q > \nu + 2m + \epsilon$, which gives a contradiction. □

The Correspondence Table

Nevanlinna Theory	**Diophantine Approximation**
Theorem A3.1.1	Theorem B3.1.2
Theorem A3.1.3	Theorem B3.1.3
Lemma A3.1.6	Lemma B3.1.4
Theorem A3.1.7	Theorem B3.1.5
Theorem A3.2.6	Conjecture B3.2.2
Theorem A3.2.8	Conjecture B3.2.3
Corollary A3.2.9	Conjecture B3.2.4
Theorem A3.3.2	Theorem B3.3.1
Theorem A3.3.4	Theorem B3.3.4
Theorem A3.4.3	Theorem B3.4.3
Theorem A3.4.4	Theorem B3.4.6
Corollary A3.4.5	Theorem B3.4.8

Chapter 4

The Moving Target Problems

While R. Nevanlinna established his Second Main Theorem(see Theorem A1.3.1), he asked whether the theorem still remains true if the targets $a_j, 1 \leq j \leq q$, are replaced by slowly growing meromorphic functions $a_j(z), 1 \leq j \leq q$. This motivated the moving target problems in Nevanlinna theory and Diophantine approximations.

Part A: Nevanlinna Theory

A4.1 The Moving Target Problem for Meromorphic Functions

We first extend Theorem A1.3.1 by replacing a_j with slowly growing meromorphic functions $a_j(z), 1 \leq j \leq q$. The proximity function $m_f(r, a)$ and the counting function $N_f(r, a)$ defined in chapter 1 for meromorphic function f and complex number a are easily extended to meromorphic functions f and a, provided f and a are **free**, i.e., $f \not\equiv a$.

Theorem A4.1.1 (Steinmetz) *Let f be a non-constant meromorphic function. Let $a_1, \ldots, a_q$ be meromorphic functions with $a_i \not\equiv a_j$ for $i \neq j$, $1 \leq i, j \leq q$. Assume that $T_{a_j}(r) = o(T_f(r))$ for $1 \leq j \leq q$. Then, for every $\epsilon > 0$, the inequality*

$$m_f(r, \infty) + \sum_{j=1}^{q} m_f(r, a_j) \leq (2 + \epsilon)T_f(r)$$

holds for all r outside a set $E \subset (0, +\infty)$ with finite Lebesgue measure.

Proof. Let $a_1, \ldots, a_q$ be the given meromorphic functions. Let $\mathcal{L}(s)$ be the vector space generated over $\mathbf{C}$ by the functions

$$\left\{ a_1^{n_1} \cdots a_q^{n_q} \;\middle|\; n_j \in \mathbf{N},\ \sum_{j=1}^{q} n_j = s \right\}.$$

Choose a basis $\{b_1, \ldots, b_{\ell(s)}\}$ for $\mathcal{L}(s)$ and $\{\tilde{b}_1, \ldots, \tilde{b}_{\ell(s+1)}\}$ for $\mathcal{L}(s+1)$, where $\ell(s) = \dim \mathcal{L}(s)$. Let $F : \mathbf{C} \to \mathbf{P}(\mathbf{C}^{\ell(s)+\ell(s+1)})$ be the holomorphic map defined by

$$F = [b_1 f : \ldots : b_{\ell(s)} f : \tilde{b}_1 : \ldots : \tilde{b}_{\ell(s+1)}].$$

We call F the **Steinmetz's map**. Since $T_{a_j}(r) = o(T_f(r))$ for $1 \le j \le q$, the components of F are linearly independent over $\mathbf{C}$. So the image of F is not contained in any proper subspaces of $\mathbf{P}(\mathbf{C}^{\ell(s)+\ell(s+1)})$. Also by definition and using $T_{a_j}(r) = o(T_f(r))$ for $1 \le j \le q$ again, we have

$$T_F(r) = T_f(r) + o(T_f(r)).$$

Notice that, for each $1 \le j \le q$ and $1 \le t \le \ell(s)$, $b_t a_j \in \mathcal{L}(s+1)$. So $b_t a_j$ can be written as a linear combination of $\{\tilde{b}_k, 1 \le k \le \ell(s+1)\}$, that is

$$-b_t a_j = \sum_{k=1}^{\ell(s+1)} c_{tjk} \tilde{b}_k, \tag{4.1}$$

where c_{tjk} are complex numbers. For each fixed j, $1 \le j \le q$, define the hyperplanes $H_1(j), \ldots, H_{\ell(s)+\ell(s+1)}(j)$ in $\mathbf{P}(\mathbf{C}^{\ell(s)+\ell(s+1)})$ as follows: for i with $1 \le i \le \ell(s)$,

$$H_i(j) = \left\{ [x_1 : \ldots : x_{\ell(s)+\ell(s+1)}] \;\middle|\; x_i + \sum_{k=1}^{\ell(s+1)} c_{ijk} x_{\ell(s)+k} = 0 \right\};$$

and for i with $\ell(s) < i \le \ell(s) + \ell(s+1)$,

$$H_i(j) = \left\{ [x_1 : \ldots : x_{\ell(s)+\ell(s+1)}] \;\middle|\; x_i = 0 \right\}.$$

Since $b_1, \ldots, b_{\ell(s)}$ are linearly independent over $\mathbf{C}$, using (4.1), it is easy to check that the hyperplanes $H_1(j), \ldots, H_{\ell(s)+\ell(s+1)}(j)$ (or more precisely the coefficient vectors of $H_1(j), \ldots, H_{\ell(s)+\ell(s+1)}(j)$) are linearly independent for each j with $1 \leq j \leq q$. By the definition for the Weil function in (3.7) and using (4.1) again, we have, for $1 \leq i \leq \ell(s)$,

$$\begin{aligned} \lambda_{H_i(j)}(F)(z) &= \log \frac{\max\{|b_1 f|, \ldots, |b_{\ell(s)} f|, |\tilde{b}_1|, \ldots, |\tilde{b}_{\ell(s+1)}|\}}{|b_i(z)(f(z) - a_j(z))|} + O(1) \\ &= \log^+ \frac{1}{|f(z) - a_j(z)|} + C(a_1, \ldots, a_q)(z), \end{aligned} \tag{4.2}$$

and, for $\ell(s) < i \leq \ell(s) + \ell(s+1)$,

$$\lambda_{H_i(j)}(F)(z) = \log^+ |f(z)| + C(a_1, \ldots, a_q)(z) \tag{4.3}$$

where $C(a_1, \ldots, a_q)(z)$ is a term that depends only on $a_1, \ldots, a_q$. Applying Theorem A3.1.3 to F and the hyperplanes $H_1(j), \ldots, H_{\ell(s)+\ell(s+1)}(j), 1 \leq j \leq q$, and using (4.2) and (4.3), it follows that

$$\begin{aligned} &\ell(s+1) m_f(r, \infty) + \ell(s) \sum_{j=1}^{q} m_f(r, a_j) \\ \leq\ & \int_0^{2\pi} \max_{1 \leq j \leq q} \sum_{i=1}^{\ell(s)+\ell(s+1)} \lambda_{H_i(j)}(F(re^{i\theta})) \frac{d\theta}{2\pi} + o(T_f(r)) \\ \leq\ & (\ell(s) + \ell(s+1)) T_F(r) + o(T_F(r)) + o(T_f(r)) \\ \leq\ & (\ell(s) + \ell(s+1)) T_f(r) + o(T_f(r)), \end{aligned}$$

where the inequality holds for all r outside a set $E \subset (0, +\infty)$ with finite Lebesgue measure. As noted by Steinmetz, we obviously have

$$0 \leq \ell(s) \leq \binom{q(n+1) + s - 1}{s}$$

for each s and therefore

$$\liminf_{s \to \infty} \frac{\ell(s+1)}{\ell(s)} = 1.$$

We conclude that the inequality

$$m_f(r, \infty) + \sum_{j=1}^{q} m_f(r, a_j) \leq (2 + \epsilon) T_f(r)$$

holds for all r outside a set $E \subset (0, +\infty)$ with finite Lebesgue measure. Theorem A4.1.1 is thus proved. □

A4.2 The Moving Target Problem for Holomorphic Curves in Projective Spaces

In this section, we extend Theorem A4.1.1 to holomorphic curves intersecting with moving hyperplanes. By a moving hyperplane H in $\mathbf{P}^n(\mathbf{C})$, we mean

$$H = \{[x_0 : \ldots : x_n] \in \mathbf{P}^n(\mathbf{C}) \mid a_0x_0 + \ldots + a_nx_n = 0\},$$

where $a_0, \ldots, a_n$ are entire functions without common zeros. So H is associated with a holomorphic map $a = [a_0 : \cdots : a_n] : \mathbf{C} \to \mathbf{P}^n(\mathbf{C})$. Write $\mathbf{a} = (a_0, \ldots, a_n)$. Given a holomorphic map $f : \mathbf{C} \to \mathbf{P}^n(\mathbf{C})$, we say that f and H are **free** if $< \mathbf{f}, \mathbf{a} > \not\equiv 0$ where $\mathbf{f}$ is a reduced representation of f and $< \ , \ >$ is the inner production on $\mathbf{C}^{n+1}$. The definitions of the Weil function, the proximity function and the counting function can easily be extended to moving hyperplane H without any change, provided that f is free with H. So, in the rest of chapter, whenever we write $\lambda_H(f), m_f(r, H)$ or $N_f(r, H)$, we automatically assume that f is free with H.

Let $H_j, 1 \leq j \leq q$, be the moving hyperplanes in $\mathbf{P}^n(\mathbf{C})$ which are given by

$$H_j = \{[x_0 : \ldots : x_n] \mid a_{j0}x_0 + \ldots + a_{jn}x_n = 0\},$$

where $a_{j0}, \ldots, a_{jn}$ are entire functions without common zeros. Let $\mathbf{a}_j = (a_{j0}, \ldots, a_{jn}) : \mathbf{C} \to \mathbf{C}^{n+1} - \{\mathbf{0}\}$ be the moving vector associated with H_j and let $a_j = \mathbf{P}(\mathbf{a}_j)$. The following theorem generalizes Theorem A3.1.3 to the moving hyperplanes case.

Theorem A4.2.1 (Ru-Stoll) *Let $f = [f_0 : \ldots : f_n] : \mathbf{C} \to \mathbf{P}^n(\mathbf{C})$ be a holomorphic map. Let $\mathcal{G}$ be a finite set of moving hyperplanes $H_1, \ldots, H_q$ (or $\mathbf{a}_1, \ldots, \mathbf{a}_q$). Let $\mathbf{a}_j = (a_{j,0}, \ldots, a_{j,n})$ and let $\mathcal{R}_{\mathcal{G}}$ be the smallest field which contains $\mathbf{C}$ and all $a_{j\mu}/a_{j\nu}$ with $a_{j\nu} \not\equiv 0$. If f is non-degenerate over $\mathcal{R}_{\mathcal{G}}$, meaning that $f_0, \ldots, f_n$ are linearly independent over $\mathcal{R}_{\mathcal{G}}$, then for every $\epsilon > 0$, the inequality*

$$\int_0^{2\pi} \max_K \sum_{k \in K} \lambda_{H_k(re^{i\theta})}(f(re^{i\theta})) \frac{d\theta}{2\pi} \leq (n+1+\epsilon)T_f(r) + O\left(\max_{1 \leq j \leq q} T_{a_j}(r)\right)$$

holds for all r outside a set $E \subset (0, +\infty)$ with finite Lebesgue measure, and $\max_K$ is taken over all subsets $K \subset \{1, \ldots, q\}$ such that $\mathbf{a}_j(z)$, for $j \in K$, *are linearly independent for some (and hence for almost all)* $z \in \mathbf{C}$.

Definition A4.2.2 *The moving hyperplanes* $H_1, \ldots, H_q$ *(or* $\mathbf{a}_1, \ldots, \mathbf{a}_q$*) are said to be* **in general position** *if* $H_1(z), \ldots, H_q(z)$ *are in general position for some (and hence for almost all)* $z \in \mathbf{C}$.

Let $\mathcal{G}$ be a finite set of moving hyperplanes $H_1, \ldots, H_q$ (or $\mathbf{a}_1, \ldots, \mathbf{a}_q$). Assume that $\mathcal{G}$ is in general position. Define, for $z \in \mathbf{C}$,

$$\Gamma(\mathcal{G})(z) = \min\left\{ \frac{\|\mathbf{a}_{\mu(0)}(z) \wedge \cdots \wedge \mathbf{a}_{\mu(n)}(z)\|}{\|\mathbf{a}_{\mu(0)}(z)\| \cdots \|\mathbf{a}_{\mu(n)}(z)\|} \;\middle|\; \mu : \mathbf{Z}[0,n] \to \{1, \ldots, q\}, \text{ injective} \right\}.$$

By the general position assumption,

$$S = \{z \in \mathbf{C} \mid \Gamma(\mathcal{G})(z) = 0\}$$

is a closed set of isolated points.

Lemma A4.2.3 (Product to the Sum Estimate) *Let* $f = [f_0 : \ldots : f_n] : \mathbf{C} \to \mathbf{P}^n(\mathbf{C})$ *be a holomorphic map. Let* $\mathcal{G}$ *be a finite set of hyperplanes* $H_1, \ldots, H_q$ *(or* $\mathbf{a}_1, \ldots, \mathbf{a}_q$*). Assume that* $\mathcal{G}$ *is in general position, Then for every* $z \in \mathbf{C} - S$ *with* $< \mathbf{f}(z), \mathbf{a}(z) > \neq 0$ *for all* $\mathbf{a} \in \mathcal{G}$*, there exist* $i(z,0), \ldots, i(z,n)$ *among* $1, \ldots, q$ *such that*

$$\prod_{\mathbf{a} \in \mathcal{G}} \frac{\|\mathbf{f}(z)\|\|\mathbf{a}(z)\|}{| < \mathbf{f}(z), \mathbf{a}(z) > |} \leq \left(\frac{2(n+1)}{\Gamma(\mathcal{G})(z)}\right)^{q-n-1} \prod_{l=0}^{n} \frac{\|\mathbf{f}(z)\|\|\mathbf{a}_{i(z,l)}(z)\|}{| < \mathbf{f}(z), \mathbf{a}_{i(z,l)}(z) > |},$$

where $< \cdot, \cdot >$ *is the inner product on* $\mathbf{C}^{n+1}$.

Proof. Given any $z \in \mathbf{C} - S$, we have fixed hyperplanes $\{H(z) | H \in \mathcal{G}\}$, and they are in general position. For the number $c = \frac{1}{2(n+1)}\Gamma(\mathcal{G})(z)$ we claim that $\# I(f(z), c, \mathcal{G}) \leq n$, where

$$I(f(z), c, \mathcal{G}) = \left\{ j \;\middle|\; \frac{| < \mathbf{f}(z), \mathbf{a}_j(z) > |}{\|\mathbf{f}(z)\|\|\mathbf{a}_j(z)\|} \leq c \right\}.$$

To verify our claim, we recall that, for integer $1 \leq q \leq p \leq n+1$, the interior product of vectors $\xi \in \bigwedge^{p+1} \mathbf{C}^{n+1}$ and $\alpha \in \bigwedge^{q+1} \mathbf{C}^{*n+1}$, where $\mathbf{C}^{*n+1}$ is the dual space of $\mathbf{C}^{n+1}$, is defined by

$$\beta(\xi \lfloor \alpha) = (\alpha \wedge \beta)(\xi)$$

for any $\beta \in \bigwedge^{p-q} \mathbf{C}^{*n+1}$. We first show that for any linearly independent vectors $\mathbf{x}_0, \ldots, \mathbf{x}_n \in \mathbf{C}^{n+1}$ and any $\mathbf{y} \in \mathbf{C}^{*n+1}$,

$$\|\mathbf{x}_0 \wedge \cdots \wedge \mathbf{x}_n \lfloor \mathbf{y}\| = \|\mathbf{x}_0 \wedge \cdots \wedge \mathbf{x}_n\| \|\mathbf{y}\|.$$

In fact, let $\mathbf{v}_0, \cdots, \mathbf{v}_n$ be an orthonormal base of $\mathbf{C}^{n+1}$ and $\mathbf{v}_0^*, \cdots, \mathbf{v}_n^*$ be the dual base. Define $y_j =< \mathbf{v}_j, \mathbf{y} >= \mathbf{y}(\mathbf{v}_j)$. Then

$$\mathbf{y} = \sum_{j=0}^{n} y_j \mathbf{v}_j^*.$$

Let $\hat{\mathbf{v}}_j = (-1)^j \mathbf{v}_0 \wedge \cdots \wedge \mathbf{v}_{j-1} \wedge \mathbf{v}_{j+1} \wedge \cdots \wedge \mathbf{v}_n$. Then $\hat{\mathbf{v}}_0, \cdots, \hat{\mathbf{v}}_n$ is an orthonormal base of $\bigwedge^n \mathbf{C}^{n+1}$. A number $A \in \mathbf{C}$ exists such that

$$\mathbf{x}_0 \wedge \cdots \wedge \mathbf{x}_n = A\mathbf{v}_0 \wedge \cdots \wedge \mathbf{v}_n.$$

Therefore we have

$$\begin{aligned}
\|\mathbf{x}_0 \wedge \cdots \wedge \mathbf{x}_n \lfloor \mathbf{y}\| &= |A| \|\mathbf{v}_0 \wedge \cdots \wedge \mathbf{v}_n \lfloor \mathbf{y}\| \\
&= |A| \|\sum_{j=0}^{n} y_j \hat{\mathbf{v}}_j\| = |A| \left(\sum_{j=0}^{n} |y_j|^2\right)^{1/2} \\
&= \|\mathbf{x}_0 \wedge \cdots \wedge \mathbf{x}_n\| \|\mathbf{y}\|.
\end{aligned}$$

We now verify our claim. For each $H \in \mathcal{G}$ with the unit normal vector $\mathbf{a}$, we can regard $\mathbf{a} \in \mathbf{C}^{*n+1}$ defined by $\mathbf{a}(\mathbf{x}) =< \mathbf{x}, \mathbf{a} >$ for every $\mathbf{x} \in \mathbf{C}^{n+1}$. Now assume that $\# I(f(z), c, \mathcal{G}) > n$, without loss of generality, we assume that $\{1, \ldots, n+1\} \subset I(f(z), c, \mathcal{G})$. So we have

$$\begin{aligned}
0 &< \Gamma(\mathcal{G})(z) \leq \frac{\|\mathbf{a}_1(z) \wedge \cdots \wedge \mathbf{a}_{n+1}(z)\|}{\|\mathbf{a}_1(z)\| \cdots \|\mathbf{a}_{n+1}(z)\|} \\
&= \frac{\|\mathbf{a}_1(z) \wedge \cdots \wedge \mathbf{a}_{n+1}(z) \lfloor \mathbf{f}(z)\|}{\|\mathbf{f}(z)\| \|\mathbf{a}_1(z)\| \cdots \|\mathbf{a}_{n+1}(z)\|} \\
&= \left\| \sum_{j=1}^{n+1} \frac{< \mathbf{f}(z), \mathbf{a}_j(z) >}{\|\mathbf{f}(z)\| \|\mathbf{a}_j(z)\|} (-1)^j \frac{\mathbf{a}_1(z) \wedge \cdots \wedge \mathbf{a}_{j-1}(z) \wedge \mathbf{a}_{j+1}(z) \wedge \cdots \wedge \mathbf{a}_{n+1}(z)}{\|\mathbf{a}_1(z)\| \cdots \|\mathbf{a}_{j-1}(z)\| \|\mathbf{a}_{j+1}(z)\| \cdots \|\mathbf{a}_{n+1}(z)\|} \right\| \\
&\leq \sum_{j=1}^{n+1} \frac{| < \mathbf{f}(z), \mathbf{a}_j(z) > |}{\|\mathbf{f}(z)\| \|\mathbf{a}_j(z)\|} \\
&\leq (n+1)c = \frac{1}{2} \Gamma(\mathcal{G})(z)
\end{aligned}$$

which gives a contradiction. So our claim is proven.

Now choose $i(z,0),\ldots,i(z,n) \in \{1,\ldots,q\}$ such that

$$I(f(z),a,\mathcal{G}) \subset \{i(z,0),\ldots,i(z,n)\},$$

then

$$\begin{aligned}\prod_{\mathbf{a}\in\mathcal{G}} \frac{\|\mathbf{f}(z)\|\|\mathbf{a}(z)\|}{|<\mathbf{f}(z),\mathbf{a}(z)>|} &\leq \left(\frac{1}{c}\right)^{q-n-1} \prod_{l=0}^{n} \frac{\|\mathbf{f}(z)\|\|\mathbf{a}_{i(z,l)}(z)\|}{|<\mathbf{f}(z),\mathbf{a}_{i(z,l)}>|} \\ &= \left(\frac{2(n+1)}{\Gamma(\mathcal{G})(z)}\right)^{q-n-1} \prod_{l=0}^{n} \frac{\|\mathbf{f}(z)\|\|\mathbf{a}_{i(z,l)}(z)\|}{|<\mathbf{f}(z),\mathbf{a}_{i(z,l)}>|}.\end{aligned}$$

Lemma A4.2.3 is proven. □

Combining Theorem A4.2.1 and Lemma A4.2.3 gives

Theorem A4.2.4 (SMT with moving targets) *Under the same assumptions in Theorem A4.2.1, and in addition, we assume that the hyperplanes H_j (or $\mathbf{a}_j$), $1 \leq j \leq q$, are in general position. Let $a_j = \mathbf{P}(\mathbf{a}_j)$. Then for every $\epsilon > 0$, the inequality*

$$\sum_{j=1}^{q} m_f(r,H_j) \leq (n+1+\epsilon)T_f(r) + O\left(\max_{1\leq j\leq q} T_{a_j}(r)\right)$$

holds for all r outside of a set E with finite Lebesgue measure.

We now prove Theorem A4.2.1. The proof is taken from [Ru3].

Proof. Without loss of generality, we can assume $q \geq n+1$, and that at least $n+1$ of the hyperplanes are linearly independent. Let T be the set of all maps $\mu : \{0,1,\ldots,n\} \to \{1,\ldots,q\}$ such that $\mathbf{a}_{\mu(0)}(z),\ldots,\mathbf{a}_{\mu(n)}(z)$ are linearly independent for some (thus for almost all) $z \in \mathbf{C}$. Let $\mathbf{f} = (f_0,\ldots,f_n)$ be a reduced representation for f, meaning that $f_i, 0 \leq i \leq n$, are entire functions and without common zeros, and $\mathbf{P}(\mathbf{f}) = f$. For each $1 \leq j \leq q$, choose an index $\hat{j}$ with $0 \leq \hat{j} \leq n$ and $a_{j,\hat{j}} \not\equiv 0$, and define

$$\zeta_{j,l}(z) = a_{j,l}(z)/a_{j,\hat{j}}(z),\ j=1,\ldots,q;\ l=0,\ldots,n.$$

Let $\mathcal{L}(s)$ be the vector space generated over $\mathbf{C}$ by

$$\left\{\zeta_{1,0}^{n_{1,0}}\cdots\zeta_{q,0}^{n_{q,0}}\cdots\zeta_{1,n}^{n_{1,n}}\cdots\zeta_{q,n}^{n_{q,n}} \;\middle|\; n_{j,l}\in\mathbf{N},\ \sum_{j=1}^{q}\sum_{l=0}^{n} n_{j,l} = s\right\}.$$

We have $\mathcal{L}(s) \subset \mathcal{L}(s+1)$. Let $\{b_1, \ldots, b_{\ell(s+1)}\}$ be a basis of $\mathcal{L}(s+1)$ such that $\{b_1, \ldots, b_{\ell(s)}\}$ is a basis of $\mathcal{L}(s)$, where $\ell(s) = \dim \mathcal{L}(s)$. Let $F = \mathbf{P}(\tilde{F}) : \mathbf{C} \to \mathbf{P}(\mathbf{C}^{(n+1)\ell(s+1)})$ be the holomorphic map defined by

$$\begin{aligned} \tilde{F}(z) &= (b_1(z)f_0(z), \ldots, b_{\ell(s+1)}(z)f_0(z), b_1(z)f_1(z), \ldots, b_{\ell(s+1)}(z)f_n(z)) \\ &\in \mathbf{C}^{(n+1)\ell(s+1)}. \end{aligned} \tag{4.4}$$

Because f is linearly non-degenerate over $\mathcal{R}_{\mathcal{G}}$, F is linearly non-degenerate. We will apply Theorem A3.1.3 to F. The next step is to construct, for each $\mu \in T$, a set of (fixed) hyperplanes $\{\hat{H}_{l,j}(\mu) \mid l = 0, \ldots, n; j = 1, \ldots, \ell(s)\}$ in $\mathbf{P}(\mathbf{C}^{(n+1)\ell(s+1)})$ such that the Weil functions satisfy

$$\lambda_{\hat{H}_{l,j}(\mu)}(F(z)) \sim \lambda_{H_{\mu(l)}(z)}(f(z)).$$

To do so, let h_j, $1 \le j \le q$, be the meromorphic function defined by

$$h_j(z) = \zeta_{j,0} + \sum_{l=1}^{n} \zeta_{j,l}(z)\frac{f_l(z)}{f_0(z)}, \quad (j = 1, \ldots, q). \tag{4.5}$$

Noticing that $b_j\zeta_{k,l} \in \mathcal{L}(s+1)$ for $1 \le j \le \ell(s)$, $1 \le k \le q$ and $0 \le l \le n$, so it can be written as a linear combination of $b_r, 1 \le r \le \ell(s+1)$. Thus functions $b_j h_{\mu(l)}, 1 \le j \le \ell(s), 0 \le l \le n$, can be written as a combination of $b_r, 1 \le r \le \ell(s+1)$, and $b_\alpha\left(f_\beta/f_0\right), 1 \le \alpha \le \ell(s+1), 0 \le \beta \le n$. In other words there is an $(n+1)\ell(s) \times ((n+1)\ell(s+1))$ matrix $C(\mu)$ with entries in $\mathbf{C}$ such that

$$\begin{pmatrix} b_1 h_{\mu(0)} \\ \vdots \\ b_{\ell(s)} h_{\mu(0)} \\ \vdots \\ b_1 h_{\mu(n)} \\ \vdots \\ b_{\ell(s)} h_{\mu(n)} \end{pmatrix} = C(\mu) \begin{pmatrix} b_1 \\ \vdots \\ b_{\ell(s+1)} \\ b_1\left(f_1/f_0\right) \\ \vdots \\ b_{\ell(s+1)}\left(f_1/f_0\right) \\ \vdots \\ b_1\left(f_n/f_0\right) \\ \vdots \\ b_{\ell(s+1)}\left(f_n/f_0\right) \end{pmatrix}. \tag{4.6}$$

For $l = 0, \ldots, n$, and $j = 1, \ldots, \ell(s)$, let $\hat{H}_{l,j}(\mu)$ be the hyperplane in $\mathbf{P}(\mathbf{C}^{(n+1)\ell(s+1)})$ defined by the corresponding row in $C(\mu)$, i.e., if we denote

$c_{ij}(\mu)$ the elements of $C(\mu)$, then

$$\hat{H}_{l,j}(\mu) = \Big\{ [y_{1,0} : \dots : y_{\ell(s+1),0} : y_{1,1} : \dots : y_{\ell(s+1),1} : \dots : y_{1,n} : \dots : y_{\ell(s+1),n}] \in \mathbf{P}(\mathbf{C}^{(n+1)\ell(s+1)}) \;\Big|\; c_{l\ell(s)+j,1}(\mu)y_{1,0} + \dots + c_{l\ell(s)+j,\ell(s+1)}(\mu)y_{\ell(s+1),0} + c_{l\ell(s)+j,\ell(s+1)+1}(\mu)y_{1,1} + \dots + c_{l\ell(s)+j,2\ell(s+1)}(\mu)y_{\ell(s+1),1} + \dots + c_{l\ell(s)+j,(n+1)\ell(s+1)}(\mu)y_{\ell(s+1),n} = 0\Big\}. \tag{4.7}$$

Since $\mathbf{a}_{\mu(0)}(z), \dots, \mathbf{a}_{\mu(n)}(z)$ are linearly independent for some z and $f_0, \dots, f_n$ are linearly independent over $\mathcal{R}_{\mathcal{G}}$, $h_{\mu(0)}, \dots, h_{\mu(n)}$ are linearly independent over $\mathcal{R}_{\mathcal{G}}$. Thus, by the choice of $b_1, \dots, b_{\ell(s)}$, the set $\{b_j h_{\mu(l)}, j = 1, \dots, \ell(s);\ l = 0, \dots, n\}$ is linearly independent over $\mathbf{C}$. Hence, $\hat{H}_{l,j}(\mu), l = 0, \dots, n;\ j = 1, \dots, \ell(s)$ (more precisely the coefficient vectors of $\hat{H}_{l,j}(\mu), l = 0, \dots, n;\ j = 1, \dots, \ell(s)$), are linearly independent for each $\mu \in T$. Applying Theorem A3.1.3 for F, with the hyperplanes $\{\hat{H}_{l,j}(\mu) | l = 0, \dots, n; j = 1, \dots, \ell(s)\}$, yields that the inequality

$$\int_0^{2\pi} \max_{\mu \in T} \sum_{l=0}^{n} \sum_{j=1}^{\ell(s)} \lambda_{\hat{H}_{l,j}(\mu)}(F(re^{i\theta}))\frac{d\theta}{2\pi} \le ((n+1)\ell(s+1) + \epsilon/2)\, T_F(r) \tag{4.8}$$

holds for all r outside a set $E \subset (0, +\infty)$ with finite Lebesgue measure. We now compare $T_f(r)$ and $T_F(r)$. In fact, for each $z \in \mathbf{C}$, not in the set of the poles of $b_1, \dots, b_{\ell(s+1)}$, by (4.4) we have

$$\begin{aligned} &\|\tilde{F}(z)\| \\ &= \max(|b_1(z)f_0(z)|, \dots, |b_{\ell(s+1)}(z)f_0(z)|, |b_1(z)f_1(z)|, \dots, |b_{\ell(s+1)}(z)f_n(z)|) \\ &\quad +O(1) \\ &\le \max(|f_0(z)|, \dots, |f_n(z)|) \cdot \max(|b_1(z)|, \dots, |b_{\ell(s+1)}(z)|) + O(1) \\ &= \|\mathbf{f}\| \cdot \max(|b_1(z)|, \dots, |b_{\ell(s+1)}(z)|) + O(1). \end{aligned}$$

By the First Main Theorem

$$\begin{aligned} T_F(r) &= \int_0^{2\pi} \log \frac{\|\tilde{F}\|}{|b_1 f_0|}(re^{i\theta})d\theta + N(r, [b_1 f_0 = 0]) - N(r, [1/b_1 = 0]) + O(1) \\ &\le \int_0^{2\pi} \log \frac{\|\mathbf{f}\|}{|f_0|}(re^{i\theta})d\theta + N(r, [f_0 = 0]) + O\left(\max_{1 \le j \le q} T_{a_j}(r)\right) \end{aligned}$$

$$= \quad T_f(r) + O\left(\max_{1\le j\le q} T_{a_j}(r)\right). \tag{4.9}$$

Next we compare $\lambda_{H_{\mu(l)}(z)}(f(z))$ and $\lambda_{\hat{H}_{l,j}(\mu)}(F(z))$, for each $\mu \in T$. By (3.7), (4.4), (4.5), (4.6), and (4.7), for $0 \le l \le n$,

$$\begin{aligned}
&\lambda_{\hat{H}_{l,j}(\mu)}(F(z)) \\
= \quad & \log\left(\|\tilde{F}(z)\| \cdot \max_{1\le r\le (n+1)\ell(s+1)} |c_{l\ell(s)+j,r}(\mu)|\right) \\
- \quad & \log\left(|c_{l\ell(s)+j,1}(\mu)b_1(z)f_0(z) + \cdots + c_{l\ell(s)+j,(n+1)\ell(s+1)}(\mu)b_{\ell(s+1)}(z)f_n(z)|\right) \\
\le \quad & \log\left(\|\mathbf{f}\| \cdot \max_{1\le\alpha\le\ell(s+1)} |b_\alpha(z)| \cdot \max_{1\le r\le (n+1)\ell(s+1)} |c_{l\ell(s)+j,r}(\mu)|\right) \\
& -\log\left(|b_j(z)h_{\mu(l)}(z)| \cdot |\, f_0(z)\,|\right) \\
= \quad & \log\left(\|\mathbf{f}\| \cdot \max_{1\le\alpha\le\ell(s+1)} |b_\alpha(z)| \cdot \max_{1\le r\le (n+1)\ell(s+1)} |c_{l\ell(s)+j,r}(\mu)|\right) \\
& -\log\left(|b_j(z)| \cdot |\zeta_{\mu(l),0}(z)f_0(z) + \cdots + \zeta_{\mu(l),n}(z)f_n(z)|\right) \\
= \quad & \lambda_{H_{\mu(l)}(z)}(f(z)) - \log\frac{\max_{0\le r\le n}|a_{\mu(l),r}(z)|}{|a_{\mu(l),\hat{\mu(l)}}(z)|} - \log\frac{|b_j(z)|}{\max_{1\le\alpha\le\ell(s+1)}|b_\alpha(z)|} \\
& +O(1),
\end{aligned}$$

where the above equality holds for those $z \in \mathbf{C}$ such that $< \mathbf{f}(z), \mathbf{a}_{\mu(l)}(z) > \neq 0$, $f_0(z) \neq 0$, and such that z is not in the union of the sets of the zeros and poles of $\zeta_{j,l}$ and b_r, $1 \le j \le q, 0 \le l \le n, 1 \le r \le \ell(s+1)$. Combining this with (4.8) and (4.9) gives

$$\begin{aligned}
&\ell(s)\int_0^{2\pi} \max_{\mu\in T}\sum_{l=0}^{n} \lambda_{H_{\mu(l)}(re^{i\theta})}(f(re^{i\theta}))\frac{d\theta}{2\pi} \\
&= \quad \int_0^{2\pi} \max_{\mu\in T}\sum_{l=0}^{n}\sum_{j=1}^{\ell(s)} \lambda_{\hat{H}_{l,j}(\mu)}(F(re^{i\theta}))\frac{d\theta}{2\pi} + O\left(\max_{1\le j\le q} T_{a_j}(r)\right) \\
&\le \quad ((n+1)\ell(s+1) + \epsilon/2)\, T_f(r) + O\left(\max_{1\le j\le q} T_{a_j}(r)\right),
\end{aligned}$$

where the inequality holds for all r outside of a set $E \subset (0, +\infty)$ with finite Lebesgue measure. Hence, the inequality

$$\int_0^{2\pi} \max_{\mu\in T}\sum_{l=0}^{n} \lambda_{H_{\mu(l)}(re^{i\theta})}(f(re^{i\theta}))\frac{d\theta}{2\pi}$$

$$\leq \quad ((1+n)\ell(s+1)/\ell(s) + \epsilon/2\ell(s))\, T_f(r) + O\left(\max_{1\leq j\leq q} T_{a_j}(r)\right)$$

holds for all r outside of a set $E \subset (0, +\infty)$ with finite Lebesgue measure. As noted by Steinmetz, we obviously have

$$0 \leq \ell(s) \leq \binom{q(n+1)+s-1}{s}$$

for each s and therefore

$$\liminf_{s\to\infty} \frac{\ell(s+1)}{\ell(s)} = 1.$$

Thus, for the given $\epsilon > 0$ there exists a positive integer $s > 0$ such that

$$\ell(s+1)/\ell(s) < 1 + \epsilon/2(n+1).$$

Choosing such an s concludes our proof of Theorem A4.2.1 □

A4.3 Cartan's Conjecture with Moving Targets

In this section, we consider the degenerate case of Theorem A4.2.1, i.e., the case when f is degenerate over $\mathcal{R}_{\mathcal{G}}$.

Definition A4.3.1 *Let $\mathcal{R}_{\mathcal{G}}$ be the field that appears in Theorem A4.2.1. A holomorphic map $f = [f_0 : \cdots : f_n] : \mathbf{C} \to \mathbf{P}^n(\mathbf{C})$ is said to be* **m-nondegenerate over** *$\mathcal{R}_{\mathcal{G}}$ if m is the largest integer with the property that there is an injection $\eta : \{0, \ldots, m\} \to \{0, \ldots, n\}$ such that $\{f_{\eta(0)}, \ldots, f_{\eta(m)}\}$ is linearly independent over $\mathcal{R}_{\mathcal{G}}$.*

Clearly $0 \leq m \leq n$, and $m = n$ if and only if f is non-degenerate over $\mathcal{R}_{\mathcal{G}}$.

Theorem A4.3.2 (Ru-Stoll) *Let $f = [f_0 : \ldots : f_n] : \mathbf{C} \to \mathbf{P}^n(\mathbf{C})$ be a holomorphic map. Let $\mathcal{G}$ be a finite set of moving hyperplanes $H_1, \ldots, H_q$ (or $\mathbf{a}_1, \ldots, \mathbf{a}_q$) with $T_{a_j}(r) = o(T_f(r)), 1 \leq j \leq q$. Assume that $\mathcal{G}$ is in general position. Let $\mathbf{a}_j = [a_{j,0} : \ldots : a_{j,n}], 1 \leq j \leq q$. Let $\mathcal{R}_{\mathcal{G}}$ be the smallest field which contains $\mathbf{C}$ and all $a_{j\mu}/a_{j\nu}$ with $a_{j\nu} \not\equiv 0$. Assume that f is m-nondegenerate over $\mathcal{R}_{\mathcal{G}}$. We further assume that $< \mathbf{f}, \mathbf{a}_j > \not\equiv 0$ for*

$j = 1, \ldots, q$. *Then, for every $\epsilon > 0$, the inequality*

$$\sum_{j=1}^{q} m_f(r, H_j) \leq (2n - m + 1 + \epsilon) T_f(r)$$

holds for all r outside of a set $E \subset (0, +\infty)$ with finite Lebesgue measure.

Definition A4.3.3 *We say that $H_1, \ldots, H_q$ (or $\mathbf{a_1}, \ldots, \mathbf{a_q}$) are in* **$u$-subgeneral position** *if for every $1 \leq i_0 < \cdots < i_u \leq q$, the linear span of $\mathbf{a}_{i_0}(z), \ldots, \mathbf{a}_{i_u}(z)$ is $\mathbf{C}^{n+1}$ for some z, hence for almost all z.*

The significance of the condition of u-subgeneral position is that the hyperplanes $H_1, \cdots,, H_q$ are in u-subgeneral position if and only if there is an embedding $\mathbf{P}^n \hookrightarrow \mathbf{P}^u$ and hyperplanes $H_1', \ldots, H_q'$ in $\mathbf{P}^u$ such that $H_j' \cap \mathbf{P}^n = H_j$ for all j and such that $H_1', \ldots, H_q'$ are in general position.

Theorem A4.3.4 (Ru-Stoll) *Let $u \geq n$. Let $f = [f_0 : \ldots : f_n] : \mathbf{C} \to \mathbf{P}^n(\mathbf{C})$ be a holomorphic map. Let $\mathcal{G}$ be a finite set of moving hyperplanes $H_1, \ldots, H_q$ (or $\mathbf{a}_1, \ldots, \mathbf{a}_q$) with $T_{a_j}(r) = o(T_f(r)), 1 \leq j \leq q$. Assume that $\mathcal{G}$ is in u-subgeneral position. If f is nondegenerate over $\mathcal{R}_{\mathcal{G}}$, then, for every $\epsilon > 0$, the inequality*

$$\sum_{j=1}^{q} m_f(r, H_j) \leq (2u - n + 1 + \epsilon) T_f(r)$$

holds for all r outside of a set $E \subset (0, +\infty)$ with finite Lebesgue measure.

We will prove Theorem A4.3.4 first and then derive Theorem A4.3.2 from Theorem A4.3.4.

Take $0 \leq p \in \mathbf{Z}$ and let $T_p = T_p(\mathcal{G})$ be the set of all injective map $\lambda : \{0, \ldots, p\} \to \{1, \ldots, q\}$ such that $\mathbf{a}_\lambda = \mathbf{a}_{\lambda(0)} \wedge \cdots \wedge \mathbf{a}_{\lambda(p)} \not\equiv 0$. For $\lambda \in T_p$, the set $S_\lambda = \mathbf{a}_{\lambda(0)}^{-1}(0)$ is a discrete set. Then $S = \cup_{p=0}^{n} \cup_{\lambda \in T_p} S_\lambda$ is a discrete set and $\mathbf{C} - S$ is open and dense in $\mathbf{C}$. It is easy to check that for $z \in \mathbf{C} - S$, $\{\mathbf{a}_j(z), 1 \leq j \leq q\}$ is in u-subgeneral position (as fixed hyperplanes). For $\emptyset \neq P \subset \{1, \ldots, q\}$, let $L(z, P)$ be the linear subspaces of $\mathbf{C}^{n+1}$ spanned by $\{\mathbf{a}_j(z) \mid j \in P\}$. Define $d(z, P) = \dim L(z, P)$.

Lemma A4.3.5 *If $\emptyset \neq P \subset \{1, \ldots, q\}$, then $d(z, P)$ is constant for $z \in \mathbf{C} - S$.*

Proof. Take z_1 and z_2 in $\mathbf{C} - S$. Let $p = d(z_1, P) - 1$. Then $0 \leq p \leq n$. There is a bijective map $\lambda : \mathbf{Z}[0,p] \to P$ such that $\mathbf{a}_{\lambda(0)}(z_1), \ldots, \mathbf{a}_{\lambda(p)}(z_1)$ is a base of $L(z, P)$. Then $\mathbf{a}_\lambda \not\equiv 0$ and $\lambda \in T_p$. Because $z_2 \in \mathbf{C} - S$, also $\mathbf{a}_\lambda(z_2) \neq 0$. Hence $\mathbf{a}_{\lambda(0)}(z_2), \ldots, \mathbf{a}_{\lambda(p)}(z_2)$ are linearly independent. Hence $p+1 \leq d(z_2, P)$. Thus, $d(z_1, P) \leq d(z_2, P)$. By symmetry we obtain equality. □

We now use Theorem A3.4.3 to derive the product to the sum estimate. Given our $\mathcal{G} = \{H_j \mid 1 \leq j \leq q\}$ in u-subgeneral position with $q \geq 2u - n + 1$.

Theorem A4.3.6 *Given $\mathcal{G}$, a set of moving hyperplanes $H_1, \ldots, H_q$ in u-subgeneral position with $q \geq 2u - n + 1$, there exists a function $\omega : \{1, \ldots, q\} \to \mathbf{R}(0,1]$ called the Nochka weight and a real number $\theta \geq 1$ called the Nochka constant satisfying the following properties:*

(i) If $j \in \{1, \ldots, q\}$, then $0 < \omega(j)\theta \leq 1$.

(ii) $q - 2u + n - 1 = \theta(\sum_{j=1}^{q} \omega(j) - n - 1)$.

(iii) If $\emptyset \neq B \subset \{1, \ldots, q\}$ with $\#B \leq u + 1$, then

$$\sum_{j \in P} \omega(j) \leq \dim L(B).$$

(iv) $1 \leq (u+1)/(n+1) \leq \theta \leq (2u - n + 1)/(n+1)$.

(v) Let $\{E_1, \ldots, E_q\}$ be a family of functions $E_j : \mathbf{C} - S \to \mathbf{R}[1, +\infty)$. Given any $A \subset \{1, \ldots, q\}$ with $0 < \#A \leq u + 1$. Take a $z \in \mathbf{C} - S$. Then there is a subset $B(z)$ of A such that $\#B(z) = \dim L(z, A) = d(A)$ and such that $\{\mathbf{a}_j(z) \mid j \in B(z)\}$ is a base of $L(z, A)$, and such that

$$\prod_{j \in A} E_j^{\omega(j)} \leq \prod_{j \in B} E_j.$$

Proof. We select a point $z_0 \in \mathbf{C} - S$. Then $\mathcal{G}(z_0) = \{H_j(z_0) \mid 1 \leq j \leq q\}$ is a collection of fixed hyperplanes in u-subgeneral position. Now we take the Nochka weight function ω and the Nochka constant θ for this family $\mathcal{G}(z_0)$. Then Theorem A3.4.3 gives (i) to (iv). So we only need to verify (v). By (v) of Theorem A3.4.3 it follows that, for $z \in \mathbf{C} - S$, there is a subset $B(z)$

of A such that $\#B(z) = \dim L(z, A) = d(A) = p + 1$, $\{\mathbf{a}_j(z_0) \mid j \in B(z)\}$ is a base of $L(z_0, A)$, and such that

$$\prod_{j\in A} E_j^{\omega(j)}(z) \leq \prod_{j\in B} E_j(z).$$

Take a bijective map $\lambda : \mathbf{Z}[0,p] \to B(z)$. Then $\mathbf{a}_\lambda(z_0) = \mathbf{a}_{\lambda(0)}(z_0) \wedge \cdots \wedge \mathbf{a}_{\lambda(p)}(z_0) \neq 0$. Thus $\mathbf{a}_\lambda \not\equiv 0$ and $\lambda \in T_p$. Thus $\mathbf{a}_\lambda(z) \neq 0$ since $z \in \mathbf{C} - S$. Hence $\mathbf{a}_{\lambda(0)}(z), \ldots, \mathbf{a}_{\lambda(p)}(z)$ are linearly independent. Also $\mathbf{a}_{\lambda(j)}(z) \in L(z, A)$ for $j = 0, \ldots, p$. Since $z \in \mathbf{C} - S$, we have $\dim L(z, A) = d(z_0) = p + 1$. Thus $\{\mathbf{a}_j(z) \mid j \in B(z)\}$ is a base of $L(z, A)$. □

Proof of Theorem A4.3.4.

Proof. The theorem is trivial when $q \leq 2u - n + 1$, so we may assume that $q > 2u - n + 1$. Let $\mathcal{G} = \{H_j \mid 1 \leq j \leq q\}$ be a finite collection of moving hyperplanes in u-subgeneral position. Define, for $z \in \mathbf{C}$,

$$\Gamma(\mathcal{G})(z) = \min\left\{ \frac{\|\mathbf{a}_{\mu(0)}(z) \wedge \cdots \wedge \mathbf{a}_{\mu(n)}(z)\|}{\|\mathbf{a}_{\mu(0)}(z)\| \cdots \|\mathbf{a}_{\mu(n)}(z)\|} \;\middle|\; \mu : \mathbf{Z}[0,n] \to \{1, \ldots, q\},\ \text{injective} \right\}.$$

By the u-subgeneral position assumption,

$$S = \{z \in \mathbf{C} \mid \Gamma(\mathcal{G})(z) = 0\}$$

is a closed set of isolated points. Fix a $z_0 \in \mathbf{C} - S$ and let $\omega(j), 1 \leq j \leq q$, be the Nochka weights associated with the hyperplanes $\{H_j(z_0)\}$. Since $H_1, \cdots, , H_q$ are in u-subgeneral position, there is an embedding $\mathbf{P}^n \hookrightarrow \mathbf{P}^u$ and hyperplanes $H_1', \ldots, H_q'$ in $\mathbf{P}^u$ such that $H_j' \cap \mathbf{P}^n = H_j$ for all j and such that $H_1', \ldots, H_q'$ are in general position. So by Lemma A4.2.3, for every $z \in \mathbf{C} - S$, there exists $i(z, 0), \ldots, i(z, u)$ among $1, \ldots, q$ such that

$$\begin{aligned} &\prod_{\mathbf{a}\in\mathcal{G}} \left(\frac{\|\mathbf{f}(z)\|\|\mathbf{a}(z)\|}{|<\mathbf{f}(z), \mathbf{a}(z)>|} \right)^{\omega(j)} \\ &= \left(\frac{2(u+1)}{\Gamma(\mathcal{G})(z)} \right)^{q-u-1} \prod_{l=0}^{u} \left(\frac{\|\mathbf{f}(z)\|\|\mathbf{a}_{i(z,l)}(z)\|}{|<\mathbf{f}(z), \mathbf{a}_{i(z,l)}(z)>|} \right)^{\omega(i(z,l))}. \end{aligned} \tag{4.10}$$

Let $A = \{i(z, 0), \ldots, i(z, u)\}$. Then $d(A) = n + 1$. Recall that the Weil function $\lambda_{H(z)}(f(z))$ of f is defined by

$$\lambda_{H(z)}(f(z)) = \log \frac{\|\mathbf{f}(z)\|\|\mathbf{a}(z)\|}{|<\mathbf{f}(z), \mathbf{a}(z)>|}$$

and our Weil function has the property that $\lambda_H(f) \geq 0$. Applying Theorem A4.3.6 with $E_l = e^{\lambda_{H_{i(z,l)}}(f(z))}, 0 \leq l \leq u$, for $z \in \mathbf{C} - S$, there is a subset $B(z)$ of A such that $\#B(z) = \dim L(z, A) = d(A) = n + 1$ and such that $\{\mathbf{a}_j(z) \mid j \in B(z)\}$ is a base of $\mathbf{C}^{n+1}$. Moreover

$$\begin{aligned}
\prod_{l=0}^{u} \left(\frac{\|\mathbf{f}(z)\|\|\mathbf{a}_{i(z,l)}(z)\|}{|<\mathbf{f}(z), \mathbf{a}_{i(z,l)}(z)>|} \right)^{\omega(i(z,l))} &\leq \prod_{j \in B(z)} \frac{\|\mathbf{f}(z)\|\|\mathbf{a}_j(z)\|}{|<\mathbf{f}(z), \mathbf{a}_j(z)>|} \\
&\leq \max_{\gamma \in \Gamma} \prod_{t=0}^{n} \frac{\|\mathbf{f}(z)\|\|\mathbf{a}_{\gamma(t)}(z)\|}{|<\mathbf{f}(z), \mathbf{a}_{\gamma(t)}(z)>|} \qquad (4.11)
\end{aligned}$$

where Γ is the set of all maps $\gamma : \{0, \ldots, n\} \to \{1, \ldots, q\}$ such that $\mathbf{a}_{\gamma(0)}(z), \ldots, \mathbf{a}_{\gamma(n)}(z)$ are linearly independent for some (and hence for almost all) $z \in \mathbf{C}$. It follows from (4.10) and (4.11) that

$$\sum_{j=1}^{q} \omega(j) m_f(r, H_j) \leq \int_0^{2\pi} \max_{\gamma \in \Gamma} \sum_{l=0}^{n} \lambda_{H_{\gamma(l)}(re^{i\theta})}(f(re^{i\theta})) \frac{d\theta}{2\pi} + C(a_1, \ldots, a_q),$$

where $C(a_1, \ldots, a_q)$ is a term depends only on $a_1, \ldots, a_q$. Applying Theorem A4.2.1 yields, for every $\epsilon > 0$,

$$\int_0^{2\pi} \max_{\gamma \in \Gamma} \sum_{l=0}^{n} \lambda_{H_{\gamma(l)}(re^{i\theta})}(f(re^{i\theta})) \frac{d\theta}{2\pi} \leq (n + 1 + \epsilon/2) T_f(r),$$

where the inequality holds for all r outside a set $E \subset (0, +\infty)$ with finite Lebesgue measure. So the inequality

$$\sum_{j=1}^{q} \omega(j) m_f(r, H_j) \leq (n + 1 + \epsilon) T_f(r)$$

holds for all r outside a set $E \subset (0, +\infty)$ with finite Lebesgue measure. Thus

$$\sum_{j=1}^{q} m_f(r, H_j) = \sum_{j=1}^{q} (1 - \theta\omega(j)) m_f(r, H_j) + \sum_{j=1}^{q} \theta\omega(j) m_f(r, H_j)$$

$$
\begin{aligned}
&\leq \sum_{j=1}^{q}(1-\theta\omega(j))m_f(r,H_j)+\theta(n+1+\epsilon)T_f(r)\\
&\leq \sum_{j=1}^{q}(1-\theta\omega(j))T_f(r)+\theta(n+1+\epsilon))T_f(r)\\
&= \left\{q-\theta\left(\sum_{1\leq j\leq q}\omega(j)-n-1\right)+\epsilon\right\}T_f(r)\\
&= (2u-n+1+\epsilon)T_f(r),
\end{aligned}
$$

where the inequality holds for all r outside a set $E \subset (0,+\infty)$ with finite Lebesgue measure. □

We now prove Theorem A4.3.2.

Proof. The theorem trivially holds when $q \leq 2n-m+1$. So we can assume that $q > 2n-m+1$. By the assumption, f is m-nondegenerate over $\mathcal{R}_{\mathcal{G}}$, that is m is the largest integer with the property that there is an injection $\eta : \{0,\ldots,m\} \to \{0,\ldots,n\}$ such that $\{f_{\eta(0)},\ldots,f_{\eta(m)}\}$ is linearly independent over $\mathcal{R}_A$. Without loss of generality, we assume that $\{f_0,\ldots,f_m\}$ is linearly independent over $\mathcal{R}_A$. So for each $r = m+1,\ldots,n$ there are $c_{0,r},\ldots,c_{m,r} \in \mathcal{R}_A$ such that

$$f_r = c_{0,r}f_0 + \ldots + c_{m,r}f_m.$$

We now modify f and hyperplanes $H_1,\ldots,H_q$ so that Theorem A4.3.4 can be applied. Let $\hat{f} = [f_0 : \ldots : f_m] : \mathbf{C} \to \mathbf{P}^m(\mathbf{C})$. Then $\hat{f}$ is non-degenerate over $\mathcal{R}_{\mathcal{G}}$ and then

$$T_{\hat{f}}(r) = T_f(r) + o(T_f(r)). \tag{4.12}$$

We define the modified moving hyperplanes

$$\hat{H}_j = \{[x_0 : \ldots : x_m] \in \mathbf{P}^m(\mathbf{C})\ |b_{j,0}x_0 + \ldots + b_{j,m}x_m = 0\},$$

where

$$b_{j,l} = a_{j,l} + a_{j,m+1}c_{l,m+1} + \ldots + a_{j,n}c_{l,n}$$

for $j = 1, \ldots, q$, and $l = 0, \ldots, m$. These new hyperplanes are clearly in n-subgeneral position. We also have, for $1 \leq j \leq q$,

$$a_{j,0}x_0 + \ldots + a_{j,n}x_n = b_{j,0}x_0 + \ldots + b_{j,m}x_m.$$

Thus

$$\lambda_{H_j}(f) = \lambda_{\hat{H}_j}(\hat{f}). \tag{4.13}$$

By applying Theorem A4.3.4 to $\hat{f}$ and to the hyperplanes $\hat{H}_1, \ldots, \hat{H}_q$, combining with (4.12) and (4.13), we obtain Theorem A4.3.2. □

A4.4 Applications

As an application of Theorem A4.3.2, we derive the Second Main Theorem with moving targets for algebroid functions. Let f be a v-valued meromorphic algebroid function in $|z| < +\infty$, defined by an irreducible equation

$$A_v w^v + A_{v-1} w^{v-1} + \ldots + A_0 = 0,$$

where $A_v, A_{v-1}, \ldots, A_0$ are entire functions without any common zero. We prove the following SMT with moving targets for meromorphic algebroid functions.

Theorem A4.4.1 *Let f be a v-valued algebroid function on* $\mathbf{C}$*. Let $a_1, \ldots, a_q$ be meromorphic functions with $T_{a_j}(r) = o(T_f(r)), 1 \leq j \leq q$. Assume that $a_i \not\equiv a_j$ for $i \neq j, i, j = 1, 2, \ldots, q$. Then, for any $\epsilon > 0$, the inequality*

$$\sum_{j=1}^{q} m_f(r, a_j) \leq (2v + \epsilon) T_f(r)$$

holds for all $r \notin E$, where E is a set of finite measure.

Proof. Given f, a v-valued meromorphic algebroid function on $\mathbf{C}$, defined by an irreducible equation

$$A_v w^v + A_{v-1} w^{v-1} + \ldots + A_0 = 0,$$

where $A_v, A_{v-1}, \ldots, A_0$ are entire functions without any common zero. Let $F = [A_0 : A_1 : \ldots : A_v] : \mathbf{C} \to \mathbf{P}^v(\mathbf{C})$. Then, by a theorem of Valiron [Val] we have

$$|T_f(r) - T_F(r)| < C,$$

where C is a constant. For each j, $1 \leq j \leq q$, the given meromorphic functions a_j defines a moving hyperplane H_j defined by the equation

$$X_0 + a_j X_1 + \ldots + a_j^v X_v = 0.$$

Because $a_i \not\equiv a_j$, the set of hyperplanes $\{H_j\}$ lies in general position by the non-vanishing of the van der Monde determinant. Moreover, by the standard definition, $m_f(r, a_j) = m_F(r, H_j) + O(1)$. Thus the theorem follows by applying Theorem A4.3.2 to F and $H_j, 1 \leq j \leq q$. □

Part B: Diophantine Approximation

B4.1 Schmidt's Subspace Theorem with Moving Targets

We prove, in this section, the counterpart of the theorems of Part A in number theory: Schmidt's subspace theorem with moving targets. The method of proof is similar to one used in Part A. It is worth noting that it basically says that "Schmidt's subspace theorem (with fixed targets) implies Schmidt's subspace theorem with moving targets."

Let k be a number field. Let Λ be an infinite index set. A **moving hyperplane H** indexed by Λ assigns, for each $\alpha \in \Lambda$, a hyperplane $H(\alpha)$. Also a collection of points $\{\mathbf{x}(\alpha) \in \mathbf{P}^n(k) \mid \alpha \in \Lambda\}$ will be regarded as a map $\mathbf{x} : \Lambda \to \mathbf{P}^n(k)$. We note that the definitions of the Weil function, proximity function, and the counting function defined in chapter 3 can be easily extended to the moving hyperplane H. Also, if $H(\alpha) = \{[x_0 : \ldots : x_n] \mid a_0(\alpha)x_0 + \ldots + a_n(\alpha)x_n = 0\}$, we define the height, for the hyperplane $H(\alpha)$, as $h(H(\alpha)) = h([a_0(\alpha) : \cdots : a_n(\alpha)])$. We note that $h(H(\alpha))$ depends only on $H(\alpha)$, not on the choice of $a_0(\alpha), \ldots, a_n(\alpha)$.

Definition B4.1.1 *Given moving hyperplanes $H_1, \ldots, H_q$ indexed by Λ with $H_j(\alpha) = \{[x_0 : \ldots : x_n] \mid a_{j,0}(\alpha)x_0 + \ldots + a_{j,n}(\alpha)x_n = 0\}$ for $1 \leq j \leq q$ and $\alpha \in \Lambda$. An infinite index subset $A \subseteq \Lambda$ is said to be* **coherent with respect to** $(H_1, \ldots, H_q)$ *if for every homogeneous polynomial P in variables $X_{1,0}, \ldots, X_{1,n}, X_{2,0}, \ldots, X_{q,n}$ either $P(a_{1,0}(\alpha), \ldots, a_{q,n}(\alpha))$ vanishes for all $\alpha \in A$, or it vanishes for only finitely many $\alpha \in A$.*

Remark. *The above definition is independent of the choice of the coefficients $a_{j,0}(\alpha), \ldots, a_{j,n}(\alpha)$.*

Lemma B4.1.2 *There exists an infinite index subset $A \subseteq \Lambda$ which is coherent with respect to $(H_1, \ldots, H_q)$.*

Proof. Suppose the statement is false; i.e., for any infinite subset $A \subseteq \Lambda$ there is a $P \in k[X_{1,0}, \ldots, X_{1,n}, X_{2,0}, \ldots, X_{q,n}]$ such that

$$P(a_{1,0}(\alpha), \ldots, a_{q,n}(\alpha)) = 0, \quad \text{for infinitely many } \alpha \in A$$

but $P(a_{1,0}(\alpha), \ldots, a_{q,n}(\alpha)) \neq 0$ for some $\alpha \in A$. We will construct a chain of ideals in $k[X_{1,0}, \ldots, X_{1,n}, X_{2,0}, \ldots, X_{q,n}]$: for any index subset $A \subseteq \Lambda$, let $I(A)$ denote the ideal generated by multi-homogeneous polynomials $P \in k[X_{1,0}, \ldots, X_{1,n}, X_{2,0}, \ldots, X_{q,n}]$ such that $P(a_{1,0}(\alpha), \ldots, a_{q,n}(\alpha)) = 0$ for all $\alpha \in A$. Let $A_1 = \Lambda$. Then A_1 is infinite, and (by assumption) not coherent. Therefore there is a multi-homogeneous polynomial $P \in k[X_{1,0}, \ldots, X_{1,n}, X_{2,0}, \ldots, X_{q,n}]$ and an infinite index subset $A_2 \subset A_1$ such that $P(a_{1,0}(\alpha), \ldots, a_{q,n}(\alpha)) = 0$ for all $\alpha \in A_2$, but $P(a_{1,0}(\alpha), \ldots, a_{q,n}(\alpha)) \neq 0$ for all $\alpha \in A_1 \setminus A_2$. So $I(A_1) \subset I(A_2)$, and $I(A_1) \neq I(A_2)$. Continuing in this way, we obtain an infinite chain of ideals $I(A_1) \subset I(A_2) \subset \ldots \subset k[X_{1,0}, \ldots, X_{1,n}, X_{2,0}, \ldots, X_{q,n}]$, which contradicts the fact that

$$k[X_{1,0}, \ldots, X_{1,n}, X_{2,0}, \ldots, X_{q,n}]$$

is Noetherian. □

Now we define $\mathcal{R}_A$:

Definition B4.1.3 *Let $A \subseteq \Lambda$ be an infinite index subset which is coherent with respect to $(H_1, \ldots, H_q)$. Let $\mathcal{R}_A^0$ be the set of equivalence classes of pairs (C, a) where C is a subset of A with finite complement, a is a map $a : C \to k$, and (C_1, a_1) is equivalent to (C_2, a_2) if there is a subset $C \subseteq C_1 \cap C_2$ such that C has finite complement in A, and such that the restrictions of a_1 and a_2 to C coincide. $\mathcal{R}_A^0$ has an obvious ring structure. Moreover we embed k into $\mathcal{R}_A^0$ as constant functions. If $j \in \{1, \ldots, q\}$ and $\mu, \nu \in \{0, \ldots, n\}$ are such that $a_{j,\nu}(\alpha) \neq 0$ for at least one $\alpha \in A$, then the pair $(\{\alpha \in A \mid a_{j,\nu}(\alpha) \neq 0\}, \alpha \mapsto a_{j,\mu}(\alpha)/a_{j,\nu}(\alpha))$ lies in $\mathcal{R}_A^0$ by coherence. Moreover, the subring of $\mathcal{R}_A^0$ generated over k by all such pairs is entire. Therefore we define $\mathcal{R}_A$ to be the quotient field of this subring.*

Note that the field $\mathcal{R}_A$ is independent of the choice of coefficients.

Definition B4.1.4 *Let $H_1, \ldots, H_q$ be the moving hyperplanes indexed by Λ. A map $\mathbf{x} = [x_0 : \ldots : x_n] : \Lambda \to \mathbf{P}^n(k)$ is said to be* **non-degenerate over $\mathcal{R}$ (with respect to $H_1, \ldots, H_q$)** *if there is an infinite subset $A \subseteq \Lambda$, coherent with respect to $H_1, \ldots, H_q$, such that $x_0, \ldots, x_n$ is linearly independent over $\mathcal{R}_A$.*

Remark B4.1.5 *Let $B \subseteq A \subseteq \Lambda$ be two infinite index subsets. Then it is clear that if A is coherent then so is B, and if $x_0, \ldots, x_n$ is linearly independent over $\mathcal{R}_A$ then it is linearly independent over $\mathcal{R}_B$.*

The following theorem is due to Ru-Vojta.

Theorem B4.1.6 (Schmidt's subspace theorem with moving targets) *Let k be a number field, let S be a finite set of places of k, let q be a positive integer, and let $\epsilon > 0$. Let Λ be an infinite index set, let $H_1, \ldots, H_q$ be the moving hyperplanes indexed by Λ, and let $\mathbf{x} : \Lambda \to \mathbf{P}^n(k)$ be a collection of points. Assume that $h(H_j(\alpha)) = o(h(\mathbf{x}(\alpha)))$ for all $j = 1, \ldots, q$. We also assume that $\mathbf{x}$ is non-degenerate over $\mathcal{R}$ (see Definition B4.1.4). Then*

$$\frac{1}{[k:\mathbf{Q}]} \sum_{v \in S} \max_K \sum_{j \in K} \lambda_{v, H_j(\alpha)}(\mathbf{x}(\alpha)) \leq (n+1+\epsilon) h(\mathbf{x}(\alpha))$$

holds for all $\alpha \in A$, where the maximum is taken over all subsets K of $\{1, \ldots, q\}$ such that the hyperplanes $H_j(\alpha), j \in K$, (or more precisely the linear forms $L_j(\alpha)$ defining $H_j(\alpha)$), are linearly independent over k for each $\alpha \in \Lambda$.

If, in addition, we assume that $H_1(\alpha), \ldots, H_q(\alpha)$ are in general position for each $\alpha \in \Lambda$, then by the product to the sum estimate, we have the following theorem.

Theorem B4.1.7 *Let k be a number field, let S be a finite set of places of k, let q be a positive integer, and let $\epsilon > 0$. Let Λ be an infinite index set, let $H_1, \ldots, H_q$ be the moving hyperplanes indexed by Λ, and let $\mathbf{x} : \Lambda \to \mathbf{P}^n(k)$ be a collection of points such that:*

(i) for each $\alpha \in \Lambda$, $H_1(\alpha), \ldots, H_q(\alpha)$ are in general position;

(ii) $\mathbf{x}$ is non-degenerate over $\mathcal{R}$ (see Definition B4.1.4); and

(iii) $h(H_j(\alpha)) = o(h(\mathbf{x}(\alpha)))$ for all $j = 1, \ldots, q$.

Then there exists an infinite index subset $A \subseteq \Lambda$ such that

$$\sum_{j=1}^{q} m(\mathbf{x}(\alpha), H_j(\alpha)) \leq (n+1+\epsilon)h(\mathbf{x}(\alpha))$$

for all $\alpha \in A$.

We note that when $n = 1$, the non-degeneracy condition is automatically satisfied if $h(H_j(\alpha)) = o(h(x(\alpha)))$, $j = 1, \ldots, q$.

We now prove Theorem B4.1.6.

Proof. Without loss of generality, we may assume that $q \geq n+1$ and $\#K = n+1$. Let T be the set of all injective maps $\mu : \{0, \ldots, n\} \to \{1, \ldots, q\}$ such that $H_{\mu(j)}(\alpha), j = 0, 1, \ldots, n$ (or more precisely the linear forms defining $H_{\mu(j)}(\alpha)$, $j = 0, 1, \ldots, n$) are linearly independent over k for each $\alpha \in \Lambda$. Let $\mathbf{x} = [x_0 : \ldots : x_n] : \Lambda \to \mathbf{P}^n(k)$ be the map given in Theorem B4.1.6. Since $\mathbf{x}$ is nondegenerate over $\mathcal{R}$, there is a coherent infinite index subset $A \subseteq \Lambda$ such that $x_0, \ldots, x_n$ are linearly independent over $\mathcal{R}_A$. By Remark B4.1.5, if B is any infinite subset of A, then B is still coherent and $x_0, \ldots, x_n$ are still linearly independent over $\mathcal{R}_B$. Therefore we may freely pass to infinite subsequences.

Choose $a_{j,0}, \ldots, a_{j,n}$ such that, for each $\alpha \in A$, $H_j(\alpha)$ is the hyperplane determined by the equation $a_{j,0}(\alpha)x_0 + \ldots + a_{j,n}(\alpha)x_n = 0$. By coherence, for each $j = 1, \ldots, q$, there is j_0 such that $0 \leq j_0 \leq n$ and $a_{j,j_0} \neq 0$ for all but finitely many $\alpha \in A$. Therefore, for $j = 1, \ldots, q$ and $l = 0, \ldots, n$ we may let $\zeta_{j,l} \in \mathcal{R}_A$ be defined by

$$\zeta_{j,l} = a_{j,l}/a_{j,j_0}. \tag{4.14}$$

Let $\mathcal{L}(s) \subset \mathcal{R}_A$ be the vector space generated over k by

$$\left\{ \zeta_{1,0}^{n_{1,0}} \cdots \zeta_{q,0}^{n_{q,0}} \cdots \zeta_{1,n}^{n_{1,n}} \cdots \zeta_{q,n}^{n_{q,n}} \;\middle|\; n_{i,j} \in \mathbf{N}, \sum_{i=1}^{q}\sum_{j=0}^{n} n_{i,j} = s \right\}.$$

We have $\mathcal{L}(s) \subset \mathcal{L}(s+1)$. For each s, let $\ell(s) = \dim \mathcal{L}(s)$. As noted by Steinmetz, we obviously have

$$0 \leq \ell(s) \leq \binom{q(n+1)+s-1}{s}$$

for each s and therefore

$$\liminf_{s\to\infty} \frac{\ell(s+1)}{\ell(s)} = 1.$$

Thus, given any $\delta > 0$ we may find a positive integer s such that

$$\ell(s+1) \leq (1+\delta)\ell(s). \tag{4.15}$$

Fix such an s.

Let $\{b_1, \dots, b_{\ell(s+1)}\}$ be a basis of $\mathcal{L}(s+1)$ such that $\{b_1, \dots, b_{\ell(s)}\}$ is a basis of $\mathcal{L}(s)$. Then $(b_\mu x_\nu)_{\mu=1,\dots,\ell(s+1);\ \nu=0,\dots,n}$ are, by condition (ii), linearly independent over k. For $j = 1, \dots, q$ let $h_j \in \mathcal{R}_A$ be defined by

$$h_j = \sum_{l=0}^{n} \zeta_{j,l} x_l. \tag{4.16}$$

For each $\mu \in T$, the $b_j h_{\mu(l)}$ (for $j = 1, \dots, \ell(s)$ and $l = 0, \dots, n$) can be written uniquely as k-linear combinations of the products $b_\mu x_\nu$ for $\mu = 1, \dots, \ell(s+1)$ and $\nu = 0, \dots, n$. In other words there is an $(n+1)\ell(s) \times (n+1)\ell(s+1)$ matrix $C(\mu)$, with entries in k (not just in $\mathcal{R}_A$) such that

$$\begin{pmatrix} b_1 h_{\mu(0)} \\ \vdots \\ b_{\ell(s)} h_{\mu(0)} \\ \vdots \\ b_1 h_{\mu(n)} \\ \vdots \\ b_{\ell(s)} h_{\mu(n)} \end{pmatrix} = C(\mu) \begin{pmatrix} b_1 x_0 \\ \vdots \\ b_{\ell(s+1)} x_0 \\ \vdots \\ b_1 x_n \\ \vdots \\ b_{\ell(s+1)} x_n \end{pmatrix}. \tag{4.17}$$

For $\mu \in T$, $l = 0, \dots, n$, and $j = 1, \dots, \ell(s)$, let $\hat{H}_{l,j}(\mu)$ be the hyperplane in $\mathbf{P}^{(n+1)\ell(s+1)-1}$ defined by the corresponding row in $C(\mu)$; i.e., if $c_{ij}(\mu)$ denote the elements of $C(\mu)$, then

$$\begin{aligned} &\hat{H}_{l,j}(\mu) \\ &= \{[y_{1,0} : \dots : y_{\ell(s+1),0} : \dots : y_{1,n} : \dots : y_{\ell(s+1),n}] \in \mathbf{P}^{(n+1)\ell(s+1)-1}(k) \\ &\quad \Big|\ c_{l\ell(s)+j,1}(v) y_{1,0} + \dots + c_{l\ell(s)+j,(n+1)\ell(s+1)}(v) y_{\ell(s+1),n} = 0\}. \end{aligned} \tag{4.18}$$

Since $H_{\mu(0)}(\alpha), \ldots, H_{\mu(n)}(\alpha)$ are in general position for each α, and since $x_0, \ldots, x_n$ are linearly independent over $\mathcal{R}_A$, it follows that $h_{\mu(0)}, \ldots, h_{\mu(n)}$ are linearly independent over $\mathcal{R}_A$. Thus, by the choice of $b_1, \ldots, b_{\ell(s)}$, $b_j h_{\mu(l)}, j = 1, \ldots, \ell(s); l = 0, \ldots, n$, are linearly independent over k. Hence, for each $\mu \in T$, $\hat{H}_{l,j}(\mu), 0 \le l \le n, 1 \le j \le \ell(s)$, (or more precisely, the linear forms defining $\hat{H}_{l,j}(\mu)$) are linearly independent.

Let k^A denote, as is standard, the set of (set-theoretic) maps $A \to k$. This is a ring; as before we embed k into k^A as constant functions. After deleting finitely many elements from A, we may assume that $a_{j,j_0}(\alpha) \neq 0$ for all $\alpha \in A$. Then (4.14) defines elements $\zeta_{j,l}^\natural \in k^A$ for all j and all l. Since $\mathcal{L}(s)$ is generated by monomials in the $\zeta_{j,l}$, the b_i are all polynomials in the $\zeta_{j,l}$; these polynomials define corresponding elements $b_i^\natural \in k^A$ for all $i = 1, \ldots, \ell(s+1)$. Let $h_1^\natural, \ldots, h_q^\natural \in k^A$ be defined by (4.16). Then, after deleting finitely many elements from A, (4.17) holds in k^A.

From now on we work entirely in k^A instead of $\mathcal{R}_A$, and omit the superscripts $\natural$.

For each $\alpha \in A$ let $P(\alpha) \in \mathbf{P}^{(n+1)\ell(s+1)-1}(k)$ be the point defined by homogeneous coordinates

$$\begin{aligned} &P(\alpha) \\ &= [b_1(\alpha)x_0(\alpha) : \ldots : b_{\ell(s+1)}(\alpha)x_0(\alpha), b_1(\alpha)x_1(\alpha) : \ldots : b_{\ell(s+1)}(\alpha)x_n(\alpha)] \\ &\quad \in \mathbf{P}^{(n+1)\ell(s+1)-1}(k). \end{aligned} \tag{4.19}$$

By applying Theorem B3.1.3 to the points

$$P(\alpha) \in \mathbf{P}^{(n+1)\ell(s+1)-1}(k)$$

with hyperplanes $\{\hat{H}_{l,j}(\mu) \mid l = 0, \ldots, n;\ j = 1, \ldots, \ell(s)\}$, there is a finite collection $\mathcal{L}$ of proper linear subspaces of $\mathbf{P}^{(n+1)\ell(s+1)-1}(k)$ such that

$$\frac{1}{[k:\mathbf{Q}]} \sum_{v \in S} \max_{\mu \in T} \sum_{l=0}^{n} \sum_{j=1}^{\ell(s)} \lambda_{v,\hat{H}_{l,j}(\mu)}(P(\alpha)) \le ((n+1)\ell(s+1) + \delta) h(P(\alpha))$$

for all α such that $P(\alpha) \notin \bigcup_{L \in \mathcal{L}} L$. By the condition that $\mathbf{x}$ is nondegenerate over $\mathcal{R}$, we may pass to an infinite subsequence satisfying $P(\alpha) \notin$

$\bigcup_{L\in\mathcal{L}} L$ for all α in the subsequence. Therefore,

$$\frac{1}{[k:\mathbf{Q}]}\sum_{v\in S}\max_{\mu\in T}\sum_{l=0}^{n}\sum_{j=1}^{\ell(s)}\lambda_{v,\hat{H}_{l,j}(\mu)}(P(\alpha)) \leq ((n+1)\ell(s+1)+\delta)h(P(\alpha)) \tag{4.20}$$

for all $\alpha \in A$.

We now translate the various terms above in terms of Schmidt's subspace theorem with moving targets. First consider the height $h(P_\alpha)$. We have

$$\begin{aligned} &\max(\|b_1(\alpha)x_0(\alpha)\|_v,\ldots,\|b_{\ell(s+1)}(\alpha)x_0(\alpha)\|_v,\ldots,\|b_{\ell(s+1)}(\alpha)x_n(\alpha)\|_v) \\ &\quad\leq \max(\|x_0(\alpha)\|_v,\ldots,\|x_n(\alpha)\|_v)\cdot\max(\|b_1(\alpha)\|_v,\ldots,\|b_{\ell(s+1)}(\alpha)\|_v) \\ &\quad= \|\mathbf{x}\|_v\cdot\max(\|b_1(\alpha)\|_v,\ldots,\|b_{\ell(s+1)}(\alpha)\|_v) \end{aligned} \tag{4.21}$$

and therefore

$$h(P(\alpha)) = h(\mathbf{x}(\alpha)) + o(h(\mathbf{x}(\alpha))).$$

Next consider the Weil functions $\lambda_{\hat{H}_{l,j}(v),v}(P(\alpha))$. By (3.74), (4.14), (4.16), (4.17), (4.18), (4.19), and (4.21)

$$\begin{aligned} &\lambda_{v,\hat{H}_{l,j}(\mu)}(P(\alpha)) \\ &\quad\leq \log\left(\|\mathbf{x}\|_v\cdot\max_{1\leq r\leq\ell(s+1)}\|b_r(\alpha)\|_v\cdot\max_{1\leq r\leq(n+1)\ell(s+1)}\|c_{l\ell(s)+j,r}(\mu)\|_v\right) \\ &\quad- \log\left(\|c_{l\ell(s)+j,1}(\mu)b_1(\alpha)x_0(\alpha)+\ldots+c_{l\ell(s)+j,(n+1)\ell(s+1)}(\mu)b_{\ell(s+1)}(\alpha)x_n(\alpha)\|_v\right) \\ &\quad= \log\left(\|\mathbf{x}\|_v\cdot\max_{1\leq r\leq\ell(s+1)}\|b_r(\alpha)\|_v\cdot\max_{1\leq r\leq(n+1)\ell(s+1)}\|c_{l\ell(s)+j,r}(\mu)\|_v\right) \\ &\qquad -\log\left(|b_j(\alpha)h_{\mu(l)}(\alpha)\|_v\right) \\ &\quad= \log\left(\|\mathbf{x}\|_v\cdot\max_{1\leq r\leq\ell(s+1)}\|b_r(\alpha)\|_v\cdot\max_{1\leq r\leq(n+1)\ell(s+1)}\|c_{l\ell(s)+j,r}(\mu)\|_v\right) \\ &\qquad -\log\left(\|b_j(\alpha)\|_v\cdot\|\zeta_{\mu(l),0}(\alpha)x_0(\alpha)+\ldots+\zeta_{\mu(l),n}(\alpha)x_n(\alpha)\|_v\right) \\ &\quad= \lambda_{v,H_{\mu(l)}(\alpha)}(\mathbf{x}(\alpha)) - \log\max_{0\leq r\leq n}\|\zeta_{\mu(l),r}(\alpha)\|_v - \log\frac{\|b_j(\alpha)\|_v}{\max_{1\leq t\leq\ell(s+1)}\|b_t(\alpha)\|_v} \\ &\qquad +O(1). \end{aligned}$$

But

$$-o(h(\mathbf{x}(\alpha))) \leq -\log\max_{0\leq r\leq n}\|\zeta_{\mu(l),r}(\alpha)\|_v \leq 0$$

by the condition that $h(H_j(\alpha)) = o(h(\mathbf{x}(\alpha)))$ for all $j = 1, \ldots, q$, and likewise

$$0 \leq -\log \frac{||b_j(\alpha)||_v}{\max_{1\leq t\leq \ell(s+1)} ||b_t(\alpha)||_v} \leq o(h(\mathbf{x}(\alpha)))$$

since the b_μ are polynomials in the $\zeta_{j,l}$. Thus

$$\lambda_{v,\hat{H}_{l,j}(\mu)}(P(\alpha)) = \lambda_{v,H_{\mu(l)}(\alpha)}(\mathbf{x}(\alpha)) + o(h(\mathbf{x}(\alpha))).$$

Combining this with (4.20) then gives

$$\begin{aligned}
&\frac{1}{[k:\mathbf{Q}]}\sum_{v\in S}\ell(s)\max_{\mu\in T}\sum_{l=0}^{n}\lambda_{v,H_{\mu(l)}(\alpha)}(\mathbf{x}(\alpha)) \\
&= \frac{1}{[k:\mathbf{Q}]}\sum_{v\in S}\max_{\mu\in T}\sum_{l=0}^{n}\sum_{j=1}^{\ell(s)}\lambda_{v,\hat{H}_{l,j}(\mu)}(P(\alpha)) + o(h(\mathbf{x}(\alpha))) \\
&\leq ((n+1)\ell(s+1)+\delta)h(P(\alpha)) + o(h(\mathbf{x}(\alpha))) \\
&\leq ((n+1)\ell(s+1)+\delta)h(\mathbf{x}(\alpha)) + o(h(\mathbf{x}(\alpha))),
\end{aligned}$$

for all $\alpha \in A$. Therefore

$$\begin{aligned}
&\frac{1}{[k:\mathbf{Q}]}\sum_{v\in S}\max_{\mu\in T}\sum_{l=0}^{n}\lambda_{v,H_{\mu(l)}(\alpha)}(\mathbf{x}(\alpha)) \\
&\leq \left((n+1)\frac{\ell(s+1)}{\ell(s)} + \frac{\delta}{\ell(s)}\right)h(\mathbf{x}(\alpha)) + o(h(\mathbf{x}(\alpha))) \\
&\leq (n+1+\epsilon)h(\mathbf{x}(\alpha))
\end{aligned}$$

for all $\alpha \in A$. Thus Theorem B4.1.6 is proven. □

B4.2 The Degenerate Case

In this section, we consider the degenerate case of Theorem B4.1.6; i.e., the case where condition (ii) in Theorem 1.1 fails. We will use the Nochka weights, similar to the one appearing in section A4.3.

Definition B4.2.1 *Let $H_1, \ldots, H_q$ be moving hyperplanes indexed by Λ. The map $\mathbf{x} = [x_0 : \ldots : x_n] : \Lambda \to \mathbf{P}^n(k)$ is said to be m-nondegenerate over $\mathcal{R}$ if m is the largest integer with the property that there is an infinite subset $A \subseteq \Lambda$, coherent with respect to $H_1, \ldots, H_q$, and an injection η :*

$\{0,\ldots,m\} \to \{0,\ldots,n\}$ such that $\{x_{\eta(0)},\ldots,x_{\eta(m)}\}$ is linearly independent over $\mathcal{R}_A$.

Clearly $0 \le m \le n$, and $m = n$ if and only if $\mathbf{x}$ is non-degenerate over $\mathcal{R}$ with respect to $H_1,\ldots,H_q$.

Theorem B4.2.2 (Ru-Vojta) *Let k be a number field, let S be a finite set of places of k, let q be a positive integer, and let $\epsilon > 0$. Let Λ be an infinite index set, let $H_1,\ldots,H_q$ be the moving hyperplanes in $\mathbf{P}^n(k)$ indexed by Λ, and let $\mathbf{x} : \Lambda \to \mathbf{P}^n(k)$ be the moving points, such that*

(i) for each $\alpha \in \Lambda$, $H_1(\alpha),\ldots,H_q(\alpha)$ are in general position;
(ii) $\mathbf{x}$ is m-nondegenerate over $\mathcal{R}$ (see Definition B4.2.1);
(iii) $h(H_j(\alpha)) = o(h(\mathbf{x}(\alpha)))$ for all $j = 1,\ldots,q$; and
(iv) $\mathbf{x}(\alpha) \notin H_j(\alpha)$ for all $j = 1,\ldots,q$ and $\alpha \in \Lambda$.

Then there exists an infinite index subset $A \subseteq \Lambda$ such that

$$\sum_{j=1}^{q} m(\mathbf{x}(\alpha), H_j(\alpha)) \le (2n - m + 1 + \epsilon) h(\mathbf{x}(\alpha))$$

for all $\alpha \in A$.

In particular, if only (i), (iii), and (iv) are satisfied, then there exists an infinite index subset $A \subseteq \Lambda$ such that

$$\sum_{j=1}^{q} m(\mathbf{x}(\alpha), H_j(\alpha)) \le (2n + \epsilon) h(\mathbf{x}(\alpha))$$

for all $\alpha \in A$.

Now we introduce the concept of u-subgeneral position.

Definition B4.2.3 *Let n, u, and q be positive integers with $u \ge n$ and $q \ge 2u - n + 1$. We say that hyperplanes $H_1,\ldots,H_q \subseteq \mathbf{P}^n(k)$ are in u-subgeneral position if for every set $P \subseteq \{1,\ldots,q\}$ with $\#P = u + 1$, there are $\mu(0),\ldots,\mu(n) \in P$ such that $H_{\mu(0)},\ldots,H_{\mu(n)}$ are in general position.*

The significance of the condition of u-subgeneral position is that $H_1,\ldots,H_q$ are in u-subgeneral position if and only if there is an embedding $\mathbf{P}^n \hookrightarrow \mathbf{P}^u$ and hyperplanes $H_1',\ldots,H_q'$ in $\mathbf{P}^u$ such that $H_j' \cap \mathbf{P}^n = H_j$ for all j and such that $H_1',\ldots,H_q'$ are in general position.

The following theorem is a trivial consequence of Theorem B4.2.2.

Theorem B4.2.4 (Ru-Vojta) *Let k be a number field, let S be a finite set of places of k, let n be a positive integer, let u be an integer with $u \geq n$, let q be an integer with $q \geq 2u - n + 1$, and let $\epsilon > 0$. Let Λ be an infinite index set, let $H_1, \ldots, H_q$ be the moving hyperplanes in $\mathbf{P}^n(k)$ indexed by Λ, and let $\mathbf{x} : \Lambda \to \mathbf{P}^n(k)$ be moving points, such that*

(i) $H_1(\alpha), \ldots, H_q(\alpha)$ are in u-subgeneral position for each $\alpha \in \Lambda$;
(ii) $\mathbf{x}$ is non-degenerate over $\mathcal{R}$ (Definition 4.1.4); and
(iii) $h(H_j(\alpha)) = o(h(\mathbf{x}(\alpha)))$ for all $j = 1, \ldots, q$.

Then there exists an infinite subset $A \subseteq \Lambda$ such that

$$\sum_{j=1}^{q} m(\mathbf{x}(\alpha), H_j(\alpha)) \leq (2u - n + 1 + \epsilon) h(\mathbf{x}(\alpha))$$

for all $\alpha \in A$.

We will prove Theorem B4.2.4 first and then derive Theorem B4.2.2 from Theorem B4.2.4.

Proof of Theorem B4.2.4.

Proof. The theorem is trivial when $q \leq 2u - n + 1$, so we may assume that $q > 2u - n + 1$.

Since $\mathbf{x}$ is nondegenerate over $\mathcal{R}$, there is a coherent infinite subset $A \subseteq \Lambda$ such that $x_0, \ldots, x_n$ are linearly independent over $\mathcal{R}_A$. Let $H_j, 1 \leq j \leq q$, be the given moving hyperplanes $\mathbf{P}^n(k)$ indexed by Λ. For every $\alpha \in A$, let $\omega_1(\alpha), \ldots, \omega_q(\alpha)$ be the Nochka weights and let $\theta(\alpha)$ be the Nochka constant associated by Lemma B3.4.3 to the hyperplanes $H_1(\alpha), \ldots, H_q(\alpha)$. The u-subgeneral position condition implies that, by the product to the sum formula(Lemma B3.4.4), after passing to an infinite index subset (which we still denote by A), for each $v \in S$, there exist distinct $i(v, 0), \ldots, i(v, u)$ among $\{1, \ldots, q\}$ such that

$$\begin{aligned} &\frac{1}{[k:\mathbf{Q}]} \sum_{v \in S} \sum_{j=1}^{q} \omega_j(\alpha) \lambda_{v, H_j(\alpha)}(\mathbf{x}(\alpha)) \\ \leq \quad &\frac{1}{[k:\mathbf{Q}]} \sum_{v \in S} \sum_{l=0}^{u} \omega_{i(v,l)}(\alpha) \lambda_{v, H_{i(v,l)}(\alpha)}(\mathbf{x}(\alpha)) + O(1) \qquad (4.22) \end{aligned}$$

for all $\alpha \in \Lambda$ (after passing to an infinite index subset, which we still denote by Λ). Since our Weil functions have the property that $\lambda_{v,H}(\mathbf{x}) \geq O(1)$ with a constant depending only on n and v, we can apply Theorem B3.4.3 with $E_j(\alpha) = (n+1)^{N_v} \exp(\lambda_{v,H_j(\alpha)}(\mathbf{x}(\alpha)))$, and $Y = \{i(v,0), \ldots, i(v,u)\}$, where $N_v = [k_v : \mathbf{Q}_p]$. Then there is a subset $M(v,\alpha)$ of Y such that:

(i) $\# M(v,\alpha) = \dim L(\alpha, Y) = d(Y) = n+1$,

(ii) $\{H_j(\alpha) \mid j \in M(v,\alpha)\}$ lie in general position, and

(iii) $\prod_{l=0}^{u} e^{\omega_{i(v,l)}(\alpha)\lambda_{v,H_{i(v,l)}(\alpha)}(\mathbf{x}(\alpha))} \leq C \prod_{j \in M(v,\alpha)} e^{\lambda_{v,H_j(\alpha)}(\mathbf{x}(\alpha))}$, where C is a constant depending only on n and v.

Since there are only finitely many subsets $M(v,\alpha) \subset Y$, after passing to an infinite index subset, we may assume that $M(v,\alpha) = M(v)$ is independent of α. So

$$\begin{aligned} & \sum_{l=0}^{u} \omega_{i(v,l)}(\alpha)\lambda_{v,H_{i(v,l)}(\alpha)}(\mathbf{x}(\alpha)) \\ & \quad \leq \sum_{j \in M(v)} \lambda_{v,H_j(\alpha)}(\mathbf{x}(\alpha)) + O(1) \\ & \quad \leq \max_{\mu \in T} \sum_{l=0}^{n} \lambda_{v,H_{\mu(l)}(\alpha)}(\mathbf{x}(\alpha)) + O(1) \end{aligned} \tag{4.23}$$

where T is the set of all maps $\mu : \mathbf{Z}[0,n] \to \{i(v,0), \ldots, i(v,u)\}$ such that $H_{\mu(0)}(\alpha), \ldots, H_{\mu(n)}(\alpha)$ are in general position for all $\alpha \in A$. Applying Theorem B4.1.6, for each $\mu \in T$ (note that μ depends on v),

$$\begin{aligned} & \frac{1}{[k : \mathbf{Q}]} \sum_{v \in S} \max_{\mu \in T} \sum_{l=0}^{n} \lambda_{v,H_{\mu(l)}(\alpha)}(\mathbf{x}(\alpha)) \\ & \leq (n+1+\epsilon)h(\mathbf{x}(\alpha)) + o(h(\mathbf{x}(\alpha))). \end{aligned} \tag{4.24}$$

Therefore, (4.22), (4.23), and (4.24) imply that

$$\begin{aligned} \sum_{j=1}^{q} \omega_j(\alpha) m(\mathbf{x}(\alpha), H_j(\alpha)) &= \frac{1}{[k : \mathbf{Q}]} \sum_{v \in S} \sum_{j=1}^{q} \omega_j(\alpha)\lambda_{v,H_j(\alpha)}(\mathbf{x}(\alpha)) \\ &\leq (n+1+\epsilon)h(\mathbf{x}(\alpha)) + o(h(\mathbf{x}(\alpha))) \end{aligned}$$

for all $\alpha \in A$. Combining with Lemma B3.4.3, we have

$$\begin{aligned}
\sum_{j=1}^{q} m(\mathbf{x}(\alpha), H_j(\alpha)) &= \sum_{j=1}^{q}(1-\theta(\alpha)\omega_j(\alpha))m(\mathbf{x}(\alpha), H_j(\alpha)) \\
&\quad + \sum_{j=1}^{q}\theta(\alpha)\omega_j(\alpha)m(\mathbf{x}(\alpha), H_j(\alpha)) \\
&\leq \sum_{j=1}^{q}(1-\theta(\alpha)\omega_j(\alpha))h(\mathbf{x}(\alpha)) \\
&\quad + \theta(\alpha)(n+1+\epsilon')h(\mathbf{x}(\alpha)) \\
&\leq \left\{q-\theta(\alpha)\left(\sum_{j=1}^{q}\omega_j(\alpha)-n-1-\epsilon'\right)\right\}h(\mathbf{x}(\alpha)) \\
&= (q-q+2u-n+1+\theta(\alpha)\epsilon')h(\mathbf{x}(\alpha)) \\
&\leq (2u-n+1+\epsilon)h(\mathbf{x}(\alpha)),
\end{aligned}$$

for all $\alpha \in A$. □

We now prove Theorem B4.2.2.

Proof. The theorem trivially holds when $q \leq 2n-m+1$. So we can assume that $q > 2n-m+1$. Let $H_1, \ldots, H_q$ be the given moving hyperplanes in $\mathbf{P}^n(k)$ indexed by Λ. Let $\mathbf{x} = [x_0 : \ldots : x_n] : \Lambda \to \mathbf{P}^n(k)$ be the given points. Let $\eta : \{0, \ldots, m\} \to \{0, \ldots, n\}$ be an injection and A a coherent subset of Λ such that $x_{\eta(0)}, \ldots, x_{\eta(m)}$ are linearly independent over $\mathcal{R}_A$ (these exist by the definition of m-nondegeneracy). Without loss of generality we may assume that $\mu(i) = i$ for all $i = 0, \ldots, m$. By Remark B4.1.5, for any infinite $B \subseteq A$, B is coherent and $\{x_0, \ldots, x_m\}$ is linearly independent over $\mathcal{R}_B$, so we may freely pass to infinite subsequences.

Since m is chosen to be maximal, for each $r = m+1, \ldots, n$ there are $c_{0,r}, \ldots, c_{m,r} \in \mathcal{R}_A$ such that

$$x_r = c_{0,r}x_0 + \ldots + c_{m,r}x_m.$$

For $j = 1, \ldots, q$ and $l = 0, \ldots, n$ let $\zeta_{j,l} \in \mathcal{R}_A$ be defined by (4.15) as before, and let $\zeta_{j,l}^{\natural} \in k^A$ be their liftings to k^A. Since each $c_{i,r}$ can be written as a rational function of the $\zeta_{j,l}$, there are liftings $c_{i,r}^{\natural}$ to k^A such that

$$x_r(\alpha) = c_{0,r}^{\natural}(\alpha)x_0(\alpha) + \ldots + c_{m,r}^{\natural}(\alpha)x_m(\alpha) \tag{4.25}$$

for all but finitely many $\alpha \in A$. After throwing out finitely many α, we may assume that this equation holds for all $\alpha \in A$. As before, from now on we work entirely in k^A and omit the superscripts $\natural$.

We now modify $\mathbf{x}$ and $H_1, \ldots, H_q$, so that Theorem B4.2.4 applies. The basic idea is that each $\mathbf{x}(\alpha)$ lies in a linear subspace of dimension m, which depends on α, but whose height does not grow quickly. Let

$$\mathbf{y} = [x_{\mu(0)} : \ldots : x_{\mu(m)}] : A \to \mathbf{P}^m(k).$$

We claim that

$$h(\mathbf{x}(\alpha)) = h(\mathbf{y}(\alpha)) + o(h(\mathbf{x}(\alpha))). \tag{4.26}$$

In fact, it is obvious that $h(\mathbf{y}(\alpha)) \leq h(\mathbf{x}(\alpha))$. On the other hand, (4.25) holds for all $\alpha \in A$, so the opposite inequality follows since $h(c_{i,r}(\alpha)) = o(h(\mathbf{x}(\alpha)))$. So the claim is true.

We now define the modified moving hyperplanes

$$H_j'(\alpha) = \{[x_0 : \ldots : x_m] \in \mathbf{P}^m(k) \mid b_{j,0}(\alpha)x_0 + \ldots + b_{j,m}(\alpha)x_m = 0\},$$

where $b_{j,l}(\alpha) = a_{j,l}(\alpha) + a_{j,m+1}(\alpha)c_{l,m+1}(\alpha) + \ldots + a_{j,n}(\alpha)c_{l,n}(\alpha)$ for all $j = 1, \ldots, q$, all $l = 0, \ldots, m$, and all $\alpha \in A$. These new hyperplanes are clearly the restrictions of the original hyperplanes to the linear subspaces mentioned above, so $H_1', \ldots, H_q'$ are in n-subgeneral position. We also have

$$a_{j,0}x_0 + \ldots + a_{j,n}x_n = b_{j,0}x_0 + \ldots + b_{j,m}x_m$$

in k^A for all $j = 1, \ldots, q$. Thus

$$\lambda_{v,H_j(\alpha)}(\mathbf{x}(\alpha)) = \lambda_{v,H_j'(\alpha)}(\mathbf{y}(\alpha)) + o(h(\mathbf{x}(\alpha))), \tag{4.27}$$

for all j and all v. By applying Theorem B4.2.4 to $\mathbf{y}$ and to the hyperplanes $H_1', \ldots, H_q'$, and combining with (4.26) and (4.27), we obtain Theorem B4.2.4. □

B4.3 Applications of Schmidt's Subspace Theorem with Moving Targets

As an application of Theorem B4.2.2, we first give a proof of Wirsing's theorem with moving targets which generalizes Roth's theorem with moving targets.

Theorem B4.3.1 (Wirsing's Theorem with Moving Targets) *Let k be a number field and let S be a finite set of places of k. Let r be a positive integer. Let Λ be an infinite index set, let $a_v(\alpha) \in k$ for each $v \in S$ and $\alpha \in \Lambda$, and let $x(\alpha) \in \bar{k}$ with $[k(x(\alpha)) : k] \leq r$ for all $\alpha \in \Lambda$, such that*

(i) $h(x(\alpha)) \to \infty$ *and*
(ii) $h(a_v(\alpha)) = o(x(\alpha))$ *for each* $v \in S$.

Then there exists an infinite subset $A \subseteq \Lambda$ such that

$$\prod_{v \in S} \prod_{w \in M_{k(x(\alpha))},\ w|v} \frac{\|x(\alpha) - a_v(\alpha)\|_w}{\max(1, \|x(\alpha)\|_w) \cdot \max(1, \|a_v(\alpha)\|_w)} \geq \frac{1}{H_{k(x(\alpha))}(x(\alpha))^{2r+\epsilon}} \tag{4.28}$$

for all $\alpha \in A$, where $H_k(x)$ is the (relative, multiplicative) height of x over the number field k.

Proof. It will suffice to prove the statement under the additional assumption that $[k(x(\alpha)) : k] = r$. For those points let

$$f(X) = A_r(\alpha)X^r + \ldots + A_0(\alpha)$$

be the minimal polynomial of $x(\alpha)$ over k. This defines a collection of points $P_{x(\alpha)} := [A_0(\alpha) : \ldots : A_r(\alpha)]$ in $\mathbf{P}^r(k)$. It is easy to verify that

$$\begin{aligned} C_{1,v} \cdot \frac{\max(\|A_0\|_v, \ldots, \|A_r\|_v)}{\|A_r\|_v} \leq & \prod_{w \in M_{k(x(\alpha))},\ w|v} \max(1, \|x(\alpha)\|_w) \\ \leq \quad & C_{2,v} \cdot \frac{\max(\|A_0\|_v, \ldots, \|A_r\|_v)}{\|A_r\|_v} \end{aligned} \tag{4.29}$$

for all $v \in M_k$, where $C_{1,v}$ and $C_{2,v}$ are positive constants independent of α. Moreover, $C_{1,v} = C_{2,v} = 1$ if v is non-Archimedean. Hence the heights of $x(\alpha)$ and $P_{x(\alpha)}$ are related by

$$C_1 H_k(P_{x(\alpha)}) \leq H_{k(x(\alpha))}(x(\alpha)) \leq C_2 H_k(P_{x(\alpha)}) \tag{4.30}$$

where C_1 and C_2 are positive constants independent of α.

For each $v \in S$ and $\alpha \in \Lambda$, $a_v(\alpha)$ determines a hyperplane $H_v(\alpha)$ defined by the equation

$$X_0 + a_v(\alpha)X_1 + \ldots + a_v^r(\alpha)X_r = 0.$$

The hyperplanes $H_v(\alpha)$ are defined over k, and the set $\{H_v(\alpha)\}_{v \in S}$ (ignoring duplicates) lies in general position for each $\alpha \in \Lambda$ by the non-vanishing of the van der Monde determinant. Moreover, it is easy to check that the left-hand side of (4.28) is related to the Weil functions of $P_{x(\alpha)}$ with respect to H_v, as follows. First of all,

$$\begin{aligned} \prod_{w\in M_{k(x(\alpha))},\, w|v} \max(1, \|a_v(\alpha)\|_w) &= \max(1, \|a_v(\alpha)\|_v)^r \\ &= \max(1, \|a_v(\alpha)\|_v, \|a_v(\alpha)\|_v^2, \ldots, \|a_v(\alpha)\|_v^r). \end{aligned} \tag{4.31}$$

Also, we have

$$\prod_{w\in M_{k(x(\alpha))},\, w|v} \|x(\alpha)-a_v(\alpha)\|_w = \|N_k^{k(x(\alpha))}(x(\alpha)-a_v(\alpha))\|_v = \|f(a_v(\alpha))\|_v. \tag{4.32}$$

Combining (4.29), (4.31), and (4.32) then gives

$$\begin{aligned} &-\log \prod_{w\in M_{k(x(\alpha))},\, w|v} \frac{\|x(\alpha)-a_v(\alpha)\|_w}{\max(1,\|x(\alpha)\|_w)\cdot\max(1,\|a_v(\alpha)\|_w)} \\ &= \lambda_{v,H_v(\alpha)}(P_{x(\alpha)}) + O(1) \end{aligned} \tag{4.33}$$

where the implicit constant in $O(1)$ is independent of α. By (4.30) and (4.33), (4.28) is equivalent to

$$\frac{1}{[k:\mathbf{Q}]}\sum_v \lambda_{v,H_v(\alpha)}(P_{x(\alpha)}) \le (2r+\epsilon)h(P_{x(\alpha)}) + O(1).$$

This follows immediately from the last assertion of Theorem B4.2.2. □

Theorem B4.2.2 also gives the finiteness of the number of integer solutions of decomposable form equations.

For any $x \in k - \{0\}$, let $N_S(x) = \prod_{v\in S}\|x\|_v$ denote the S-norm of x. Also for $\mathbf{x} = (x_0, \ldots, x_m) \in k^{m+1}$, define the S-height as $H_S(\mathbf{x}) = \prod_{v\in S}\|\mathbf{x}\|_v$.

Theorem B4.3.2 (Györy-Ru) *Let q, m be positive integers with $q > 2m$. Let c, ν be real numbers with $c > 0, \nu < q - 2m$ and $\mathbf{G}$ a finite extension of k. For $n = 1, 2, \ldots$, let $F_n(\mathbf{X}) = F_n(X_0, \ldots, X_m) \in \mathcal{O}_S[\mathbf{X}]$ denote a decomposable form of degree q which factors into linear factors over $\mathbf{G}$, and suppose that these factors are in general position for each n. Then there* *does not exist an infinite sequence of* $\mathcal{O}_S^*$*-non-proportional* $\mathbf{x}_n \in \mathcal{O}_S^{m+1}, n = 1, 2, \dots,$ *for which*

$$0 < N_S(F_n(\mathbf{x}_n)) \leq cH_S(\mathbf{x}_n)^\nu \text{ for } n = 1, 2, \dots, \tag{4.34}$$

and

$$h(F_n) = o(h(\mathbf{x}_n)) \quad \text{if} \quad h(\mathbf{x_n}) \to \infty \text{ as } n \to \infty \tag{4.35}$$

hold.

Proof. Assume that there is an infinite sequence $\mathbf{x}_n = (x_{0,n}, \dots, x_{m,n}) \in O_S^{m+1}$ which satisfies (4.34) and (4.35). First consider the case when the values $h(\mathbf{x}_n)$ are bounded. We may assume without loss of generality that $x_{0,n} \neq 0$ for each n. Then the $h(\mathbf{x}_n/x_{0,n})$ are bounded and this implies that $\mathbf{x}_n/x_{0,n}$ may assume only finitely many values in k^{m+1}. Hence there are infinitely many n such that $\mathbf{x}_n = x_{0,n}\mathbf{x}_0$ for some $\mathbf{x}_0 \in k^{m+1}$. For these n we deduce from (4.34) that

$$0 < N_S(x_{0,n})^q N_S(F_n(\mathbf{x}_0)) \leq cN_S(x_{0,n})^\nu H_S(\mathbf{x}_0)^\nu$$

and hence $N_S(x_{0,n})$ are bounded. Since $x_{0,n} \in \mathcal{O}_S$, it follows that there are infinitely many n for which $x_{0,n} = \eta_n x_0'$ with some $\eta_n \in \mathcal{O}_S^*$ and fixed $x_0' \in \mathcal{O}_S$. This implies that for these n the $\mathbf{x}_n$ considered above are $\mathcal{O}_S^*$-proportional, which is a contradiction.

Next consider the case when $h(\mathbf{x}_n)$ are not bounded. We may assume that $h(\mathbf{x}_n) \to \infty$ as $n \to \infty$. Then, by assumption, (4.35) also holds. Further it follows that $H_S(\mathbf{x}_n) \to \infty$ as $n \to \infty$. For $n = 1, 2, \dots$, let $F_n = L_{1,n} \dots L_{q,n}$ be a factorization of F_n over $\mathbf{G}$ into linear factors. Then

$$\max_j h(L_{j,n}) \leq h(F_n) + c_1$$

where c_1 is a positive constant which depends only on q, m and $\mathbf{G}$. Together with (4.35) this gives

$$\max_j h(L_{j,n}) = o(h(\mathbf{x}_n)) \text{ as } n \to \infty. \tag{4.36}$$

Let $M(\mathbf{G})$ denote the set of places of $\mathbf{G}$. For $v \in M(\mathbf{G})$, define and normalize $\| \, \|_v$ in a similar manner as over k above. Further, let T denote the set of extensions to $\mathbf{G}$ of the places in S. Then we deduce from (4.34) that

$$0 < N_T((F_n(\mathbf{x}_n)) = N_S(F_n(\mathbf{x}_n))^{[\mathbf{G}:k]} \leq (cH_S(\mathbf{x}_n)^\nu)^{[\mathbf{G}:k]} = c_2 H_T(\mathbf{x}_n)^\nu, \tag{4.37}$$

where $c_2 = c^{[\mathbf{G}:k]}$. Here $N_T(\ )$, $H_T(\ )$ are defined over $\mathbf{G}$ in the same way as $N_S(\ )$, $H_S(\ )$ over k.

Let $\epsilon > 0$ with $0 < \epsilon < q - 2m - \nu$. Then by Theorem B4.2.2, there is an infinite subsequence $\mathbf{x}_{n_k} \in \mathcal{O}_S^{m+1}, k = 1, 2, \ldots$, of $\{\mathbf{x}_n\}$, without loss of generality, we assume $\{\mathbf{x}_n\}$ itself, such that

$$\frac{1}{[\mathbf{G}:\mathbf{Q}]} \sum_{v \in T} \sum_{j=1}^{q} \log \frac{\|\mathbf{x}_n\|_v \cdot \|L_{j,n}\|_v}{\|L_{j,n}(\mathbf{x}_n)\|_v} \leq (2m+\epsilon) h(\mathbf{x}_n).$$

However, $F_n(\mathbf{x}_n) = \prod_{j=1}^q L_{j,n}(\mathbf{x}_n)$. Furthermore, in view of $\mathbf{x}_n \in \mathcal{O}_S^{m+1}$, we have

$$h(\mathbf{x}_n) \leq \frac{1}{[\mathbf{G}:\mathbf{Q}]} \log H_T(\mathbf{x}_n).$$

Hence it follows that

$$\prod_{v \in T} \frac{\|\mathbf{x}_n\|_v^q \cdot \prod_{j=1}^q \|L_{j,n}\|_v}{\|F_n(\mathbf{x}_n)\|_v} \leq H_T(\mathbf{x}_n)^{2m+\epsilon}, \tag{4.38}$$

whence

$$\frac{H_T^q(\mathbf{x}_n) \cdot \prod_{v \in T} \prod_{j=1}^q \|L_{j,n}\|_v}{N_T(F_n(\mathbf{x}_n))} \leq H_T(\mathbf{x}_n)^{2m+\epsilon}. \tag{4.39}$$

Since the coefficients of $L_{j,n}$ are T-integers,

$$\prod_{v \in T} \prod_{j=1}^{q} \|L_{j,n}\|_v \geq 1, \text{ for } n = 1, 2 \ldots. \tag{4.40}$$

Furthermore, it follows from (4.37) that

$$N_T(F_n(\mathbf{x}_n)) \leq c_2 H_T(\mathbf{x}_n)^\nu \text{ for } n = 1, 2 \ldots. \tag{4.41}$$

Combining (4.39), (4.40) and (4.41) gives

$$H_T(\mathbf{x}_n)^q \leq c_2 H_T(\mathbf{x}_n)^{\nu + 2m + \epsilon}.$$

Since $H_T(\mathbf{x}_n) \to \infty$ as $n \to \infty$, and $q > \nu + 2m + \epsilon$, we derives a contradiction. □

Theorem B4.3.2 implies the following theorem.

Theorem B4.3.3 (Győry-Ru) *Given positive integers q, m with $q > 2m$, a finite extension* $\mathbf{G}$ *of k, and a sequence of polynomials $G_n(\mathbf{X}) \in \mathcal{O}_S[\mathbf{X}]$ in $\mathbf{X} = (X_0, \ldots, X_m)$ such that $\deg(G_n(\mathbf{X})) < q - 2m$ for $n = 1, 2, \ldots$. Let $F_n(\mathbf{X}) = F_n(X_0, ..., X_m) \in \mathcal{O}_S[\mathbf{X}]$ be a sequence of decomposable forms of degree q such that F_n factors into linear forms over* $\mathbf{G}$*, and suppose that these factors are in general position for each n. Then there does not exist an infinite sequence of $\mathcal{O}_S^*$-non-proportional $\mathbf{x}_n \in \mathcal{O}_S^{m+1}$ satisfying (i) $F_n(\mathbf{x}_n) = G_n(\mathbf{x}_n) \neq 0$, $n = 1, 2, \ldots$, (ii) $\log H_S(G_n) = o(\log H_S(\mathbf{x}_n))$, if $H_S(\mathbf{x}_n) \to \infty$ and (iii) $h(F_n) = o(h(\mathbf{x}_n))$, if $h(\mathbf{x}_n) \to \infty$ as $n \to \infty$.*

Note that Theorem B3.4.8 is a special case of the theorem above when $F_n(\mathbf{X}) = F(\mathbf{X})$, which is independent of the index n.

The Correspondence Table

Nevanlinna Theory	**Diophantine Approximation**
Theorem A4.2.1	Theorem B4.1.6
Theorem A4.2.4	Theorem B4.1.7
Theorem A4.3.2	Theorem B4.2.2
Lemma A4.3.4	Lemma B4.2.4
Theorem A4.4.1	Theorem B4.3.1

Chapter 5

Equi-dimensional Nevanlinna Theory and Vojta's Conjecture

Part A: Nevanlinna Theory

To generalize the Nevanlinna theory to the case of higher dimensional complex manifolds, we note that if $f : A \to M$ is holomorphic, where A, M are complex manifolds with $\dim A < \dim M$, then $f(A)$ never covers a non-empty open set of M, and moreover, even in the equidimensional case $\dim A = \dim M$, we know that the Fatou-Bieberbach map $f : \mathbf{C}^m \to \mathbf{C}^m$ satisfies $|df| \equiv 1$ and $\overline{f(\mathbf{C}^m)} \neq \mathbf{C}^m$. So to get a reasonable generalization, we should regard $a \in \mathbf{P}^1(\mathbf{C})$ not simply as a point, but as a sub-variety of $\mathbf{P}^1(\mathbf{C})$. In this reason, a higher dimensional Nevanlinna theory usually deals with holomorphic maps intersecting divisors in M. However it is, in general, very hard to establish the Second Main Theorem in the higher dimensional case. P. Griffiths et al. [Ca-G] in the 1970's successfully proved the S.M.T. for the equi-dimensional holomorphic mappings, i.e., holomorphic maps from $\mathbf{C}^n$ to a projective compact complex manifold M with dimension n. In this section, we will introduce their results. However, here we will not follow their method. Instead, we will use the logarithmic derivative lemma to prove the main theorem. We note that this chapter can be regarded as a generalization of the results in Chapter 2.

A5.1 Logarithmic Derivative Lemma for Meromorphic Functions on $\mathbf{C}^n$

In this section, we extend the Logarithmic Derivative Lemma to meromorphic functions on $\mathbf{C}^n$, following the methods of A. Biancofiore and

W. Stoll [Bia-S], and Z. Ye[Ye]. First we introduce some notations. For $z = (z_1, \ldots, z_n) \in \mathbf{C}^n$, we define, for any positive number r, $|z| = (|z_1|^2 + \ldots + |z_n|^2)^{1/2}$, $\mathbf{C}_n(r) = \{z \in \mathbf{C}^n \mid |z| < r\}$, $\mathbf{C}_n[r] = \{z \in \mathbf{C}^n \mid |z| \leq r\}$, and $\mathbf{S}_n(r) = \{z \in \mathbf{C}^n \mid |z| = r\}$. The sphere $\mathbf{S}_n(r)$ is considered to be an analytic manifold oriented to the exterior of $\mathbf{C}_n(r)$. Define $d^c = \frac{\sqrt{-1}}{4\pi}(\bar{\partial} - \partial)$. The pull-back of the form

$$\sigma_n = d^c \log |z|^2 \wedge (dd^c \log |z|^2)^{n-1} \quad \text{on} \quad \mathbf{C}^n - \{0\}$$

defines a positive measure on $\mathbf{S}_n(r)$ with total measure 1. For $z \in \mathbf{C}^n - \{0\}$, let $\omega_n(z) = dd^c \log |z|^2$. For $z \in \mathbf{C}^n$ let $v_n(z) = dd^c|z|^2$ and $\rho_n(z) = (dd^c|z|^2)^n$. Then ρ_n is a Lebesgue measure on $\mathbf{C}^n$ such that $\mathbf{C}_n(r)$ has measure r^{2n}.

A **divisor** on $\mathbf{C}^n$ is a formal finite sum $D = \sum_j n_j D_j$, where $n_j \in \mathbf{Z}$ and D_j is a sub-variety of codimension 1. Each n_j is called the order of D, while each D_j is called the support of D. If $n_j \geq 0$ for all j, then we call D an **effective divisor.**

To each meromorphic function f on $\mathbf{C}^n$, we can assign a divisor D_f associated to f, while the supports consist of the union of the set N_f of zeros of f and the set P_f of its poles. The order of D_f is equal to the order of zero on every component of N_f and the order of the pole with a minus sign on every component of P_f. For $a \in \mathbf{C} \cup \{\infty\}$, let D_f^a be the divisor associated to $f - a$ if $a \in \mathbf{C}$ and the divisor associated to $1/f$ if $a = \infty$. Let

$$n_f(t, a) = t^{2-2n} \int_{D_f^a \cap \mathbf{C}_n[t]} v_n^{n-1}.$$

The **counting function of f with respect to** a is defined by

$$N_f(r, a) = \int_0^r [n_f(t, a) - n_f(0, a)] \frac{dt}{t} + n_f(0, a) \log r, \tag{5.1}$$

where $n_f(0, a) = \lim_{t \to 0} n_f(t, a)$ and the integration $\int_{D_f^a \cap \mathbf{C}_n[t]} v_n^{n-1}$ is in the following sense: write $D_f^a = \sum m_j M_j$, where M_j are the irreducible components, then

$$\int_{D_f^a \cap \mathbf{C}_n[t]} v_n^{n-1} = \sum m_j \int_{M_j \cap \mathbf{C}_n[t]} v_n^{n-1}.$$

The proximity function $m_f(r,a)$ is defined by

$$m_f(r,a) = \int_{\mathbf{S}_n(r)} \log^+ \frac{1}{|f-a|}\sigma_n, \quad \text{if } a \neq \infty \tag{5.2}$$

and

$$m_f(r,\infty) = \int_{\mathbf{S}_n(r)} \log^+ |f|\, \sigma_n. \tag{5.3}$$

Let $T_f(r) = m_f(r,\infty) + N_f(r,\infty)$. Similar to Theorem A1.1.5, the First Main Theorem states

$$T_f(r) = m_f(r,a) + N_f(r,a) + O(1).$$

Lemma A5.1.1 *Let $r > 0$ and let h be a function on $S_n(r)$ such that $h\sigma_n$ is integrable over $\mathbf{S}_n(r)$. Let $p(w) = \sqrt{r^2 - |w|^2}$. Then*

$$\int_{\mathbf{S}_n(r)} h\sigma_n = r^{2-2n} \int_{\mathbf{C}_{n-1}[r]} \left(\int_{\mathbf{S}_1(p(w))} h(w,\zeta)\sigma_1(\zeta) \right) \rho_{n-1}(w).$$

Proof. Define $E = \{(z_1,\ldots,z_n) \in \mathbf{S}_n(r) \mid 0 \le z_n \in \mathbf{R}\}$ and $F = \mathbf{S}_n(r) - E$. Then E is a closed subset of $S_n(r)$ with zero $(2n-1)$-dimensional Hausdorff measure. A bijective map

$$g : \mathbf{R}(0,2\pi) \times \mathbf{C}_{n-1}(r) \to F$$

of class C^∞ is defined by $g(\phi, w) = (w, p(w)e^{i\phi})$ for all $\phi \in \mathbf{R}(0,2\pi)$ and all $w \in \mathbf{C}_{n-1}(r)$. Let ϕ be the variable in $\mathbf{R}(0,2\pi)$ and let $w_1,\ldots,w_{n-1}$ be the complex variables on $\mathbf{C}^{n-1}$ with $w_j = x_j + y_j\sqrt{-1}$. The partial derivatives $g_\phi, g_{x_1}, g_{y_1}, \cdots, g_{x_{n-1}}, g_{y_{n-1}}$ are pointwise linearly independent over $\mathbf{R}$ and g is perpendicular to these derivatives. Here $g(\phi,w)$ points in the direction of $\mathbf{S}_{n-1}(r)$ at $g(\phi,w)$. Since

$$\det(g, g_\phi, g_{x_1}, g_{y_1}, \cdots, g_{x_{n-1}}, g_{y_{n-1}}) = r^2 > 0$$

the map g is an orientation preserving diffeomorphism. We shall compute $g^*(\sigma_n)$. Write $z = (z_1,\ldots,z_n) \in \mathbf{C}^n$ and $w = (w_1,\ldots,w_{n-1}) \in \mathbf{C}^{n-1}$ where actually $w_1 = z_1,\ldots,w_{n-1} = z_{n-1}$, hence $z = (w, z_n)$. We have

$$g^*(d^c|z|^2) = d^c|w|^2 + (1/2\pi)(r^2 - |w|^2)d\phi$$

$$g^*(dd^c|z|^2) = dd^c|w|^2 - (1/2\pi)d|w|^2 \wedge d\phi$$

$$g^*(dd^c|z|^2)^{n-1} = (dd^c|w|^2)^{n-1} - \frac{n-1}{2\pi}d|w|^2 \wedge d\phi \wedge (dd^c|w|^2)^{n-2}$$

$$\begin{aligned} g^*(d^c|z|^2 \wedge (dd^c|z|^2)^{n-1}) &= (1/2\pi)(r^2 - |w|^2)d\phi \wedge (dd^c|w|^2)^{n-1} \\ &+ \frac{n-1}{2\pi}d\phi \wedge d|w|^2 \wedge d^c|w|^2 \wedge (dd^c|w|^2)^{n-2}. \end{aligned}$$

It is easy to check that $|w|^2(dd^c|w|^2)^{n-1} = (n-1)d|w|^2 \wedge d^c|w|^2 \wedge (dd^c|w|^2)^{n-2}$, hence

$$g^*(d^c|z| \wedge (dd^c|z|^2)^{n-1}) = (1/2\pi)r^2 d\phi \wedge (dd^c|w|^2)^{n-1}.$$

Now $|g|^2 = r^2$ and $(dd^c|w|^2)^{n-1} = \rho_{n-1}$, we obtain

$$\begin{aligned} g^*(\sigma_n) &= g^*(d^c \log|z|^2 \wedge (dd^c \log|z|^2)^{n-1}) = g^*(r^{-2n}d^c|z|^2 \wedge (dd^c|z|^2)^{n-1}) \\ &= (1/2\pi)r^{2-2n}d\phi \wedge \rho_{n-1}. \end{aligned}$$

Fubini's theorem implies that

$$\int_{\mathbf{S}_n(r)} h\sigma_n = r^{2-2n}\int_{\mathbf{C}_{n-1}[r]} \frac{1}{2\pi}\int_0^{2\pi} h(w, p(w)e^{i\phi})d\phi \wedge \rho_{n-1}(w).$$

□

Let f be a meromorphic function on $\mathbf{C}^n$. Take $w \in \mathbf{C}^{n-1}$ and define, for $z \in \mathbf{C}$, $f_{[w]}(z) = f(w, z)$.

Lemma A5.1.2 *Let f be a non-constant meromorphic function on $\mathbf{C}^n$, and let $a \in \mathbf{C} \cup \{\infty\}$. Then, for $r > 0$,*

$$\frac{1}{r^{2n-2}}\int_{\mathbf{C}_{n-1}[r]} n_{f_{[w]}}(\sqrt{r^2 - |w|^2}, a)\rho_{n-1}(w) \le n_f(r, a),$$

where $f_{[w]} = f(w, z)$ for $w \in \mathbf{C}^{n-1}$ and $z \in \mathbf{C}$.

Proof. For each $j \in \mathbf{N}[1, n]$ define $\pi_j : \mathbf{C}^n \to \mathbf{C}^{n-1}$ by

$$\pi(z_1, \ldots, z_n) = (z_1, \ldots, z_{j-1}, z_{j+1}, \ldots, z_n).$$

Then

$$\upsilon_n^{n-1} = \sum_{j=1}^{n} \pi_j^*(\upsilon_{n-1}^{n-1}) \ge \pi_n^*(\rho_{n-1}).$$

Define $A = \mathbf{C}_n(r) \cap \text{supp } D_f^a$. If A is empty, then the lemma is trivial. Assume that A is not empty. Then A is a pure $(n-1)$-dimensional subset of $\mathbf{C}_n(r)$. Define $\pi = \pi_n|_A$. Then $\pi(A) \subset \mathbf{C}_{n-1}(r)$. The set $E = \{z \in A \mid \text{rank}_z \pi < n-1\}$ is analytic in $\mathbf{C}_n(r)$. Let E_0 be the union of all $(n-1)$-dimensional branches of E and let E_1 be the union of all other branches of E. Then E_0 and E_1 are analytic in $\mathbf{C}_n(r)$ with $\dim E_1 \leq n-2$ and $\dim E_0 = n-1$ if $E_0 \neq \emptyset$. Also $E_0' = \pi(E_0)$ is analytic in $\mathbf{C}_{n-1}(r)$ with $E_0 = \pi_n^{-1}(E_0') \cap \mathbf{C}_n(r)$. Therefore $\pi^*(\rho_{n-1}) = 0$ on E_0. The complement $A_0 = A - E_0$ is open in A and $A_0 \cap E_1$ is thin analytic in A_0 if $A_0 \neq \emptyset$. Then $A_1 = A_0 - E_1 = A - E$ is open in A. A thin analytic subset E_2 of A_1 exists such that π is locally biholomorphic on $A_2 = A_1 - E_2$. Then $F = \pi(E_0 \cup E_1 \cup E_2)$ has a measure of zero in $\mathbf{C}^{n-1}$. Also $B_2 = \pi(A_2)$ is open in $\mathbf{C}^{n-1}$. Here $B = \pi(A) = B_2 \cup F$ and $B_3 = B_2 - F$ differ by sets of measure zero from B_2.

The intersection $\mathbf{S}_n(r) \cap \text{supp } D_f^a$ has a $(2n-2)$-dimensional Hasudorff measure of zero. If $w \in \mathbf{C}_{n-1}(r)$, then

$$\pi^{-1}(w) = (\{w\} \times \mathbf{C}_1(\sqrt{r^2 - |w|^2})) \cap A$$

and for almost all $w \in \mathbf{C}_{n-1}(r)$ we have

$$\pi^{-1}(w) = (\{w\} \times \mathbf{C}_1[\sqrt{r^2 - |w|^2}]) \cap \text{supp} D_f^a.$$

For all $w \in \mathbf{C}_{n-1}(r) - B$, we have $\pi^{-1}(w) = \emptyset$, hence $n_{f_{[w]}}(\sqrt{r^2 - |w|^2}, a) = 0$ for almost all $w \in \mathbf{C}_{n-1}(r) - B$. If $w \in B_3$, then

$$\pi^{-1}(w) = (\{w\} \times \mathbf{C}_1(\sqrt{r^2 - |w|^2})) \cap A.$$

Take $P = (b, c) \in A_2$ with $b = \pi(P) \in \mathbf{C}_{n-1}(r)$ and $c \in \mathbf{C}$. There exist open, connected neighborhoods V of b in $\mathbf{C}_{n-1}(r)$ and W of c and $U = V \times W$ of P in $\mathbf{C}_n(r)$ with $U \cap A_2 = U \cap A$ such that $\pi : U \cap A \to V$ is biholomorphic. Let λ be the inverse map. Then a holomorphic function $h : V \to W$ exists such that $\lambda(w) = (w, h(w))$ for all $w \in V$. The multiplicity $\nu_f^a(w, z) = q$ is constant for all $(w, z) \in U \cap A$. A holomorphic function $H : U \to \mathbf{C} - \{0\}$ exists such that

$$f(w, z) = a + (z - h(w))^q H(w, z)$$

for all $(w, z) \in U$. Therefore $\nu_f^a(w, z) = q = \nu_{f_{[w]}}^a(z)$ for all $(w, z) \in U \cap A$.

All together we obtain

$$\begin{aligned}
r^{2n-2}n_f(r,a) &= \int_A \nu_f^a v_n^{n-1} \geq \int_A \nu_f^a \pi^*(\rho_{n-1}) = \int_{A_0} \nu_f^a \pi^*(\rho_{n-1}) \\
&= \int_{A_2} \nu \pi^*(\rho_{n-1}) = \int_{B_2} \sum_{(w,z)\in A_2} \nu_f^a(w,z)\rho_{n-1}(w) \\
&= \int_{B_3} \left(\sum_{z\in \mathbf{C}_1(\sqrt{r^2-|w|^2})} \nu_{f_{[w]}}^a(z) \right) \rho_{n-1}(w) \\
&= \int_{B_3} n_{f_{[w]}}(\sqrt{r^2-|w|^2}, a)\rho_{n-1}(w) \\
&= \int_{B} n_{f_{[w]}}(\sqrt{r^2-|w|^2}, a)\rho_{n-1}(w) \\
&= \int_{\mathbf{C}_{n-1}[r]} n_{f_{[w]}}(\sqrt{r^2-|w|^2}, a)\rho_{n-1}(w).
\end{aligned}$$

□

Lemma A5.1.3 *Let f be a non-constant meromorphic function on $\mathbf{C}^n$. Then for any $0 < \alpha < 1/2$ there is a constant $C > 1$ such that for any real numbers r and R with $r < R$ and $1 \leq j \leq n$, we have*

$$\int_{\mathbf{S}_n(r)} \left| \frac{f_{z_j}}{f} \right|^\alpha \sigma_n \leq C \left(\frac{R}{r} \right)^{\alpha(2n-2)} \left(\frac{R}{r(R-r)} \right)^\alpha T_f^\alpha(R).$$

Proof. Without loss of generality, we take $j = n$. Let $s = (R+r)/2$ and write $p(w) = \sqrt{r^2-|w|^2}$ and $P(w) = \sqrt{s^2-|w|^2}$. Then Lemma A5.1.1 implies that

$$\int_{\mathbf{S}_n(r)} \left| \frac{f_{z_n}}{f} \right|^\alpha \sigma_n = r^{2-2n} \int_{\mathbf{C}_{n-1}[r]} \left(\int_{\mathbf{S}_1(p(w))} \left| \frac{f'_{[w]}(z)}{f_{[w]}(z)} \right|^\alpha \sigma_1(z) \right) \rho_{n-1}(w).$$

We write $n_f(t,0,\infty) = n_f(t,0) + n_f(t,\infty)$ and $m_f(t,0,\infty) = m_f(t,0) + m_f(t,\infty)$. By Lemma A1.2.2,

$$\begin{aligned}
&\int_{\mathbf{S}_1(p(w))} \left| \frac{f'_{[w]}(z)}{f_{[w]}(z)} \right|^\alpha \sigma_1(z) \\
&\leq C\sec(\alpha\pi/2) \left(\frac{P(w)}{p(w)(P(w)-p(w))} \right)^\alpha m_{f_{[w]}}^\alpha(P(w),0,\infty)
\end{aligned}$$

$$+\frac{C\sec(\alpha\pi/2)}{p^{\alpha}(w)}n^{\alpha}_{f_{[w]}}(P(w),0,\infty).$$

So, by Lemma A5.1.1,

$$\begin{aligned}\int_{\mathbf{S}_n(r)}\left|\frac{f_{z_n}}{f}\right|^{\alpha}\sigma_n &= r^{2-2n}\int_{\mathbf{C}_{n-1}[r]}\left(\int_{\mathbf{S}_1(p(w))}\left|\frac{f'_{[w]}(z)}{f_{[w]}(z)}\right|^{\alpha}\sigma_1(z)\right)\rho_{n-1}(w)\\ &\le r^{2-2n}\int_{\mathbf{C}_{n-1}[r]}C\sec(\alpha\pi/2)\left[\left(\frac{P(w)m_{f_{[w]}}(P(w),0,\infty)}{p(w)(P(w)-p(w))}\right)^{\alpha}\right.\\ &\quad\left.+\frac{1}{p^{\alpha}(w)}n^{\alpha}_{f_{[w]}}(P(w),0,\infty)\right]\rho_{n-1}(w). \qquad (5.4)\end{aligned}$$

Clearly, for any $w\in\mathbf{C}_{n-1}[r]$, since $p(w)\le P(w)r/s$,

$$\frac{P(w)}{P(w)-p(w)}\le\frac{s}{s-r}. \qquad (5.5)$$

For any $0<\beta<1$, set $C=\int_{\mathbf{C}_{n-1}[1]}\frac{1}{(1-|\tau|^2)^{\beta/2}}\rho_{n-1}(\tau)$, then

$$\begin{aligned}\int_{\mathbf{C}_{n-1}[r]}\frac{1}{p^{\beta}(w)}\rho_{n-1}(w) &= r^{2-2n}\int_{\mathbf{C}_{n-1}[1]}\frac{1}{r^{\beta}(1-|\tau|^2)^{\beta/2}}\rho_{n-1}(\tau)\\ &= Cr^{2-2n-\beta}. \qquad (5.6)\end{aligned}$$

For any $a\in\mathbf{C}\cup\{\infty\}$, applying lemma A5.1.1 gives

$$\begin{aligned}s^{2n-2}m_f(s,a) &= s^{2n-2}\int_{\mathbf{S}_n(s)}\log^+|1/(f(z)-a)|\sigma_n(z)\\ &= \int_{\mathbf{C}_{n-1}[s]}\left(\int_{\mathbf{S}(P(w))}\log^+|1/(f_{[w]}(z)-a)|\sigma_1(z)\right)\rho_{n-1}(w)\\ &= \int_{\mathbf{C}_{n-1}[s]}m_{f_{[w]}}(P(w),a)\rho_{n-1}(w)\\ &\ge \int_{\mathbf{C}_{n-1}[r]}m_{f_{[w]}}(P(w),a)\rho_{n-1}(w). \qquad (5.7)\end{aligned}$$

Therefore, using (5.5), the Hölder inequality, (5.6) for $\beta=\alpha/(1-\alpha)$, and (5.7), we obtain

$$r^{2-2n}\int_{\mathbf{C}_{n-1}[r]}\left(\frac{P(w)}{p(w)(P(w)-p(w))}\right)^{\alpha}m^{\alpha}_{f_{[w]}}(P(w),0,\infty)\rho_{n-1}(w)$$

$$
\begin{aligned}
&\leq \left(\frac{s}{s-r}\right)^{\alpha} r^{2-2n} \int_{\mathbf{C}_{n-1}[r]} \frac{m^{\alpha}_{f_{[w]}}(P(w),0,\infty)}{p^{\alpha}(w)} \rho_{n-1}(w) \\
&\leq \left(\frac{s}{s-r}\right)^{\alpha} r^{2-2n} \left(\int_{\mathbf{C}_{n-1}[r]} m_{f_{[w]}}(P(w),0,\infty)\rho_{n-1}(w)\right)^{\alpha} \\
&\quad \times \left(\int_{\mathbf{C}_{n-1}[r]} (p(w))^{\frac{\alpha}{\alpha-1}} \rho_{n-1}(w)\right)^{1-\alpha} \\
&\leq C\left(\frac{s}{r(s-r)}\right)^{\alpha} \left(\frac{s}{r}\right)^{\alpha(2n-2)} m^{\alpha}_{f}(s,0,\infty). \qquad (5.8)
\end{aligned}
$$

Similarly, using the Hölder inequality, (5.6) for $\beta = \alpha/(1-\alpha)$, and Lemma A5.1.2, we obtain

$$
\begin{aligned}
&r^{2-2n} \int_{\mathbf{C}_{n-1}[r]} \frac{1}{p^{\alpha}(w)} n^{\alpha}_{f_{[w]}}(P(w),0,\infty)\rho_{n-1}(w) \\
&\leq r^{2-2n} \left(\int_{\mathbf{C}_{n-1}[r]} n_{f_{[w]}}(P(w),0,\infty)\rho_{n-1}(w)\right)^{\alpha} \\
&\quad \times \left(\int_{\mathbf{C}_{n-1}[r]} (p(w))^{\alpha/(\alpha-1)} \rho_{n-1}(w)\right)^{1-\alpha} \\
&\leq r^{2-2n}(s^{2n-2} n_f(s,0,\infty))^{\alpha} \left(\frac{Cr^{2n-2}}{r^{\alpha/(1-\alpha)}}\right)^{1-\alpha} \\
&\leq \frac{C}{r^{\alpha}} \left(\frac{s}{r}\right)^{\alpha(2n-2)} n^{\alpha}_{f}(s,0,\infty). \qquad (5.9)
\end{aligned}
$$

Noting that $s = (R+r)/2$ we have $m_f(s,\infty) \leq T_f(R)+O(1)$ and $m_f(s,0) \leq T_f(R)+O(1)$, $s/r \leq R/r$, and $s/r(s-r) \leq 2R/r(R-r)$. Also

$$
n_f(s,\infty) \leq \frac{R}{R-s} N_f(s,\infty) \leq \frac{R}{R-s} T_f(R) = \frac{2R}{R-r} T_f(R),
$$

and $n_f(s,0) \leq \frac{2R}{R-r}(T_f(R)+O(1))$. Combining these estimates with (5.4), (5.8) and (5.9) proves the Lemma. □

Let f be a meromorphic function on $\mathbf{C}^n$. Let $I = (i_1, \ldots, i_n)$ be a multi-index with $i_j \in \mathbf{Z}^+ \cup \{0\}$ with $1 \leq j \leq n$. We denote the length of I

by $|I| = \sum_{j=1}^{n} i_j$, and define

$$\partial^I f = \frac{\partial^{|I|} f}{\partial z_1^{i_1} \cdots \partial z_n^{i_n}}.$$

Then Lemma A5.1.3 is extended to the following theorem.

Theorem A5.1.4 *Let f be a meromorphic function on $\mathbf{C}^n$ and let $I = (i_1, \ldots, i_n)$ be a multi-index with length $l = \sum_{j=1}^{n} i_j$. For any α with $0 < l\alpha < 1/2$, there are positive constants C, C_1, C_2 such that for any $r < \rho < R$, we have*

$$\int_{\mathbf{S}_n(r)} \left|\frac{\partial^I f}{f}\right|^\alpha \sigma_n \leq C\left(\frac{\rho}{r}\right)^{l\alpha(2n-2)} \left(\frac{\rho}{r(\rho - r)}\right)^{l\alpha} \left[C_1 T_f(\rho) + C_2 \log\left\{\left(\frac{R}{\rho}\right)^{\alpha(2n-2)} \frac{R}{\rho(R-\rho)} T_f(R)\right\}\right]^{l\alpha}.$$

Proof. We prove the theorem by induction on the number of non-zero elements in I. First we assume that there is only one non-zero element in I, say $I = (l, 0, \ldots, 0)$. Then

$$\frac{\partial^I f}{f} = \frac{f_{z_1^l}}{f} = \frac{f_{z_1^l}}{f_{z_1^{l-1}}} \cdot \frac{f_{z_1^{l-1}}}{f_{z_1^{l-2}}} \cdots \frac{f_{z_1}}{f},$$

where $f_{z_1^l}$ means $\partial^l f / \partial z_1^l$. It turns out from the Hölder inequality and Lemma A5.1.3 that, for $r < \rho$,

$$\begin{aligned}\int_{\mathbf{S}_n(r)} \left|\frac{\partial^I f}{f}\right|^\alpha \sigma_n &= \int_{\mathbf{S}_n(r)} \left|\frac{f_{z_1^l}}{f_{z_1^{l-1}}}\right|^\alpha \cdot \left|\frac{f_{z_1^{l-1}}}{f_{z_1^{l-2}}}\right|^\alpha \cdots \left|\frac{f_{z_1}}{f}\right|^\alpha \sigma_n \\ &\leq \left(\int_{\mathbf{S}_n(r)} \left|\frac{f_{z_1^l}}{f_{z_1^{l-1}}}\right|^{l\alpha} \sigma_n\right)^{1/l} \cdots \left(\int_{\mathbf{S}_n(r)} \left|\frac{f_{z_1}}{f}\right|^{l\alpha} \sigma_n\right)^{1/l} \\ &\leq C\left(\frac{\rho}{r}\right)^{l\alpha(2n-2)} \left(\frac{\rho}{r(\rho - r)}\right)^{l\alpha} T^{l\alpha}_{f_{z_1^{l-1}}}(\rho) \cdots T^{l\alpha}_{f_{z_1}}(\rho) \cdot T_f^{l\alpha}(\rho) \qquad (5.10)\end{aligned}$$

By Lemma A5.1.3, we have, for any $\rho \leq \rho' < \rho'' \leq R$,

$$\begin{aligned}T_{f_{z_1^j}}(\rho') &= m_{f_{z_1^j}}(\rho', \infty) + N_{f_{z_1^j}}(\rho', \infty) \\ &\leq m_{f_{z_1^j}/f_{z_1^{j-1}}}(\rho', \infty) + m_{f_{z_1^{j-1}}}(\rho', \infty) + 2N_{f_{z_1^{j-1}}}(\rho', \infty)\end{aligned}$$

$$\leq \frac{1}{\alpha}\int_{\mathbf{S}_n(r)}\log^+\left|\frac{f_{z_1^j}}{f_{z_1^{j-1}}}\right|^\alpha \sigma_n + 2T_{f_{z_1^{j-1}}}(\rho')$$

$$\leq \frac{1}{\alpha}\log^+\int_{\mathbf{S}_n(r)}\left|\frac{f_{z_1^j}}{f_{z_1^{j-1}}}\right|^\alpha \sigma_n + 2T_{f_{z_1^{j-1}}}(\rho')$$

$$\leq 2T_{f_{z_1^{j-1}}}(\rho') + C\left\{\left(\frac{\rho''}{\rho'}\right)^{\alpha(2n-2)}\frac{\rho''}{\rho'(\rho''-\rho')}T_{f_{z_1}^{j-1}}(\rho'')\right\}, \quad (5.11)$$

where $C > 1$ is a constant. Using (5.11) with $j = l-1$, $\rho' = \rho$ and $\rho'' = (\rho+R)/2$, we have

$$T_{f_{z_1^{l-1}}}(\rho) \leq 2T_{f_{z_1^{l-2}}}(\rho) + C\log\left\{\left(\frac{R}{\rho}\right)^{\alpha(2n-2)}\frac{2R}{\rho(R-\rho)}T_{f_{z_1^{l-2}}}((R+\rho)/2)\right\}. \quad (5.12)$$

Using (5.11) with $j = l-2$, $\rho' = (\rho+R)/2$ and $\rho'' = (\rho+3R)/4$, we have

$$T_{f_{z_1^{l-2}}}((R+\rho)/2) \leq 2T_{f_{z_1^{l-3}}}((R+\rho)/2) + C\log\left\{\left(\frac{R}{\rho}\right)^{\alpha(2n-2)}\frac{8R}{\rho(R-\rho)}T_{f_{z_1^{l-3}}}((3R+\rho)/4)\right\}.$$

In this way, using (5.11) to $f_{z_1^{l-1}}, \cdots, f_{z_1}$ consecutively, and combining it with (5.10) proves the theorem in this case.

Now suppose the theorem is true when the number of non-zero elements in I is $l-1$. Set $I_{l-1} = (i_1, \ldots, i_{n-1}, 0)$. Then

$$\frac{\partial^I f}{f} = \frac{(\partial^{I_{n-1}}f)_{z_n^{i_n}}}{\partial^{I_{n-1}}f}\cdot\frac{\partial^{I_{n-1}}f}{f} \quad \text{and} \quad l = i_n + |I_{n-1}|.$$

Therefore, from the Hölder inequality and the induction hypothesis,

$$\int_{\mathbf{S}_n(r)}\left|\frac{\partial^I f}{f}\right|^\alpha \sigma_n$$

$$\leq \left(\int_{\mathbf{S}_n(r)}\left|\frac{(\partial^{I_{n-1}}f)_{z_n^{i_n}}}{\partial^{I_{n-1}}f}\right|^{2\alpha}\sigma_n\right)^{1/2}\cdot\left(\int_{\mathbf{S}_n(r)}\left|\frac{\partial^{I_{n-1}}f}{f}\right|^{2\alpha}\sigma_n\right)^{1/2}$$

$$\leq C\left(\frac{\rho}{r}\right)^{(i_n+|I_{n-1}|)\alpha(2n-2)}\left(\frac{\rho}{r(\rho-r)}\right)^{(i_n+|I_{n-1}|)\alpha}$$

$$\cdot T^{i_n\alpha}_{\partial^{I_{n-1}}f}(\rho)\cdot T_f^{|I_{n-1}|\alpha}(\rho). \quad (5.13)$$

Again making use of (5.11), we get

$$T_{\partial^{I_{n-1}}f}(\rho) \le C_1 T_f(\rho) + C_2 \log\left\{\left(\frac{R}{\rho}\right)^{\alpha(2n-2)} \frac{R}{\rho(R-\rho)} T_f(R)\right\}.$$

It turns out from (5.13) that

$$\begin{aligned}\int_{\mathbf{S}_n(r)} \left|\frac{\partial^I f}{f}\right|^\alpha \sigma_n \le\ & C\left(\frac{\rho}{r}\right)^{l\alpha(2n-2)} \left(\frac{\rho}{r(\rho-r)}\right)^{l\alpha} \Big[C_1 T_f(\rho) \\ & + C_2 \log\left\{\left(\frac{R}{\rho}\right)^{\alpha(2n-2)} \frac{R}{\rho(R-\rho)} T_f(R)\right\}\Big]^{l\alpha}.\end{aligned}$$

This completes the proof. □

The following Theorem is due to A. Biancofiore and Stoll (see also [Ye]).

Theorem A5.1.5 (Lemma on the Logarithmic Derivative) *Let f be a non-constant meromorphic function on $\mathbf{C}^n$ and let $I = (i_1, \ldots, i_n)$ be a multi-index with length $l = \sum_{j=1}^n i_j$. Assume that $T_f(r_0) \ge e$ for some r_0. Then for any $\epsilon > 0$, the inequality*

$$\int_{\mathbf{S}_n(r)} \log^+ \left|\frac{\partial^I f}{f}\right| \sigma_n \le l \log T_f(r) + (1+\epsilon)\log^+ \log T_f(r) + C$$

holds for all $r \ge r_0$ outside a set $E \subset (0, +\infty)$ with $\int_E dr < \infty$, where C is a constant which depends only on f.

Proof. Using the con-cavity of $\log^+$ to pull the $\log^+$ outside the integral we get, for any small positive α,

$$\begin{aligned}\int_{\mathbf{S}_n(r)} \log^+ \left|\frac{\partial^I f}{f}\right| \sigma_n &= \frac{1}{\alpha}\int_{\mathbf{S}_n(r)} \log^+ \left|\frac{\partial^I f}{f}\right|^\alpha \sigma_n \\ &\le \frac{1}{\alpha}\log^+\left(\int_{\mathbf{S}_n(r)} \left|\frac{\partial^I f}{f}\right|^\alpha \sigma_n\right) + O(1) \\ &\le \frac{1}{\alpha}\log^+\Big\{C\left(\frac{\rho}{r}\right)^{l\alpha(2n-2)} \left(\frac{\rho}{r(\rho-r)}\right)^{l\alpha}\Big[c_1 T_f(\rho) \\ &\quad + C_2 \log\left\{\left(\frac{R}{\rho}\right)^{\alpha(2n-2)} \frac{R}{\rho(R-\rho)} T_f(R)\right\}\Big]^{l\alpha}\Big\}. \qquad (5.14)\end{aligned}$$

Take

$$R = r + \frac{1}{\log^{1+\epsilon} T_f(r)}$$

and

$$\rho = \frac{R+r}{2} = r + \frac{1}{2\log^{1+\epsilon} T_f(r)}.$$

Applying Lemma A1.2.4, we have

$$\log T(\rho) \leq \log T_f(r) + 1 \tag{5.15}$$

for all r outside a set $E \subset (0, +\infty)$ with finite Lebesgue measure. In addition, for all large r,

$$\rho/r \leq 2,\ 1/(\rho - r) \leq 2\log^{1+\epsilon} T_f(r),\ R/\rho \leq 2, 1/(R-\rho) \leq 4\log^{1+\epsilon} T_f(r). \tag{5.16}$$

The theorem follows by (5.14), (5.15), (5.16) and the inequality

$$\log^+(x+y) \leq \log^+ x + \log^+ y + \log 2.$$

□

A5.2 Equi-dimensional Nevanlinna Theory

We first extend the concepts about divisors, line bundles, etc., to a compact complex manifold.

Definition A5.2.1 *Let M be a complex manifold. A* **divisor** *on M is a formal finite sum $D = \sum_j n_j D_j$, where $n_j \in \mathbf{Z}$ and D_j is a sub-variety of codimension 1. Each n_j is called the order of D, while each D_j is called the support of D. If $n_j \geq 0$ for all j, then we call D an* **effective divisor**.

To each function f which is meromorphic on the manifold M, we can assign a divisor D_f associated to f, while the supports consist of the union of the set N_f of zeros of the function and the set P_f of its poles. The order of D_g is equal to the order of zero on every component of N_f and the order of the pole with a minus sign on every component of P_f.

Definition A5.2.2 *By a* **line bundle** *L over a complex manifold M, we mean a collection $\{U_\alpha, g_{\alpha\beta}\}$ where $\{U_\alpha\}$ is a finite open cover of M and*

$g_{\alpha\beta}$ is a nowhere-zero holomorphic function on $U_\alpha \cap U_\beta$ satisfying the compatibility condition $g_{\alpha\gamma} = g_{\alpha\beta}g_{\beta\gamma}$ on $U_\alpha \cap U_\beta \cap U_\gamma$. The functions $\{g_{\alpha\beta}\}$ are called **transition functions**.

To every divisor D on a complex manifold M one can associate a line bundle $\mathcal{O}(D)$ called the **line bundle of the divisor** D. This is how it is defined: since every divisor is locally solvable, the covering $\{U_\alpha\}$ can be assumed so fine that the restriction $D|_{U_\alpha} = D_{f_\alpha}$ for some f_α meromorphic on U_α. If $D \cap U_\alpha = \emptyset$, we set $f_\alpha \equiv 1$. Let $g_{\alpha\beta} = f_\beta / f_\alpha$ on $U_\alpha \cap U_\beta$. Then $\mathcal{O}(D)$ is the line bundle with the above defined transition functions.

Definition A5.2.3 *Let $L = \{U_\alpha, g_\alpha\}$ be a line bundle. A* **holomorphic section** *s of L is a collection $\{s_\alpha\}$ where each s_α is a holomorphic function defined on U_α and satisfies $s_\alpha = g_{\alpha\beta}s_\beta$ on $U_\alpha \cap U_\beta$.*

Definition A5.2.4 *Let $L = \{U_\alpha, g_{\alpha\beta}\}$ be a line bundle. A* **metric** *on L is a collection of positive smooth functions*

$$h_\alpha : U_\alpha \to \mathbf{R}_{>0}$$

such that on $U_\alpha \cap U_\beta$ we have

$$h_\beta = |g_{\alpha\beta}|^2 h_\alpha.$$

Definition A5.2.5 *Let $L = \{U_\alpha, g_{\alpha\beta}\}$ be a metrized line bundle with metric $\{h_\alpha\}$. The form $\theta_L = -\frac{\sqrt{-1}}{2\pi}\partial\bar{\partial}\log h_\alpha$ on U_α is called the* **Chern form** *of L with respect to the metric $\{h_\alpha\}$. Denoted by $c_1(L, h)$, or just $c_1(L)$. A holomorphic line bundle L with a metric is called* **positive** *if the Chern form θ_L for the metric of L is positive definite everywhere on M.*

A line bundle L (resp. a divisor D) is said to be **ample** if there is a positive metric (i.e., Chern form θ_L for the metric of L is positive definite everywhere on M) on L (resp. on $\mathcal{O}(D)$). A complex manifold M is said to be a **projective complex manifold** if it can be embedded into some projective space $\mathbf{P}^N$. Then, by pulling back the hyperplane line bundle to M, we have that there always exists an ample bundle L over M. We also define the **Zariski topology** on M, it is the topology such that the closed sets are those algebraic subsets of M.

Definition A5.2.6 *For a metrized line bundle L with metric $\{h_\alpha\}$. Given any two sections, s_i, s_j, we define the* **inner product**

$$< s_i, s_j >= s_{i\alpha}\bar{s}_{j\alpha}h_\alpha.$$

In particular, $||s|| = |s_\alpha|^2 h_\alpha$. By the transition properties of s_α and h_α, it is well-defined.

We now introduce briefly the concept of currents. Let M be a complex manifold of dimension n. A **current** of dimension (r,s) is a linear functional T on the space $\mathcal{F}_c^{r,s}$, where $\mathcal{F}_c^{r,s}$ is the space of forms of bidegree (r,s) with coefficients of smooth functions on M with compact support. In our case, only two types of currents are involved: the currents induced by differential forms, and the currents induced by submanifolds $N \subset M$. For an $(n-r, n-s)$ form ω, it induces a current $[\omega]$ of dimension (r,s), which, as a linear functional on the space $\mathcal{F}_c^{r,s}$, is defined by

$$[\omega](\phi) = \int_M \omega \wedge \phi,$$

for every $\phi \in \mathcal{F}_c^{r,s}$. In exactly the same way, a k-dimensional submanifold $N \subset M$ induces a current $[N]$ given by

$$[N](\phi) = \int_N \phi,$$

for every $\phi \in \mathcal{F}_c^{k,k}$. In particular, for a divisor D, write $D = \sum_j n_j D_j$ where D_j are the irreducible components. Then D determines a current $[D]$ as

$$[D](\phi) = \sum_j n_j \int_{D_j} \phi,$$

for every $\phi \in \mathcal{F}_c^{n-1,n-1}$. The multiplication of current $[D]$ with ω can be defined by

$$[D] \wedge \omega(\phi) = \int_D \omega \wedge \phi.$$

Currents can be differentiated by the rule

$$\partial T(\phi) = (-1)^{r+s+1} T(\partial \phi),$$

for a current T of dimension (r,s). Thus $dd^c T(\phi) = T(dd^c \phi)$.

Similar to Theorem A2.1.2, we have the following Poincaré-Lelong Formula.

Theorem A5.2.7 (Poincaré-Lelong Formula) *Let f be a holomorphic function on an n-dimensional manifold M and let D_f be its divisor. Then the following equality, in the sense of currents, is true:*

$$dd^c[\log |f|^2] = [D_f].$$

We also have, similar to Theorem A2.1.3, the following Green-Jensen's formula.

Theorem A5.2.8 (Green-Jensen's Formula) *Let g be a function of class C^2 on $\mathbf{C}^n$ or a pluri-subharmonic (resp. pluri-superharmonic) function on $\mathbf{C}^n$. Then, for any $0 \leq r < R$,*

$$\int_r^R \frac{dt}{t} \int_{\mathbf{C}_n[t]} dd^c[g] \wedge (dd^c \log |z|^2)^{n-1} = \frac{1}{2} \int_{\mathbf{S}_n(R)} g\sigma_n - \frac{1}{2} \int_{\mathbf{S}_n(r)} g\sigma_n.$$

Let M be a compact projective complex manifold of dimension n. Let $f : \mathbf{C}^n \to M$ be a holomorphic map. Let D be an effective divisor on M. Take a metric on $\mathcal{O}(D)$, where $\mathcal{O}(D)$ is the line bundle associated with D. The **proximity function of f with respect to D** is defined by, under the assumption that $f(\mathbf{C}^n) \not\subset D$,

$$m_f(r, D) = \int_{\mathbf{S}_n(r)} \log \frac{1}{||s_D \circ f||} \sigma_n, \tag{5.17}$$

where s_D is a canonical meromorphic section associated with $\mathcal{O}(D)$. We note here, since M is compact, $m_f(r, D)$ is independent, up to a bounded term, of the choice of the section s defining D and also of the choice of the metric on $\mathcal{O}(D)$. Define

$$n_f(t, D) = t^{2-2n} \int_{f^{-1}(D) \cap \mathbf{C}_n[t]} (dd^c |z|^2)^{n-1}.$$

The **counting function of f with respect to D** is defined as, under the assumption that $f(\mathbf{C}^n) \not\subset D$,

$$N_f(r, D) = \int_0^r [n_f(t, D) - n_f(0, D)] \frac{dt}{t} + n_f(0, D) \log r$$

where $n(0, D) = \lim_{t \to 0} n_f(t, D)$ is called the Lelong number. Here, the integration

$$\int_{f^{-1}(D) \cap \mathbf{C}_n[t]} (dd^c |z|^2)^{n-1}$$

is in the following sense: write $f^{-1}(D) = \sum_j m_j M_j$ where M_j are the irreducible components, then

$$\int_{f^{-1}(D) \cap \mathbf{C}_n[t]} (dd^c |z|^2)^{n-1} = \sum_j m_j \int_{M_j \cap \mathbf{C}_n[t]} (dd^c |z|^2)^{n-1}.$$

Note that we also have

$$N_f(r, D) = \int_0^r \frac{dt}{t} \int_{f^{-1}(D) \cap \mathbf{C}_n[t]} (dd^c \log |z|^2)^{n-1} + n_f(0, D) \log r.$$

For any fixed real number $r_0 > 0$,

$$N_f(r, D) = \int_{r_0}^r \frac{dt}{t} \int_{f^{-1}(D) \cap \mathbf{C}_n[t]} (dd^c \log |z|^2)^{n-1} + O(1).$$

Using the notation of currents, we can write

$$N_f(r, D) = \int_0^r \frac{dt}{t} \int_{\mathbf{C}_n[t]} [f^{-1} D] \wedge (dd^c \log |z|^2)^{n-1} + n(0, D) \log r.$$

Note that $m_f(r, D)$ and $N_f(r, D)$ are defined under the assumption that $f(\mathbf{C}^n) \not\subset D$, so whenever one of these functions appears, we automatically assume that $f(\mathbf{C}^n) \not\subset D$.

Let ω be a $(1, 1)$ form on M. The **characteristic function $T_{f,\omega}(r)$** of f with respect to ω is defined by

$$T_{f,\omega}(r) = \int_0^r \frac{dt}{t} \int_{\mathbf{C}_n[t]} f^* \omega \wedge (dd^c \log |z|^2)^{n-1}. \tag{5.18}$$

Note that $(dd^c \log |z|^2)^{n-1}$ has an infinite of order $2(n-1)$ at the point $z = 0$, this is integrable over the $2n$-dimensional ball $\mathbf{C}_n[t]$, and $f^*\omega$ is smooth. Therefore the inner integral in (5.18) exists, has a zero of no less than first order at $t = 0$. Hence the outer integral also exists. If ω is positive, then $T_{f,\omega}(r)$ measures the growth of f. Also, since M is compact, for any two positive $(1,1)$ forms ω_1, ω_2, $cT_{f,\omega_1}(r) \leq T_{f,\omega_2}(r) \leq c'T_{f,\omega_1}(r)$ for some positive constants c and c'.

Let L be a Hermitian line bundle with metric h, then the characteristic function $T_f(r, L)$ is defined by $T_f(r, L) = T_{f,c_1(L)}(r)$ where $c_1(L)$ is the Chern form associated with h. For a divisor D, take a metric on $\mathcal{O}(D)$, where $\mathcal{O}(D)$ is the line bundle associated with D, we define the characteristic function of $T_f(r, \mathcal{O}(D))$ by

$$T_f(r, \mathcal{O}(D)) = \int_0^r \frac{dt}{t} \int_{\mathbf{C}_n[t]} f^* c_1(\mathcal{O}(D)) \wedge (dd^c \log |z|^2)^{n-1},$$

where $c_1(\mathcal{O}(D))$ is the Chern form of $\mathcal{O}(D)$. Note that, since M is compact, $T_f(r, \mathcal{O}(D))$ does not, up to the addition of a bounded term, depend on the choice of the metric on $\mathcal{O}(D)$. If $\mathcal{O}(D)$ is ample, then $T_f(r, \mathcal{O}(D))$ is positive and it measures the growth of f.

Theorem A5.2.9 (First Main Theorem)

$$T_f(r, \mathcal{O}(D)) = m_f(r, D) + N_f(r, D) + O(1).$$

Proof. By the definition,

$$N_f(r, D) = \int_{r_0}^r \frac{dt}{t} \int_{\mathbf{C}_n[t]} [f^{-1}D] \wedge (dd^c \log |z|^2)^{n-1} + O(1)$$

for any $r_0 > 0$. Fix a $r_0 > 0$. Let s_D be a canonical meromorphic section associated with D. Let U_α be a finite cover of M. Write $s_D = \{s_\alpha\}$. Then the square of its Hermitian norm

$$||s_D||^2 = h_\alpha |s_\alpha|^2$$

on U_α. So

$$dd^c[\log ||s_D||^2] = dd^c \log h_\alpha + dd^c[\log |s_\alpha|^2].$$

By the Poincaré-Lelong formula, $dd^c[\log |s_\alpha|^2] = [D]$. Noticing $dd^c \log h_\alpha = -c_1(\mathcal{O}(D))$, we have

$$-dd^c[\log ||s_D||^2] + [D] = c_1(\mathcal{O}(D)).$$

We pass now to pull-backs by the map f:

$$-dd^c[\log \|f^*s_D\|^2] + [f^{-1}D] = f^*c_1(\mathcal{O}(D)).$$

Thus

$$(-dd^c[\log \|f^*s_D\|^2]+[f^{-1}D])\wedge(dd^c \log|z|^2)^{n-1} = f^*c_1(\mathcal{O}(D))\wedge(dd^c \log|z|^2)^{n-1}.$$

So

$$\begin{aligned}(-dd^c[\log \|f^*s_D\|^2] \wedge (dd^c \log|z|^2)^{n-1} + [f^{-1}D] \wedge (dd^c \log|z|^2)^{n-1})(\chi_{\mathbf{C}_n[t]}) \\ = \left(f^*c_1(\mathcal{O}(D)) \wedge (dd^c \log|z|^2)^{n-1}\right)(\chi_{\mathbf{C}_n[t]}),\end{aligned}$$

where $\chi_{\mathbf{C}_n[t]}$ is the characteristic function of the ball $\mathbf{C}_n[t]$. This simply means

$$\begin{aligned}\int_{\mathbf{C}_n[t]} f^*c_1(\mathcal{O}(D)) \wedge (dd^c \log|z|^2)^{n-1} &= \int_{\mathbf{C}_n[t]} [f^{-1}D] \wedge (dd^c \log|z|^2)^{n-1} \\ &\quad - \int_{\mathbf{C}_n[t]} dd^c[\log \|f^*s_D\|^2] \wedge (dd^c \log|z|^2)^{n-1}.\end{aligned}$$

Taking integration yields

$$\begin{aligned}&\int_{r_0}^r \frac{dt}{t} \int_{\mathbf{C}_n[t]} f^*c_1(\mathcal{O}(D)) \wedge (dd^c \log|z|^2)^{n-1} \\ &= \int_{r_0}^r \frac{dt}{t} \int_{\mathbf{C}_n[t]} [f^{-1}D] \wedge (dd^c \log|z|^2)^{n-1} \\ &\quad - \int_{r_0}^r \frac{dt}{t} \int_{\mathbf{C}_n[t]} dd^c[\log \|f^*s_D\|^2] \wedge (dd^c \log|z|^2)^{n-1} \\ &= N_f(r,D) - \int_{r_0}^r \frac{dt}{t} \int_{\mathbf{C}_n[t]} dd^c[\log \|f^*s_D\|^2] \wedge (dd^c \log|z|^2)^{n-1} + O(1).\end{aligned}$$

By Theorem A5.2.8 and the definition of $m_f(r,D)$,

$$-\int_{r_0}^r \frac{dt}{t} \int_{\mathbf{C}_n[t]} dd^c[\log \|f^*s_D\|^2] \wedge (dd^c \log|z|^2)^{n-1} = m_f(r,D) + O(1).$$

So we have $T_f(r,\mathcal{O}(D)) = m_f(r,D) + N_f(r,D) + O(1)$. □

Definition A5.2.10 *Let M be a compact complex manifold of dimension n. A divisor D on M is said to be of simple normal crossing if for every point $P \in M$, there is a neighborhood U_P of P such that $D \cap U_P$ is given by an equation $w_1 \dots w_k = 0$ in local coordinates, where $k \le n$.*

Let M be a compact projective complex manifold of dimension n. Fix a positive (1,1)-form ω on M and write $T_f(r) = T_{f,\omega}(r)$. We note such ω exists because M is a projective manifold.

Theorem A5.2.11 *Let M be a compact projective complex manifold of dimension n. Let (L, h) be a positive hermitian holomorphic line bundle over M. Let $s_1, \ldots, s_q$ be holomorphic sections of L and $D = D_1 + \cdots + D_q$ be the divisor defined by the s_j's, i.e., $D_j = [s_j^{-1}(0)]$. Assume that D is of simple normal crossing. Let $f : \mathbf{C}^n \to M$ be a non-degenerate (i.e., the image contains an open set) holomorphic map. Then, for every $\epsilon > 0$, the inequality*

$$\sum_{j=1}^{q} m_f(r, D_j) + T_f(r, K) + N(r, S_f) \leq O(\log^+ T_f(r))$$

holds for all large r outside a set $E \subset (0, +\infty)$ with finite Lebesgue measure, where K is the canonical line bundle over M and S_f is the stationary divisor.

Proof. Let Ω be a volume form on M, i.e., an (n, n)-form on M which is locally written as, in terms of local coordinates $w_1, \ldots, w_n$,

$$\Omega = h(w) \prod_{i=1}^{n} \frac{\sqrt{-1}}{2\pi} dw_i \wedge d\bar{w}_i,$$

where $h > 0$ everywhere. We note that a volume form Ω as above defines a metric on the canonical line bundle K such that $dd^c \log h = c_1(K)$.

We consider a singular volume form with singularity on D:

$$\Psi = \frac{\Omega}{\prod_{j=1}^{q} \|s_j\|^2}. \tag{5.19}$$

We write $f^*(\Psi) = \xi \Phi_z$ where Φ_z is the Euclidean volume form in $\mathbf{C}^n$, i.e. $\Phi_z = \prod_{i=1}^{n} \frac{\sqrt{-1}}{2\pi} dz_i \wedge d\bar{z}_i$. We use the logarithmic derivative lemma to prove the following claim.

Claim *The inequality*

$$\int_{\mathbf{S}_n(r)} \log^+ \xi \sigma_n \leq O(\log^+ T_f(r))$$

holds for all large r outside a set $E \subset (0, +\infty)$ with finite Lebesgue measure.

To prove the Claim, we first prove that there exists a finite set $\mathcal{H} = A \cup B$ of set of rational functions on M such that

(a) For every point $P \in M$, there is a Zariski open neighborhood U of P and rational functions $\phi_1, \ldots, \phi_n \in A$ such that $\phi_1, \ldots, \phi_n$ are holomorphic on U and form a local coordinate system around every point of U.

(b) For every point $P \in D$, there is a Zariski open neighborhood U of P and rational functions $\phi_1, \ldots, \phi_n \in B$ such that $\phi_1, \ldots, \phi_n$ are holomorphic on U, $\phi_1 = 0, \ldots, \phi_k = 0$ are the defining equation of D on U, and $\phi_1, \ldots, \phi_n$ form a local coordinate system around every point of U.

We now construct $\mathcal{H}$. We may assume without loss of generality that D is irreducible, since otherwise we will just consider each component D_i because D has only finitely many components and a finite union of a finite number of rational functions is still finite. We also assume that D is ample; otherwise we may replace D by $D+D'$ so that $D+D'$ is ample. Observe that if s is a function holomorphic on a neighborhood U such that $[s=0] = D \cap U$ then $[s^\tau = 0] = \tau D \cap U$ where τ is a real constant. This means that we may assume without loss of generality that D is very ample by replacing D with τD for some τ so that τD is very ample.

Let $u \in H^0(M, \mathcal{O}(D))$ be a section such that $D = [u = 0]$. At a point $x \in D$ choose a section $v_1 \in H^0(X, \mathcal{O}(D))$ so that $E_1 = [v_1 = 0]$ is smooth, $D + E_1$ is of simple normal crossings and v_1 is non-vanishing at x (this is possible because $\mathcal{O}(D)$ is very ample). The rational function $t_1 = u/v_1$ is regular on the affine open neighborhood $M - E_1$ of x and $(M - E_1) \cap [t_1 = 0] = (M - E_1) \cap D$. Since M is a projective variety, there exists a very ample bundle L on M. Choose rational functions $t_2 = u_2/v_2, ..., t_n = u_n/v_n$ where u_i and v_i are sections of a very ample bundle L on M (such line bundle exists because M is projective) so that $t_2, ..., t_n$ are regular at x, the divisors $D_i = [u_i = 0], E_i = [v_i = 0]$ are smooth and $D + D_2 + \cdots + D_n + E_1 + \cdots + E_n$ is of simple normal crossings. Moreover, since the bundles involved are very ample the sections can be chosen so that $dt_1 \wedge \cdots \wedge dt_n$ is non-vanishing at x (the complete system of sections provides an embedding, hence at each point there are $n+1$ sections with the property that n of the quotients of these $n+1$ sections forms a local coordinate system on some open neighborhood U_x of x). In this way, we get a set of rational functions $t_1, \ldots, t_n$ which depends on $x \in M$. More precisely, it depends on an open neighborhood of U_x of x. However, since D is compact it is covered by a finite number of such open neighborhoods, say $U_1, ..., U_m$. Moreover, there exist relatively compact open subsets U_i' of U_i $(1 \le i \le m)$ such that $\cap_{1 \le i \le m} U_i'$ still covers D. For each $U_i, 1 \le i \le m$ we get a finite set of rational functions as described above. Let B be the collections of all rational functions obtained.

Next we consider a point x in the compact set $M - \cap_{1 \le \le m} U_i'$. Repeating the procedure as above we can find rational functions $s_1 = a_1/b_1, ..., s_n = a_n/b_n$ where a_i and b_i are sections of some very ample line bundle L so that $s_1, ..., s_n$ from a holomorphic local coordinate on some open neighborhood V_x of x. We must also choose these sections so that the divisor $H = [s_1 \cdots s_n = 0]$ together with those divisors (finite in number), which had already been constructed, is still a divisor with simple normal crossings(this is possible by the very ampleness of the line bundle L). Since $M - \cap_{1 \le i \le m} U_i'$ is compact, it is covered by a finite number of such coordinate neighborhoods. Let A be the collections of such rational functions. It is a finite set. Let $\mathcal{H} = A \cup B$. Then $\mathcal{H}$ has the desired properties.

Let $U_1, ..., U_m$ be an open covering of D, as presented above. There exist relatively compact open subsets U_i' of U_i $(1 \le i \le m)$ such that $\cup_{1 \le i \le m} U_i'$ still covers D. Since Ω is a volume form on M, on each U_i,

$$\Omega = a_{U_i}(w)\Phi(w), \tag{5.20}$$

where $a_{U_i} > 0$ on U_i, and hence is bounded on U_i'. Let $\phi_1, \ldots, \phi_n \in B$ such that $\phi_1, \ldots, \phi_n$ are holomorphic on U_i, $\phi_1 = 0, \ldots, \phi_k = 0$ are defining equations of D on U_i and $\phi_1, \ldots, \phi_n$ form a local coordinate system around every point of U_i. Then

$$\Phi(w) = \prod_{j=1}^{n} \frac{\sqrt{-1}}{2\pi} d\phi_j \wedge d\bar{\phi}_j.$$

Also, after rearranging the indices, if necessary,

$$||s_j||^2 = |\phi_j|^2 h_{U_i} \quad \text{for} \quad 1 \le j \le k \tag{5.21}$$

and $||s_j|| > 0$ for $j \ge k+1$ on U_i, hence

$$\log^+ \frac{1}{\prod_{j=k+1}^{q} ||s_j||^2} \quad \text{is bounded on} \quad U_i'. \tag{5.22}$$

Also, since $h_{U_i} > 0$ and $a_{U_i} > 0$ on U_i,

$$\log^+ \frac{a_{U_i}}{h_{U_i}^k} \quad \text{is bounded on} \quad U_i'. \tag{5.23}$$

We now consider $M - \cup_{1\leq i\leq m} U_i'$. Since $M - \cup_{1\leq i\leq m} U_i'$ is compact and $\|s_j\|^2 > 0$, $\|s_j\|^2$ is bounded for $1 \leq j \leq q$ on $M - \cup_{1\leq i\leq m} U_i'$. This, together with (5.20), (5.21), (5.22), (5.23), (5.24) and the inequality $\ln^+(a+b) \leq \ln^+ a + \ln^+ b$, gives us

$$\begin{aligned}\log^+ \xi &\leq C \sum_{\phi_1,\ldots,\phi_n \in A\cup B} \log^+ \frac{\left|\det\left(\frac{\partial(\phi_i\circ f)}{\partial z^j}\right)_{1\leq i,j\leq n}\right|^2}{|\phi_1 \circ f|^2 \cdots |\phi_n \circ f|^2} \\ &= C\sum_{h\in\mathcal{H}} \log^+ \left|\frac{(h\circ f)_{z_j}}{h\circ f}\right|. \end{aligned} \tag{5.24}$$

By Theorem A5.1.5(the Logarithmic derivative lemma),

$$\int_{\mathbf{S}_n(r)} \log^+ \xi\sigma_n \leq C\sum_{h\in\mathcal{H}} \int_{\mathbf{S}_n(r)} \log^+ \left|\frac{(h\circ f)_{z_j}}{h\circ f}\right| \sigma_n \leq O(\log^+ T_{h\circ f}(r))$$

holds for all large r outside a set $E \subset (0, +\infty)$ with finite Lebesgue measure. Since h is a rational function on M,

$$T_{h\circ f}(r) \leq O(T_f(r)).$$

This proves the claim.

We proceed with the proof of Theorem A5.2.9. ξ is equal to zero on the divisor of stationary S_f and equal to infinity on $f^{-1}(D)$. So we have, by the definition of ξ,

$$dd^c[\log \xi] = f^*c_1(K) - \sum_{j=1}^{q} dd^c[\log \|f^*s_j\|^2] + S_f.$$

Since, by the Poincare-Lelong formula, $dd^c[\log\|s_j\|^2] = -c_1(L) + [D_j]$,

$$dd^c[\log \xi] = f^*c_1(K) + qf^*c_1(L) - \sum_{j=1}^{q}[f^{-1}D_j] + S_f.$$

So, for a fixed real number $r_0 > 0$,

$$\begin{aligned}&\int_{r_0}^{r} \frac{dt}{t} \int_{\mathbf{C}_n[t]} dd^c[\log \xi] \wedge (dd^c \log |z|^2)^{n-1} \\ &= T_f(r,K) + qT_f(r) + N(r,S_f) - \sum_{j=1}^{q} N_f(r,D_j) + O(1).\end{aligned}$$

On the other hand, by Green-Jensen's formula and the Claim,

$$\begin{aligned}\int_{r_0}^{r}\frac{dt}{t}\int_{\mathbf{C}_n[t]} dd^c[\log \xi]\wedge(dd^c\log|z|^2)^{n-1} &= \frac{1}{2}\int_{\mathbf{S}_n(r)}\log\xi\sigma_n+O(1)\\ &\leq O(\log^+ T_f(r))\end{aligned}$$

holds for all large r outside a set $E\subset(0,+\infty)$ with finite Lebesgue measure. Thus,

$$qT_f(r)+T_f(r,K)+N(r,S_f)\leq\sum_{j=1}^{q}N_f(r,D_j)+O(\log^+T_f(r))$$

holds for all large r outside a set $E\subset(0,+\infty)$ with finite Lebesgue measure. By the First Main Theorem,

$$\sum_{j=1}^{q}m_f(r,D_j)+\sum_{j=1}^{q}N_f(r,D_j)=qT_f(r)+O(1).$$

So

$$\sum_{j=1}^{q}m_f(r,D_j)+T_f(r,K)+N(r,S_f)\leq O(\log^+T_f(r))$$

holds for all large r outside a set $E\subset(0,+\infty)$ with finite Lebesgue measure. This proves the theorem. □

A5.3 Griffiths' Conjecture

Theorem A5.2.11 provides a "good model" of what results can be expected in the higher dimensional Nevanlinna theory. P. Griffiths made the following conjecture for holomorphic curves into compact projective complex manifolds.

Conjecture A5.3.1 (Griffiths) *Let D be a divisor of simple normal crossings on a compact projective complex manifold M. Then there exists a proper algebraic subset Z_D having the following property. Let $f:\mathbf{C}\to M$ be a holomorphic curve such that $f(\mathbf{C})\not\subset Z_D$. Let K be the canonical divisor, and let A be an ample divisor, then*

$$m_f(r,D)+T_f(r,\mathcal{O}(K))\leq O(\log^+T_f(r,\mathcal{O}(A)))$$

holds for all r outside a set $E\subset(0,+\infty)$ with finite Lebesgue measure.

We consider the case that $M = \mathbf{P}^n(\mathbf{C})$. To determine the canonical divisor we consider the volume form $\Omega = dx_1 \wedge \cdots \wedge dx_n$ in the affine coordinates $(1, x_1, \ldots, x_n)$ on $U_0 = \{[z_0 : \cdots : z_n] \in \mathbf{P}^n(\mathbf{C}) | z_0 \neq 0\}$. If we rewrite Ω with respect to $(x_0, \ldots, 1, \ldots, x_n)$ on $U_i = \{[z_0 : \cdots : z_n] \in \mathbf{P}^n(\mathbf{C}) | z_i \neq 0\}$, we find

$$\Omega = \frac{1}{x_0^{n+1}} dx_0 \wedge \cdots \wedge d\hat{x}_i \wedge \cdots \wedge dx_n.$$

Hence Ω has a pole of order $n+1$ along $x_0 = 0$. Hence the canonical divisor $K = -(n+1)H$ where H is the hyperplane at infinity. So $T_f(r, \mathcal{O}(K)) = -(n+1)T_f(r)$, where $T_f(r)$ is the Nevanlinna's characteristic function defined in Chapter 3. Take $D = H_1 + \cdots + H_q$, where $H_1, \ldots, H_q$ are hyperplanes in general position, then Conjecture A5.3.1 in this case is just Theorem A3.1.6. If M is an abelian variety, then K_M is trivial. So Griffiths' conjecture implies that if the image f omits a divisor on an Abelian variety, then f is degenerate. Furthermore, if D is ample, then f must be constant. This is known as Lang's conjecture. We shall discuss, in Chapter 6, the results for holomorphic curves in Abelian varieties.

Part B: Diophantine Approximation

B5.1 Vojta's Conjecture in Diophantine Approximation

Motivated by Conjecture A5.3.1, Paul Vojta made the following conjecture.

Conjecture B5.1.1 *Let V be a smooth projective variety defined over a number field k and let A be a pseudo ample divisor. Let D be a normal crossing divisor defined over a finite extension of k. Let K be a canonical divisor of V. Let S be a finite set of valuation on k and for each $v \in S$ let $\lambda_{v,D}$ be a Weil function for D. Let $\epsilon > 0$. Then there exists a Zariski closed variety Z of V such that for all $P \in V(k)$, $P \notin Z$ we have*

$$\sum_{v \in S} \lambda_{v,D}(P) + h_K(P) < \epsilon h_A(P) + O(1).$$

As an example consider $V = \mathbf{P}^n$. Then $K = -(n+1)H$ where H is the hyperplane at infinity. So by linearity of heights we find that $h_K =$

$-(n+1)h$ where h is the ordinary projective height. Thus Vojta's conjecture for $V = \mathbf{P}^n$ reads as follows: for all $P \in V(k)$, $P \notin Z$ we have

$$\sum_{v \in S} \lambda_{v,D}(P) < (n+1+\epsilon)\epsilon h(P) + O(1).$$

In the case when D is a union of hyperplanes in general position, this is just Schmidt's subspace theorem.

We take for D any hypersurface, and for $k = \mathbf{Q}$, the field of rational numbers, and we obtain the following conjecture.

Conjecture B5.1.2 *Let D be a hypersurface in $\mathbf{P}^n$ defined over $\mathbf{Q}$ of degree $d \geq n+2$ with at most normally crossing singularities. Suppose that it is given by the homogeneous equation $Q(x_0, \ldots, x_n)$ where Q has coefficient in $\mathbf{Z}$. Let S be a finite set of rational primes. Then the set of points $(x_0, \ldots, x_n) \in \mathbf{Z}^{n+1}$, with $gcd(x_0, \ldots, x_n) = 1$ such that $Q(x_0, \ldots, x_n)$ only contains primes from S, lies in a Zariski closed subset of $\mathbf{P}^n$.*

Proof. In fact, we apply Vojta's conjecture with the proximity functions

$$\lambda_{v,D} = \log \max_i \left\| \frac{x_i^d}{Q(x_0, \ldots, x_n)} \right\|_v$$

and the set of valuations $S \cup \{\infty\}$. Vojta's conjecture implies that for any $\epsilon > 0$ the set of projective $n+1$-tuples $(x_0, \ldots, x_n) \in \mathbf{Z}^{n+1}$ with $gcd(x_0, \ldots, x_n) = 1$ which satisfy

$$\sum_{v \in S \cup \{\infty\}} \log \max_i \left\| \frac{x_i^d}{Q(x_0, \ldots, x_n)} \right\|_v > (n+1+\epsilon)h(x_0, \ldots, x_n)$$

lie in a Zariski closed subset of $\mathbf{P}^n$. The inequality can be restated as

$$\begin{aligned} \sum_{v \in S \cup \{\infty\}} \log \|Q(x_0, \ldots, x_n)\|_v &< -(n+1+\epsilon)h(x_0, \ldots, x_n) \\ &+ \sum_{v \in S \cup \{\infty\}} \log \max_i \|x_i^d\|_v. \end{aligned}$$

The sum on the right is precisely $dh(x_0, \ldots, x_n)$ since the sum includes the infinite valuation and the $\max_i$ is 1 for all finite v, so the set of solutions to

$$\sum_{v \in S \cup \{\infty\}} \log \|Q(x_0, \ldots, x_n)\|_v < (d - n - 1 - \epsilon) h(x_0, \ldots, x_n)$$

lies in a Zariski closed subset of $\mathbf{P}^n$. Suppose $Q(x_0, \ldots, x_n)$ is composed of primes only from S. Then $\log \|Q(x_0, \ldots, x_n)\|_v = 0$ for $v \notin S$. Thus, by the product formula, $\sum_{v \in S \cup \{\infty\}} \log \|Q(x_0, \ldots, x_n)\|_v = 0$. So

$$0 = \sum_{v \in S \cup \{\infty\}} \log \|Q(x_0, \ldots, x_n)\|_v < (d - n - 1 - \epsilon) h(x_0, \ldots, x_n)$$

holds since $d \geq n + 1$. Hence the set of points $(x_0, \ldots, x_n) \in \mathbf{Z}^{n+1}$, with $gcd(x_0, \ldots, x_n) = 1$ such that $Q(x_0, \ldots, x_n)$ only contains primes from S, lies in a Zariski closed subset of $\mathbf{P}^n$. □

Note that we know that Conjecture B5.1.2 holds if Q is decomposable, i.e. $Q = L_1 L_2 \cdots L_d$ is the product of d linear forms in general position(see Chapter 3, Part B). But it in general is still an open question. Indeed, it does not appear to be known whether the integer solutions to the specific equation

$$x^5 + y^5 + z^5 + w^5 = 1$$

are Zariski dense in $\mathbf{P}^3(\mathbf{Q})$.

Recall that V is said to be of general type if K is pseudo-ample. So taking $D = 0$, Vojta's conjecture gives us the following conjecture.

Conjecture B5.1.3 (Bombieri) *Let V be a projective variety over a number field k and suppose that V is of general type. Then $V(k)$ is contained in a Zariski closed subset of V.*

Finally, if V is an Abelian variety, then Vojta's conjecture gives the following result (now known as Faltings' theorem).

Theorem B5.1.4 (Faltings) *Let A be an Abelian variety over k and E a sub-variety, also defined over F. Let h be a height on A and v a valuation on F. Let $\epsilon > 0$. Then we have*

$$\lambda_{v,E}(P) < \epsilon h(P)$$

for almost every point $P \in A(k) - E$.

The Correspondence Table

Nevanlinna Theory	**Diophantine Approximation**
Conjecture A5.3.1	Conjecture B5.1.1

Chapter 6

Holomorphic Curves in Abelian Varieties and the Theorem of Faltings

Part A: Nevanlinna Theory

A6.1 Bloch's Theorem for Holomorphic Curves in Abelian Varieties

In this section, we prove the following theorem of Bloch.

Theorem A6.1.1 (Bloch) *Let A be an Abelian variety and X be a sub-variety of A which is not a translate of an Abelian sub-variety of A. Then there exists no non-constant holomorphic map from $\mathbf{C}$ into X whose image in X is Zariski dense in X.*

The proof of Bloch's Theorem presented here is based on Bloch's original proof, along the lines of the papers of Ochiai [Och], Kawamata [Ka], Wong [Wong2], and Siu [Siu3].

There are two steps in this proof. The first one is an algebraic statement about the map from the projectivization of a jet bundle of X defined by jet differentials from the Abelian variety to a complex projective space. The second one is an argument from Nevanlinna theory.

One of the tools which we use is the k-jet bundle $J_k(M)$ of k-jets for a complex manifold M of complex dimension n.

Definition A6.1.2 *Let M be a complex manifold of dimension n. The k-jet bundle $J_k(M) = \cup_{P \in M} J_k(M)_P$ is a bundle over M, where, for each point $P \in M$, the fiber $J_k(M)_P$ is defined as follows: every element $v \in J_k(M)_P$*

is a set of complex numbers $(\xi_{j\alpha})_{1\le j\le k, 1\le\alpha\le n}$ *with respect to a local coordinates* z_α $(1 \le \alpha \le n)$ *of* M *around* P*; when another coordinate system* w_α $(1 \le \alpha \le n)$ *is used,* $v = (\eta_{j\alpha})_{1\le j\le k, 1\le\alpha\le n}$ *and the relation between* $(\xi_{j\alpha})_{1\le j\le k, 1\le\alpha\le n}$ *and* $(\eta_{j\alpha})_{1\le j\le k, 1\le\alpha\le n}$ *is as follows; let* g *be a holomorphic map to* M *from an open neighborhood* U *of* 0 *in* $\mathbf{C}$ *with coordinate* ζ *such that* $g(0) = P$ *and* $\frac{d^j}{d\zeta^j}(z_\alpha \circ g)(0) = \xi_{j\alpha}$ $(1 \le j \le k,\ 1 \le \alpha \le n)$*; then* $\frac{d^j}{d\zeta^j}(w_\alpha \circ g)(0) = \eta_{j\alpha}$ $(1 \le j \le k,\ 1 \le \alpha \le n)$.

From the above definition, with respect to a local coordinate system z_α $(1 \le \alpha \le n)$ of $P \in M$, every element $v \in J_k(M)_P$ is represented by $(\frac{d^j}{d\zeta^j}(z_\alpha \circ g)(0))_{1\le j\le k, 1\le\alpha\le n}$ for some holomorphic map g from an open neighborhood U of 0 in $\mathbf{C}$ to M such that $g(0) = P$. The complex dimension of $J_k(M)_P$ is kn. The 1-jet bundle $J_1(M)$ is simply the tangent bundle of M.

Definition A6.1.3 *For a holomorphic map* $f : \mathbf{C} \to X$*. The map* $d^k f : \mathbf{C} \to J_k X$ *sending each point* $\zeta_0 \in \mathbf{C}$ *to* $d^k f(\zeta_0) = (\frac{d^j}{d\zeta^j}(z_\alpha \circ f)(\zeta_0))_{1\le j\le k, 1\le\alpha\le n}$ *is called the* k**-jet lifting** *of* f.

An Abelian variety A is a complex tori $A = \mathbf{C}^m/\Lambda$ which can be embedded into a projective space. For an Abelian variety A, let $(z_1, \ldots, z_m)$ be the coordinates of $\mathbf{C}^m$, then $dz_1, \ldots, dz_m$ are global differential forms on A. So we can identify $J_k(A)$ with the trivial product bundle $A \times \mathbf{C}^{km}$ using the global differentials $dz_1, \ldots, dz_m$. We write $J_k(A) = A \times \mathbf{C}^{km}$ with this identification. Let

$$p_k : J_k(A) = A \times \mathbf{C}^{km} \to \mathbf{C}^{km}$$

be the natural projection.

Let $f : \mathbf{C} \to A$ be a holomorphic map. Let X be the Zariski closure of $f(\mathbf{C})$ in A. Let n be the complex dimension of X. Since X is a sub-variety of A, the inclusion $\iota : X \to A$ naturally induces a holomorphic bundle morphism $\iota_* : J_k(X) \to J_k(A)$. Let

$$\Phi_k = p_k \circ \iota_* : J_k(X) \to \mathbf{C}^{km}, \tag{6.1}$$

and write

$$\Phi = \Phi_n. \tag{6.2}$$

The algebraic statement concerning the jet bundles of X and A is the following.

Lemma A6.1.4 *Let $f : \mathbf{C} \to A$ be a holomorphic map. Denote by X the Zariski closure of $f(\mathbf{C})$. Assume that for every $\zeta \in \mathbf{C}$ the differential $d\Phi|_{d^n f(\zeta)}$ of Φ at $d^n f(\zeta)$ is not injective. Then there exists a one-parameter subgroup of A which leaves X invariant under the translation.*

To prove Lemma A6.1.4, we need a lemma concerning the properties of Wronskian.

Lemma A6.1.5(Generalized Lemma on Wronskian) *Let $n \geq 2$ be an integer. Let U be a domain in $\mathbf{C}$ and $(f_{1,\lambda}, \ldots, f_{n,\lambda})$ be an n-tuple of holomorphic functions on U for $\lambda \in \Lambda$, where Λ is a finite index set. Suppose for every point $\zeta_0 \in U$ there exist complex numbers $\hat{c}_1(\zeta_0), \ldots, \hat{c}_n(\zeta_0)$ such that $\sum_{\alpha=0}^{n} \hat{c}_\alpha(\zeta_0) f_{\alpha,\lambda}^{(\beta)}(\zeta_0) = 0$ for $0 \leq \beta \leq n-1$ and $\lambda \in \Lambda$, where $f_{\alpha,\lambda}^{(\beta)}$ denotes the β-th order derivative of $f_{\alpha,\lambda}$. Then there exist complex numbers $c_1, \ldots, c_n$ not all zero such that $\sum_{\alpha=0}^{n} c_\alpha f_{\alpha,\lambda} \equiv 0$ on U for every $\lambda \in \Lambda$.*

Proof. By replacing U by the complement in U of the common zero-set of $f_{\alpha,\lambda}(1 \leq \alpha \leq n)$ we can assume without loss of generality that some vector $(f_{1,\lambda}(\zeta), \ldots, f_{n,\lambda}(\zeta))$ is nonzero for every $\zeta \in U$. For $0 \leq \beta \leq n-1$, $\lambda \in \Lambda$, and $\zeta \in U$ let $\mathbf{v}(\beta, \lambda, \zeta) = (f_{1,\lambda}^{(\beta)}(\zeta), \ldots, f_{n,\lambda}^{(\beta)}(\zeta)) \in \mathbf{C}^n$. For $0 \leq k \leq n-1$ let $V_k(\zeta)$ denote the vector subspace of $\mathbf{C}^n$ spanned by $\mathbf{v}(\beta, \lambda, \zeta)$ for $0 \leq \beta \leq k$ and $\lambda \in \Lambda$. Let $d_k(\zeta) = \dim V_k(\zeta)$. By the assumption in the Lemma, we have $d_{n-1}(\zeta) < n$. On the other hand, since some vector $(f_{1,\lambda}(\zeta), \ldots, f_{n,\lambda}(\zeta))$ is nonzero for every $\zeta \in U$, we have $d_0(\zeta) \geq 1$ for every $\zeta \in U$. Thus we get a sequence

$$1 \leq d_0(\zeta) \leq d_1(\zeta) \leq \ldots \leq d_{n-1}(\zeta) < n.$$

It follows that there exists $\ell(\zeta) < n-1$ such that $d_{\ell(\zeta)}(\zeta) = d_{\ell(\zeta)+1}(\zeta)$. Moreover, there actually exists a subset U_0 of positive measure in U such that both $\ell(\zeta)$ and $d_{\ell(\zeta)}(\zeta)$ are constant on U_0. Denote the constant value of $\ell(\zeta)$ on U_0 by ℓ, and the constant value of $d_{\ell(\zeta)}(\zeta)$ on U_0 by m.

There exists a point $\zeta_0 \in U$ such that for every open neighborhood W of ζ_0 the measure of $W \cap U_0$ is positive, otherwise we can cover U by a countable number of sets of measure zero each of which has measure-zero intersection with U_0, contradicting the positivity of the measure of

U_0. Therefore we can find $0 \leq \beta_j \leq \ell$, $\lambda_j \in \Lambda$, and $1 \leq \alpha_j \leq n$ for $1 \leq j \leq m$ such that the rank of the $m \times m$ matrix $(f^{(\beta_j)}_{\alpha_k,\lambda_j}(\zeta_0))_{1\leq j,k\leq m}$ is m, and for some open neighborhood W of ζ_0 the rank of the $m \times m$ matrix $(f^{(\beta_j)}_{\alpha_k,\lambda_j}(\zeta))_{1\leq j,k\leq m}$ is also m for $\zeta \in W$. Thus

$$\operatorname{rank}(f^{(\beta_j)}_{\alpha_k,\lambda_j}(\zeta))_{1\leq j\leq m, 1\leq k\leq n} = m \quad \text{for} \quad \zeta \in W. \tag{6.3}$$

For $\zeta \in W$ the m-vectors $(f^{(\beta_j)}_{\alpha_k,\lambda_j}(\zeta))_{1\leq j\leq m}$ for $1 \leq k \leq m$ form a basis of the vector space $\mathbf{C}^m$. Pick $\alpha_{m+1} \in \{1, \ldots, n\} - \{\alpha_1, \ldots, \alpha_m\}$. Then for $\zeta \in W$ we can find unique complex numbers $c_{\alpha_1}(\zeta), \ldots, c_{\alpha_m}(\zeta)$ such that $f^{(\beta_j)}_{\alpha_{m+1},\lambda_j}(\zeta) = \sum_{k=1}^m c_{\alpha_k}(\zeta) f^{(\beta_j)}_{\alpha_k,\lambda_j}(\zeta)$ for $1 \leq j \leq m$. Set

$$c_{\alpha_{m+1}}(\zeta) \equiv -1 \quad \text{and} \quad c_{\alpha_k}(\zeta) \equiv 0 \quad \text{for} \quad k \in \{1, \ldots, n\} - \{\alpha_1, \ldots, \alpha_m\}. \tag{6.4}$$

Then we have holomorphic functions $c_1(\zeta), \ldots, c_n(\zeta)$ not all zero for $\zeta \in W$ such that

$$\sum_{k=1}^n c_k(\zeta) f^{(\beta_j)}_{k,\lambda_j}(\zeta) = 0 \quad \text{for} \quad \zeta \in W \quad \text{and} \quad 1 \leq j \leq m. \tag{6.5}$$

Since for $\beta \leq \ell + 1$ the n-vector $\mathbf{v}(\beta, \lambda, \zeta)$ is a linear combination of the vectors $\mathbf{v}(\beta_1, \lambda_1, \zeta), \ldots, \mathbf{v}(\beta_m, \lambda_m, \zeta)$ for $\zeta \in W \cap U_0$, it follows from the positivity of the measure of $W \cap U_0$ that for $\beta \leq \ell+1$ the n-vector $\mathbf{v}(\beta, \lambda, \zeta)$ is a linear combination of the vectors $\mathbf{v}(\beta_1, \lambda_1, \zeta), \ldots, \mathbf{v}(\beta_m, \lambda_m, \zeta)$ for $\zeta \in W$. Thus by (6.5) we have

$$\sum_{k=1}^n c_k(\zeta) f^{(\beta)}_{k,\lambda_j}(\zeta) = 0 \quad \text{for} \quad \zeta \in W \quad \text{and} \quad \beta \leq \ell + 1. \tag{6.6}$$

By differentiating (6.6) with respect to $\zeta \in W$ for $\beta \leq \ell$ and by using (6.4) and (6.6), we conclude that $\sum_{k=1}^m c'_{\alpha_k}(\zeta) f^{(\beta)}_{\alpha_k,\lambda_j}(\zeta) = 0$ for $\beta \leq \ell$. Since $\beta_j \leq \ell$ $(1 \leq j \leq m)$, it follows from (6.3) that $c'_{\alpha_k}(\zeta) \equiv 0$ on W and thus $c_{\alpha_k}(\zeta)$ is constant on W for $1 \leq k \leq m$. Finally, the identical vanishing of $\sum_{k=1}^n c_k f_{k,\lambda}(\zeta)$ on U for $\lambda \in \Lambda$ follows from (6.6). □

Proof of Lemma A6.1.4:

Proof. Let $z_1, \cdots, z_m$ be the coordinates of the universal cover $\mathbf{C}^m$ of A. Then the global differentials $dz_1, \ldots, dz_m$ are well-defined on A. Let ι :

$X \to A$ be the inclusion. Then there is a simply connected neighborhood U of $x_0 = f(0) \in X$ such that for suitable indices $1 \le i_1 < i_2 < \ldots < i_n \le m$

$$\iota^*(dz_{i_1} \wedge \ldots \wedge dz_{i_n}) \neq 0 \quad \text{on} \quad U.$$

Without loss of generality we can assume that $i_j = j$ for $1 \le j \le n$. Taking U smaller if necessary, we have a simple connected neighborhood U' of x_0 in A such that $U' \cap X = U$ and if we put $x^i(P) = \int_{x_0}^P dz_i, 1 \le i \le m$ in U' then they form a holomorphic coordinate system in U' and there are holomorphic functions $F_\alpha(x^1, \ldots, x^n), n+1 \le \alpha \le m$, such that

$$U' \cap X = \{(x^1, \ldots, x^m) \in U';\ x^\alpha = F_\alpha(x^1, \ldots, x^n),\ n+1 \le \alpha \le m\}.$$

Therefore $(x^1, \ldots, x^n)$ restricted over U is a holomorphic local coordinate system in U and

$$\iota^*(dz_\alpha) = d(x^\alpha|_U) = \sum_{i=1}^n \frac{\partial F_\alpha}{\partial x^i} dx^i, \qquad n+1 \le \alpha \le m. \tag{6.7}$$

Using this coordinate system, we identify $J_n(U)$ with $U \times \mathbf{C}^{n^2}$. By (6.1), $\Phi : U \times \mathbf{C}^{n^2} \to \mathbf{C}^{mn}$ is given by

$$\Phi : (P, (\xi_{i\alpha})_{1 \le i \le n, 1 \le \alpha \le n}) \to (\xi_{i\alpha})_{1 \le i \le n, 1 \le \alpha \le m}, \tag{6.8}$$

where, for $\alpha \ge n+1$, $\xi_{i\alpha}$ is a polynomial in $\xi_{i\beta}, 1 \le i \le n, 1 \le \beta \le \alpha$, whose coefficients are partial derivatives of F_α of order $\le \alpha$ in $x^1, \ldots, x^n$. Take $\zeta_0 \in \mathbf{C}$ close to the origin. By the assumption there exists a nonzero tangent vector $\mathbf{T}$ to $J_n(X)$ at $d^n f(\zeta_0)$ such that $d\Phi(\mathbf{T}) = 0$. The tangent vector $\mathbf{T}$ of $J_n(X)$ at $d^n f(\zeta_0)$ is represented by an integral curve, i.e., there exists a neighborhood W of ζ_0, a positive number ϵ, and a holomorphic curve $G : W \times (-\epsilon, \epsilon) \to X$ such that

$$g_0(\zeta) = f(\zeta), \quad \frac{d}{dt}(d^n g_t(\zeta_0))\bigg|_{t=0} = \mathbf{T},$$

where $g_t(\zeta) = G(\zeta, t)$. So

$$0 = d\Phi(\mathbf{T}) = \frac{d}{dt}\Phi(d^n g_t(\zeta_0))\bigg|_{t=0}. \tag{6.9}$$

By definition A6.1.3,

$$d^n g_t(\zeta_0) = \left(\frac{d^j}{d\zeta^j}(z_\alpha \circ g_t)(\zeta_0)\right)_{1\le j\le n, 1\le\alpha\le n}. \tag{6.10}$$

Combining (6.8), (6.9) and (6.10) yields, for $1 \le \alpha \le m$,

$$\frac{d}{dt}\left(\frac{d^j}{d\zeta^j}(z_\alpha \circ g_t)(\zeta_0)\right)\Bigg|_{t=0} = 0, \quad 1 \le j \le n. \tag{6.11}$$

For $n+1 \le \alpha \le m$ (6.11) means, for $1 \le j \le n$,

$$0 = \frac{d}{dt}\left(\frac{d^j}{d\zeta^j}(z_\alpha \circ g_t)(\zeta_0)\right)\Bigg|_{t=0} = \frac{\partial^{j+1}}{\partial\zeta^j\partial t}(z_\alpha \circ G)\Bigg|_{(\zeta_0,0)}. \tag{6.12}$$

Let $F_\alpha^\nu = \frac{\partial F_\alpha}{\partial z_\nu}$. Then, by (6.7),

$$dz_\alpha = \sum_{\nu=1}^{n} F_\alpha^\nu dz_\nu, \quad n+1 \le \alpha \le m.$$

Thus (6.12) becomes, for $n+1 \le \alpha \le m$,

$$\begin{aligned} 0 = \frac{\partial^{j+1}}{\partial\zeta^j\partial t}(z_\alpha \circ G)\Bigg|_{(\zeta_0,0)} &= \left(\frac{\partial^j}{\partial\zeta^j}\sum_{\nu=1}^{n}(F_\alpha^\nu \circ f)\cdot\frac{\partial}{\partial t}(z_\nu \circ G)\right)\Bigg|_{(\zeta_0,0)} \\ &= \sum_{\nu=1}^{n}\sum_{\lambda=0}^{j}\binom{j}{\lambda}\frac{d^{j-\lambda}(F_\alpha^\nu \circ f)}{d\zeta^{j-\lambda}}\Bigg|_{\zeta_0}\cdot\frac{\partial^{\lambda+1}}{\partial\zeta^\lambda\partial t}(z_\nu \circ G)\Bigg|_{(\zeta_0,0)}. \end{aligned} \tag{6.13}$$

However, from (6.11), for $1 \le \lambda \le n$ and $1 \le \nu \le n$,

$$\frac{\partial^{\lambda+1}}{\partial\zeta^\lambda\partial t}(z_\nu \circ G)\Bigg|_{(\zeta_0,0)} = 0.$$

Thus (6.13) yields, for $n+1 \le \alpha \le m$,

$$0 = \sum_{\nu=1}^{n}\frac{d^j(F_\alpha^\nu \circ f)}{d\zeta^j}\Bigg|_{\zeta_0}\cdot\frac{\partial}{\partial t}(z_\nu \circ G)\Bigg|_{(\zeta_0,0)}.$$

Since the n complex numbers $\frac{\partial}{\partial t}(z_\nu \circ G)\Big|_{(\zeta_0,0)}$ $(1 \le \nu \le n)$ are not all zero due to the non-triviality of $\mathbf{T}$, it follows from Lemma A6.1.5 that there exist constants c_ν $(1 \leq \nu \leq n)$ not all zero such that $\sum_{\nu=1}^{n} c_\nu \frac{d}{d\zeta}(F_\alpha^\nu \circ f) \equiv 0$ for $n+1 \leq \alpha \leq m$. So $\sum_{\nu=1}^{n} c_\nu F_\alpha^\nu \circ f \equiv$ constant for $n+1 \leq \alpha \leq m$. Denote this constant by $-c_\alpha$. Then

$$c_\alpha + \sum_{\nu=1}^{n} c_\nu F_\alpha^\nu \circ f \equiv 0$$

on $\mathbf{C}$. Since the Zariski closure of the image of f is dense in X, we have

$$c_\alpha + \sum_{\nu=1}^{n} c_\nu F_\alpha^\nu \equiv 0$$

on X for $n+1 \leq \alpha \leq m$. Note that $F_\alpha^\nu = \frac{\partial F_\alpha}{\partial z_\nu}$, so

$$c_\alpha + \sum_{\nu=1}^{n} c_\nu \frac{\partial F_\alpha}{\partial z_\nu} \equiv 0$$

on X for $n+1 \leq \alpha \leq m$. Consider the vector field $v = \sum_{i=1}^{m} c_i \frac{\partial}{\partial z^i}$ defined on A, then, for $n+1 \leq \alpha \leq m$,

$$v(z^\alpha - F_\alpha) \equiv 0 \tag{6.14}$$

on X. Note that X is defined by

$$U' \cap X = \{(x^1, \ldots, x^m) \in U';\ x^\alpha = F_\alpha(x^1, \ldots, x^n), n+1 \leq \alpha \leq m\}.$$

(6.14) means that the vector field $v = \sum_{i=1}^{m} c_i \frac{\partial}{\partial z^i}$ is, when restricted to X, actually in the tangent space $T(X)$ of X. Thus the exponential map $\exp(vt)$ leaves X invariant by the translation for any $t \in \mathbf{C}$. Clearly $\{\exp(vt); t \in \mathbf{C}\}$ is a one-parameter subgroup of A. This proves Lemma A6.1.4. □

Next we use Nevanlinna theory to prove the following Lemma.

Lemma A6.1.6 *Let $f : \mathbf{C} \to A$ be a holomorphic map. Denote by X the Zariski closure of the image of f. If $\dim X > 0$ then for every $\zeta \in \mathbf{C}$ the differential $d\Phi$ of Φ at $d^n f(\zeta)$ is not injective.*

To prove Lemma A6.1.6, we need the following well-known theorem from algebraic geometry.

Theorem A6.1.7 *Let Y_1, Y_2 be complex quasi-projective varieties of the same dimension and $f : Y_1 \to Y_2$ be a dominate rational mapping. Then the rational function field $\mathbf{C}(Y_1)$ is a finite algebraic extension of the rational function field $\mathbf{C}(Y_2)$ with extension degree $[\mathbf{C}(Y_1); \mathbf{C}(Y_2)]$, where f is said to be dominate if $f(W)$ is Zariski dense in Y_2 for some non-empty Zariski open subset $W \subset Y_1$. Note that we regard $\mathbf{C}(Y_2)$ as a subfield of $\mathbf{C}(Y_1)$ since for every function $F \in \mathbf{C}(Y_2)$, $f^*F \in \mathbf{C}(Y_1)$.*

Proof of Lemma A6.1.6.

Proof. Suppose the contrary, that is there exists $\zeta_0 \in \mathbf{C}$ such that the differential $d\Phi$ of Φ at $d^n f(\zeta_0)$ is injective. We are going to derive a contradiction. Without loss of generality, we may assume that $\zeta_0 = 0$.

Let A be an Abelian variety whose universal cover is $\mathbf{C}^m$ with coordinates $z_1, \cdots, z_m$. Let $d^n f : \mathbf{C} \to J_n X$ be the n-th lifting map. Let Z be the Zariski closure of $d^n(f)(\mathbf{C})$ in $J_n(X)$. Let $p : J_n(X) \to X$ be the bundle projection. Take an embedding $A \subset \mathbf{P}^N$ and take a homogeneous coordinate system $[u^0 : \cdots : u^N]$ of $\mathbf{P}^N$ so that $X \not\subset \{u^0 = 0\}$. Put

$$w^i = \frac{u^i}{u^0}, 1 \leq i \leq N.$$

It follows that, by Corollary A3.1.2,

$$T_{w^i \circ f}(r) \leq T_f(r) + O(1) \leq \sum_{i=1}^{N} T_{w^i \circ f}(r) + O(1). \tag{6.15}$$

Let V be the Zariski closure of $\Phi \circ d^n(f)(\mathbf{C})$ in $\mathbf{C}^{mn}$. Since, by the assumption, the differential $d\Phi$ of Φ at $d^n f(0)$ is injective, $\dim Z = \dim V$ and $\Phi|_Z : Z \to V$ is dominate. Since w^i are rational functions on X, the pull-backs $p^* w^i, 1 \leq i \leq N$, are rational functions on Z. By Theorem A6.1.7, there are algebraic relations, for $1 \leq i \leq N$,

$$(P_{i0} \circ \Phi)(p^* w^i)^{d_i} + (P_{i1} \circ \Phi)(p^* w^i)^{d_i - 1} + \ldots + (P_{id_i} \circ \Phi) = 0, \tag{6.16}$$

where P_{ij} are homogeneous polynomials on $\mathbf{C}^{mn}$ and $P_{i0} \circ \Phi \not\equiv 0$ on Z. However, $P_{ij} \circ \Phi \circ d^n(f)$ are polynomials in $\dfrac{d^j}{d\zeta^j}(z_\alpha \circ f), 1 \leq j \leq n, 1 \leq \alpha \leq m$. By (6.16), we have

$$P_{i0}(\{\frac{d^j}{d\zeta^j}(z_\alpha \circ f)\}) \not\equiv 0$$

and

$$P_{i0}(\{\frac{d^j}{d\zeta^j}(z_\alpha \circ f)\})(f^*w^i)^{d_i} + P_{i1}(\{\frac{d^j}{d\zeta^j}(z_\alpha \circ f)\})(f^*w^i)^{d_i-1} + \dots$$
$$+P_{id_i}(\{\frac{d^j}{d\zeta^j}(z_\alpha \circ f)\}) = 0.$$

Thus, by Theorem A1.1.6,

$$T_f(r) \le C \max_{1\le\alpha\le m, 1\le j\le k} T_{\frac{d^j}{d\zeta^j}(z_\alpha\circ f)}(r) + O(1). \tag{6.17}$$

Since $dd^c \sum_{\alpha=1}^m |z_\alpha|^2$ is a Kähler form on A, it follows that, from Theorem A5.2.8 (Green-Jensen's formula),

$$\begin{aligned} T_f(r) &= \int_0^r \frac{dt}{t}\int_{|z|\le t} f^*(dd^c \sum_{\alpha=1}^m |z_\alpha|^2) \\ &= \frac{1}{2}\int_0^{2\pi}\left(\sum_{\alpha=1}^m |z_\alpha \circ f(re^{i\theta})|^2\right)\frac{d\theta}{2\pi} + O(1) \\ &= \frac{1}{2}\int_0^{2\pi}\left(\sum_{\alpha=1}^m (\log^+ \exp|z_\alpha \circ f(re^{i\theta})|)^2\right)\frac{d\theta}{2\pi} + O(1). \end{aligned} \tag{6.18}$$

On the other hand, the logarithmic derivative lemma gives us

$$T_{\frac{d^k}{d\zeta^k}(z_\alpha\circ f)}(r) \le O(\log T_{\exp(z_\alpha\circ f)}(r)). \tag{6.19}$$

Moreover, by the First Main Theorem,

$$\begin{aligned} &T_{\exp(z_\alpha\circ f)}(r) = m_{\exp(z_\alpha\circ f)}(r,\infty) + O(1) \\ &= \int_0^{2\pi} \log^+ \exp|z_\alpha \circ f(re^{i\theta})|\frac{d\theta}{2\pi} + O(1) \\ &\le \left(\int_0^{2\pi} (\log^+ \exp|z_\alpha \circ f(re^{i\theta})|)^2 \frac{d\theta}{2\pi}\right)^{1/2} + O(1). \end{aligned} \tag{6.20}$$

Hence, by (6.18), (6.19) and (6.20),

$$\begin{aligned} &\max_{1\le\alpha\le m} T_{\frac{d^k}{d\zeta^k}(z_\alpha\circ f)}(r) \\ &\le O(\log \max_{1\le\alpha\le m}\int_0^{2\pi} (\log^+ \exp|z_\alpha \circ f(re^{i\theta})|)^2 \frac{d\theta}{2\pi}) + O(1) \end{aligned}$$

$$\begin{aligned} &\leq O(\log \int_0^{2\pi} \left(\sum_{\alpha=1}^{m} (\log^+ \exp |z_\alpha \circ f(re^{i\theta})|)^2 \right) \frac{d\theta}{2\pi}) + O(1) \\ &= O(\log T_f(r)) + O(1). \end{aligned}$$

Combining this with (6.17) we get

$$T_f(r) \leq O(\log T_f(r)) + O(1),$$

which implies that $T_f(r) = O(1)$, contradicting f being non-constant. □

Combining Lemma A6.1.4 and Lemma A6.1.6 gives us the following lemma.

Lemma A6.1.8 *Let $f : \mathbf{C} \to A$ be a holomorphic map. Denote by X the Zariski closure of the image of f. Assume that* $\dim X > 0$. *Then there exists a one-parameter subgroup of A which leaves X invariant.*

Bloch's theorem now follows easily from Lemma A6.1.8. Let $f : \mathbf{C} \to A$ be the given holomorphic map. Let X be the Zariski closure of $f(\mathbf{C})$. Assume that X is not a translate of an Abelian sub-variety of A. Let A' be the quotient of the subgroup of all elements whose translates leave X invariant, i.e., if $B = \{a \in A | a + X = X\}$, then $A' = A/B$. By replacing f by its composite with the quotient map $A \to A'$, we can assume without loss of generality that X is not invariant by the translate of any subgroup of A with positive dimension. This contradicts Lemma A6.1.8. This proves Bloch's Theorem.

A6.2 The Proof of Lang's Conjecture

In this section, we prove the conjecture of Lang that the complement of an ample divisor in an Abelian variety is hyperbolic. The proof is due to Siu-Yeung [Siu-Y2].

Theorem A6.2.1 (Siu-Yeung) *Let A be an Abelian variety and D be an ample in A. Let $f : \mathbf{C} \to A$ be a non-constant holomorphic map. Then the image of f must intersect D.*

To prove Theorem A6.2.1, we assume, in contrast, that $f : \mathbf{C} \to A - D$. We note that we can assume that the image of f is Zariski dense in A. Otherwise, by Bloch's theorem, there is a positive dimensional Abelian sub-variety B of A and a point $a \in A$ such that the Zariski closure of $f(\mathbf{C})$ is $B + a$. Let $g : \mathbf{C} \to B$ be defined by $g(\zeta) = f(\zeta) - a$. Since $B \cap (D - a)$ is an ample divisor in B, the problem is reduced to study g, and the image of g is Zariski dense in B.

The proof of Theorem A6.2.1 uses essentially the same techniques of Bloch presented in the previous section. The key point here in proving Lang's conjecture is the explicit construction of a suitable log-pole higher order jet differential in terms of the theta function. To describe it we first introduce the concept of the jet differential.

Recall that $J_k(M) = \cup_{P \in M} J_k(M)_P$ is a bundle over M, where, for each point $P \in M$, every element $v \in J_k(M)_P$ is a set of complex numbers $(\xi_{j\alpha})_{1 \le j \le k, 1 \le \alpha \le n}$ with respect to a local coordinates z_α $(1 \le \alpha \le n)$ of M in a neighborhood of P. Define $d^j z_\alpha$ such that $d^j z_\alpha(v) = \xi_{j\alpha}$ for $v = (\xi_{j\alpha})_{1 \le j \le k, 1 \le \alpha \le n}$.

Definition A6.2.2 *A k-jet differential ω of total weight m (respectively meromorphic k-jet differential) on a complex manifold M assigns, at each point $P \in M$, a function $\omega(P)$ on $J_k(M)_P$ such that, with local coordinates $z_1, \cdots, z_n$, ω is locally a polynomial, with holomorphic (respectively meromorphic) functions as coefficients, in the variables $d^\ell z_j$ $(1 \le \ell \le k, 1 \le j \le n)$ and of homogeneous weight m when $d^\ell z_j$ is given the weight ℓ. A meromorphic k-jet differential ω is said to be a log-pole k-jet differential if it is locally a polynomial, with holomorphic functions as coefficients, in the variables $d^\ell z_j, d^\nu \log g_\lambda$ $(1 \le \ell \le k, 1 \le j \le n, 1 \le \nu \le k, 1 \le \lambda \le \Lambda)$, where g_λ $(1 \le \lambda \le \Lambda)$ are local holomorphic functions whose zero-divisors are contained in a finite number of global nonnegative divisors of M.*

Thus, in terms of local coordinates $z_1, \ldots, z_n$, a meromorphic k-jet differential of total weight m is expressed in the form

$$\omega = \sum_\nu \omega_{\nu_{1,1}\cdots\nu_{1,k}\cdots\nu_{n,1}\cdots\nu_{n,k}} (dz_1)^{\nu_{1,1}} \cdots (d^k z_1)^{\nu_{1,k}} \cdots (dz_n)^{\nu_{n,1}} \cdots (d^k z_n)^{\nu_{n,k}} \tag{6.21}$$

where the summation is over the kn-tuple

$$\nu = (\nu_{1,1} \cdots \nu_{1,k} \cdots \nu_{n,1} \cdots \nu_{n,k})$$

of nonnegative integers with

$$(\nu_{1,1} + 2\nu_{1,2} + \cdots + k\nu_{1,k}) + \cdots + (\nu_{n,1} + 2\nu_{n,2} + \cdots + k\nu_{n,k}) = m$$

and $\omega_{\nu_{1,1}\cdots\nu_{1,k}\cdots\nu_{n,1}\cdots\nu_{n,k}}$ is a meromorphic function locally defined. A k-jet differential can be naturally regarded as a function on $J_k(X)$ and also naturally regarded as a function on $J_l(X)$ when $k \le l$ by the composition with the forgetful projection $\pi_{k,l} : J_l(X) \to J_k(X)$.

We now describe the explicit construction of a suitable log-pole higher order jet differential in terms of the theta function. Let $f : \mathbf{C} \to A$ be the given holomorphic map whose image is Zariski dense in A. Assume that $\dim A = n$. Choose a coordinate system $(z_1, \ldots, z_n)$ of $\mathbf{C}^n$, so that the lifting $\tilde{f} : \mathbf{C} \to \mathbf{C}^n$ of f is expressed as

$$\tilde{f} = (f_0, \ldots, f_q, 0, \ldots, 0)$$

with entire functions $f_1, \ldots, f_q$ being linearly independent over $\mathbf{C}$. Let θ be a theta function defining the ample divisor D. The locally defined $q+1$-jet differential

$$\Theta = det \begin{pmatrix} d\log\theta, dz_1, dz_2, \ldots, dz_q \\ d^2\log\theta, d^2z_1, d^2z_2, \ldots, d^2z_q \\ \cdots \\ d^{q+1}\log\theta, d^{q+1}z_1, d^{q+1}z_2, \ldots, d^{q+1}z_q \end{pmatrix}$$

gives a well-defined function Θ on the Zariski closure $X_{q+1}(f)$ of $d^{q+1}f(\mathbf{C})$ in $J_{q+1}A$. We define an algebraic subbundle $J_k(A)'$ of $J_k(A)(k = 0, 1, \ldots)$ by equations

$$d^i z_j = 0, \quad 1 \le i \le k, \ q+1 \le j \le n.$$

Then $X_{q+1}(f) \subset J_{q+1}(A)'$. Recall that from the discussion in Section A5.1, $J_k(A) = A \times \mathbf{C}^{nk}$ is a trivial bundle. So $J_k(A)' = A \times \mathbf{C}^{qk}$. Let $p : J_k(A)' = A \times \mathbf{C}^{qk} \to \mathbf{C}^{qk}$ be the projection and let $p|_{X_k(f)}$ be the restriction of p to $X_k(f)$. Note that Θ is a well-defined function on $X_{q+1}(f)$. Taking the derivatives of Θ, we have

$$d^l\Theta : J_l(X_{q+1}(f)) \to \mathbf{C}, \qquad l = 0, 1, 2, \ldots.$$

We also note that $J_l(X_{q+1}(f)) = X_{q+1+l}(f)$ and denote by π_l the projection from $J_{q+n+1}(X)$ to $J_{q+n+1-l}(X)$. So we can define a new map

$$\Psi = (p|_{X_{q+1+n}(f)}, \Theta\circ\pi_n, d\Theta\circ\pi_{n-1}, \ldots, d^n\Theta\circ\pi_0) : X_{q+1+n}(f) \to \mathbf{C}^{q(q+1+n)+n+1}. \tag{6.22}$$

We also denote by $\Omega_{f,k}$ the set of all meromorphic k-jet differentials on A which vanishes identically on $X_k(f)$.

Similar to Lemma A6.1.8, we shall prove the following lemma.

Lemma A6.2.3 *If* $\dim X_{q+1}(f) > 0$, *then there is some non-zero constant vector field* $v = \sum_{\alpha=1}^n c_\alpha\partial/\partial z_\alpha$ *on* A *such that* $(\sum_{\alpha=1}^n c_\alpha\partial/\partial z_\alpha)\omega$ *vanishes identically on* $X_{q+1}(f)$ *for every* $\omega \in \Omega_{f,q+1}$ *with pole set contained in* D *and* $(\sum_{\alpha=1}^n c_\alpha\partial/\partial z_\alpha)\Theta$ *is also identically zero on* $X_{q+1}(f)$.

Proof. The proof of Lemma A6.2.3 breaks down into two steps. The first step is similar to that of Lemma A6.1.7. We prove that for every $\zeta \in \mathbf{C}$, the differential $d\Phi$ of Ψ at the point $d^{q+1+n}f(\zeta)$ is not injective. The same argument carries over. The only thing that needs to be modified since Θ now is involved, is that in carrying the estimation we need to control the growth of the theta function θ. In fact, we always have

$$|\theta(z)| \leq exp(C'|z|^2) + C''$$

for some positive constants C'' and C''. By using the above estimate, the same argument applies.

The next step is similar to the proof of Lemma A6.1.4, where X is replaced by $X_{q+1}(f)$ and Φ is replaced by Ψ. We will prove that if for every $\zeta \in \mathbf{C}$ the differential $d\Psi$ of Ψ at the point $d^{q+1+n}f(\zeta)$ is not injective, then there is some non-zero constant vector field $v = \sum_{\alpha=1}^n c_\alpha\partial/\partial z_\alpha$ on A such that $(\sum_{\alpha=1}^n c_\alpha\partial/\partial z_\alpha)\omega$ vanishes identically on $X_{q+1}(f)$ for every $\omega \in \Omega_{f,q+1}$ with pole set contained in D and $(\sum_{\alpha=1}^n c_\alpha\partial/\partial z_\alpha)\Theta$ is also identically zero on $X_{q+1}(f)$. We still use the method of integral curves to prove this statement. Assume that the statement is not true. Take $\zeta_0 \in \mathbf{C}$ close to the origin, and such that $Q_{\zeta_0} = d^{q+1+n}f(\zeta_0) \in X_{q+1+n}(f)$ is a regular point of $X_{q+1+n}(f)$. Then there exists a non-zero tangent vector T to $X_{q+1+n}(f)$ at Q_{ζ_0} such that $d\Psi(T) = 0$. We now use an integral curve to represent T.

Recall that for a complex manifold M and a point $P \in M$, a tangent vector $T \in T_P(M)$ is represented by a unique holomorphic map $g : \mathbf{C} \to M$ defined in a neighborhood of 0 with $g(0) = P, g'(0) = T$. Such a holomorphic curve is called an integral curve. The existence of g follows from solving the system of ordinary differential equations. In our case, we consider the algebraic variety $X_k(f)$. Recall that $X_k(f)$ is the Zariski closure of $d^k f(\mathbf{C})$ in $J_k(A)$. Denote by $\Omega_{f,k}$ the set of all meromorphic k-jet differentials on A which vanishes identically on the image $\operatorname{Im}(d^k f)$ of $d^k f : \mathbf{C} \to J_k(A)$. Then $X_k(f)$ is the common zero set of all elements of $\Omega_{f,k}$.

Sub-lemma A6.2.4 *Let $f : \mathbf{C} \to A$ be a holomorphic map with Zariski dense image in A. Let $k > n = \dim A$. Then there exists a proper sub-variety of $X_k(f)$ such that for any point Q in $X_k(f)$ outside that sub-variety there exists a holomorphic map $g : U \to A$ for some open neighborhood of 0 with $d^k g(0) = Q$ and $d^k g(\zeta)$ belongs to $X_k(f)$ for $\zeta \in U$.*

Proof. The existence of g follows from a standard argument involving the fundamental theorem of the ordinary differential equation. In fact, let $X_k(f)$ be defined by meromorphic k-jet differentials $\omega_{k,\nu} (1 \leq \nu \leq N_k)$. Note that the fiber $J_k(A)_P$ of $J_k(A)$ over a point P of A can be identified with $\mathbf{C}^{kn}$ with coordinates $\{x_\alpha^{(\nu)}\}_{1 \leq \alpha \leq n, 1 \leq \nu \leq k}$. There exists a proper sub-variety of $X_k(f)$ (after renumbering $\omega_{k,\nu}$) so that, for any point $Q \in X_k(f)$ outside the sub-variety, the rank of the matrix $\left(\frac{\partial}{\partial x_j^{(k)}} \omega_{k,q} \right)_{1 \leq q \leq p, 1 \leq j \leq p}$ is equal to p at Q and is no less than the rank of the matrix $\left(\frac{\partial}{\partial x_j^{(k)}} \omega_{k,q} \right)_{1 \leq q \leq N_k, 1 \leq j \leq n}$ at any point in some neighborhood of Q in $X_k(f)$. Choose local holomorphic k-jet differentials $\omega'_s (p < s \leq n)$ such that the rank of the matrix

$$\left. \begin{pmatrix} \frac{\partial}{\partial x_j^{(k)}} \omega_{k,q} \\ \frac{\partial}{\partial x_j^{(k)}} \omega'_s \end{pmatrix} \right|_{1 \leq q \leq p, p < s \leq n, 1 \leq j \leq n}$$

is equal to n at Q. Now we solve for the n-tuple valued function $g(\zeta)$ as the unknown functions in the system of n ordinary differential equations

$$g^* \omega_{k,\nu}(\{x_j^{(\alpha)}\}) = 0 \quad (0 \leq \alpha \leq k, 1 \leq j \leq n, 1 \leq \nu \leq p),$$
$$g^* \omega'_s(\{x_j^{(\alpha)}\}) = \omega'_s(Q) \quad (0 \leq \alpha \leq k, 1 \leq j \leq n, p < s \leq n),$$

with the initial conditions

$$\frac{\partial^\alpha}{\partial \zeta^\alpha}(z_j \circ g)(0) = x_j^{(\alpha)}(Q) \quad (0 \leq \alpha \leq k, 1 \leq j \leq n).$$

The vanishing of $g^*\omega_{k,\nu}$ with order $\ell_\nu < k$ $(p < \nu \leq N_k)$ follows from the uniqueness of the solution of the system of differential equations which expresses $d^{k-\ell_\nu}\omega_{k,\nu}$ in terms of $\omega_{k,1}, \ldots, \omega_{k,N_k}$. □

Sub-lemma A6.2.5 *Any tangent vector T to $X_k(f)$ at the point Q can be presented as a holomorphic map $g : U \times U \to A$, where U is a neighborhood of 0 in* $\mathbf{C}$, *such that*

(1) $T = \frac{\partial}{\partial t}(d_\zeta^k g)(0, 0)$,

(2) $Q = (d_\zeta^k g)(0, 0)$,

(3) $(d_\zeta^k g)(\zeta, t)$ *belongs to $X_k(f)$ for $\zeta \in U$ and $t \in U$.*

Moreover, if Q is given by $d^k f(0)$, then we can choose g such that $g(\zeta, 0) = f(\zeta)$ for $\zeta \in U$.

Proof. The tangent vector T is represented by a curve in $X_k(f)$ and we simply apply sub-lemma A6.2.4 to each point on the curve. From the proof of sub-lemma A6.2.4 and the holomorphic dependence of the unique solution of the system of ordinary differential equations on the holomorphically varying initial condition, we conclude that the result depends holomorphically on the parameter of the curve.

When Q is given by $d^k f(0)$, in the proof of sub-lemma A6.2.4, the condition $g(\zeta, 0) = f(\zeta)$ for $\zeta \in U$ is satisfied, because of the uniqueness of the solution of the system of ordinary differential equations. □

We now continue the proof of Lemma A6.2.3. For Q_{ζ_0} not in a proper sub-variety of $X_{q+1+n}(f)$, by sub-lemma A6.2.5, we can represent T by a holomorphic map $g : U_{\zeta_0} \times U_0 \to A$, where U_{ζ_0}(respectively U_0) is a neighborhood of ζ_0(respectively 0) in $\mathbf{C}$, such that $g(\zeta, 0) = f(\zeta)$, T is given by $\{\frac{\partial}{\partial t}(\frac{\partial^j g}{\partial \zeta^j})(\zeta_0, 0)\}_{0 \leq j \leq q+1+n}$ and for every fixed $t \in U_0$ the pull-back $g_t^*\omega \equiv 0$ on U_{ζ_0} for every $\omega \in \Omega_{f,q+n+1}$. Note that g is called **an integral curve** of T. By the definition of Ψ, the vanishing of $d\Psi(T)$ means that

$$\frac{\partial}{\partial t}\left(\frac{\partial^\lambda(z_\alpha \circ g)}{\partial \zeta^\lambda}\right)\Bigg|_{(\zeta_0, 0)} = 0 \text{ for } 1 \leq \alpha \leq n \text{ and } 1 \leq \lambda \leq q+1+n \quad (6.23)$$

and

$$\frac{\partial}{\partial t}\left(\frac{\partial^\mu(g_t^*\Theta)}{\partial\zeta^\mu}\right)\Bigg|_{(\zeta_0,0)} = 0 \text{ for } 0 \le \mu \le n-1.$$

The non-vanishing of T means that there exists some $1 \le \alpha \le n$ with $\frac{\partial(z_\alpha \circ g)}{\partial t}\Big|_{(\zeta_0,0)} \neq 0$. Let, for $1 \le \alpha \le n$,

$$\hat{c}_\alpha\Bigg|_{(\zeta_0,0)} = \frac{\partial(z_\alpha \circ g)}{\partial t}\Bigg|_{(\zeta_0,0)}. \tag{6.24}$$

For any $\omega \in \Omega_{f,q+1}$ with pole set in D, we write it in the form of equation (6.21):

$$\begin{aligned}\omega &= \sum_\nu \omega_{\nu_{1,1}\cdots\nu_{1,q+1}\cdots\nu_{n,1}\cdots\nu_{n,q+1}}(dz_1)^{\nu_{1,1}}\cdots(d^{q+1}z_1)^{\nu_{1,q+1}} \\ &\quad\cdots(dz_q)^{\nu_{q+1,1}}\cdots(d^{q+1}z_q)^{\nu_{q+1,q+1}}\end{aligned}$$

where $\omega_{\nu_{1,1}\cdots\nu_{1,q+1}\cdots\nu_{n,1}\cdots\nu_{n,q+1}}$ is a meromorphic function locally defined. Using $g_t^*\omega \equiv 0$ for $t \in U_0$, (6.23) and (6.24), we obtain

$$\begin{aligned}0 &= \frac{\partial^\beta}{\partial\zeta^\beta}\frac{\partial}{\partial t}g_t^*\omega\Bigg|_{t=0} \\ &= \sum_{\nu,\alpha}\hat{c}_\alpha(\zeta_0)\frac{\partial^\beta}{\partial\zeta^\beta}[(\frac{\partial}{z_\alpha}\omega_{\nu_{1,1}\cdots\nu_{1,q+1}\cdots\nu_{n,1}\cdots\nu_{n,q+1}})\circ f\cdot(f^*dz_1)^{\nu_{1,1}}\cdots \\ &\quad (f^*d^{q+1}z_1)^{\nu_{1,q+1}}\cdots(f^*dz_q)^{\nu_{n,1}}\cdots(f^*d^{q+1}z_q)^{\nu_{n,q+1}}]\end{aligned}$$

at $\zeta = \zeta_0$ for $0 \le \beta \le n-1$. Similarly, for Θ, since $\frac{\partial}{\partial t}\frac{\partial^\mu(g_t^*\Theta)}{\partial\zeta^\mu}\Big|_{(\zeta_0,0)} = 0$ for $0 \le \mu \le n-1$, we have

$$\begin{aligned}0 &= \frac{\partial^\beta}{\partial\zeta^\beta}\frac{\partial}{\partial t}g_t^*\Theta\Bigg|_{t=0} \\ &= \sum_{\nu,\alpha}\hat{c}_\alpha(\zeta_0)\frac{\partial^\beta}{\partial\zeta^\beta}[(\frac{\partial}{z_\alpha}\Theta_{\nu_{1,1}\cdots\nu_{1,q+1}\cdots\nu_{n,1}\cdots\nu_{n,q+1}})\circ f\cdot(f^*dz_1)^{\nu_{1,1}}\cdots \\ &\quad (f^*d^{q+1}z_1)^{\nu_{1,q+1}}\cdots(f^*dz_q)^{\nu_{n,1}}\cdots(f^*d^{q+1}z_q)^{\nu_{n,q+1}}]\end{aligned}$$

at $\zeta = \zeta_0$ for $0 \leq \beta \leq n-1$. We note that the key point here is that the same $\hat{c}_\alpha(\zeta_0)$ is used for all $\omega \in \Omega_{f,q+1}$, as well as Θ. We now apply Lemma A6.1.5 to conclude that there exist complex numbers $c_1, \ldots, c_n$ such that

$$\begin{aligned} 0 \quad = \quad & \sum_{\nu,\alpha} c_\alpha (\frac{\partial}{z_\alpha} \omega_{\nu_{1,1}\cdots\nu_{1,q+1}\cdots\nu_{n,1}\cdots\nu_{n,q+1}}) \circ f \cdot (f^* dz_1)^{\nu_{1,1}} \cdots \\ & (f^* d^{q+1} z_1)^{\nu_{1,q+1}} \cdots (f^* dz_q)^{\nu_{n,1}} \cdots (f^* d^{q+1} z_q)^{\nu_{n,q+1}} \end{aligned}$$

for every $\omega \in \Omega_{f,k}$ with pole set in D and

$$\begin{aligned} 0 \quad = \quad & \sum_{\nu,\alpha} c_\alpha (\frac{\partial}{z_\alpha} \Theta_{\nu_{1,1}\cdots\nu_{1,q+1}\cdots\nu_{n,1}\cdots\nu_{n,q+1}}) \circ f \cdot (f^* dz_1)^{\nu_{1,1}} \cdots \\ & (f^* d^{q+1} z_1)^{\nu_{1,q+1}} \cdots (f^* dz_q)^{\nu_{n,1}} \cdots (f^* d^{q+1} z_q)^{\nu_{n,q+1}}. \end{aligned}$$

This simply means $(\sum_{\alpha=1}^n c_\alpha \frac{\partial}{z_\alpha})\omega \equiv 0$ and $(\sum_{\alpha=1}^n c_\alpha \frac{\partial}{z_\alpha})\Theta \equiv 0$ on $X_{q+1}(f)$. So Lemma 6.2.3 is proven. □

We now prove Lang's conjecture:

Proof. By Lemma A6.2.3, there exist constants c_α $(1 \leq \alpha \leq n)$ not all zero such that $f^*(\sum_{\alpha=1}^n c_\alpha \frac{\partial}{\partial z_\alpha})\Theta \equiv 0$ on $\mathbf{C}$. Since Θ is defined in the form of the Wronskian determinant, by Lemma A6.1.5, there exist constants $a_0, a_1, \ldots, a_q$ with $a_0 \neq 0$ such that

$$a_0 f^* d \left((\sum_{\alpha=1}^n c_\alpha \frac{\partial}{\partial z_\alpha}) \log \theta \right) + a_1 df_1 + \ldots + a_q df_q \equiv 0$$

on $\mathbf{C}$. Therefore

$$a_0 d \left((\sum_{\alpha=1}^n c_\alpha \frac{\partial}{\partial z_\alpha}) \log \theta \right) + a_1 dz_1 + \ldots + a_q dz_q \equiv 0 \qquad (6.25)$$

on $X_{q+1}(f)$. Let $\omega = a_0 d \left((\sum_{\alpha=1}^n c_\alpha \frac{\partial}{\partial z_\alpha}) \log \theta \right) + a_1 dz_1 + \ldots + a_q dz_q$, then $\omega \in \Omega_{f,q+1}$. Thus, by Lemma A6.2.3,

$$\left(\sum_{\alpha=1}^n c_\alpha \frac{\partial}{\partial z_\alpha} \right) \omega \equiv 0$$

on $X_{q+1}(f)$. That is

$$d\left((\sum_{\alpha=1}^{n} c_\alpha \frac{\partial}{\partial z_\alpha})^2 \log\theta\right) \equiv 0. \tag{6.26}$$

Since $\left(\sum_{\alpha=1}^{n} c_\alpha \frac{\partial}{\partial z_\alpha}\right)^2 \log\theta$ is a globally well-defined meromorphic function on A and its pull-back by f is identically equal to a constant by (6.26), it follows from the Zariski density of f in A that

$$\left(\sum_{\alpha=1}^{n} c_\alpha \frac{\partial}{\partial z_\alpha}\right)^2 \log\theta \equiv \text{constant}$$

on A.

Now let B be the subgroup of A so that D is invariant under translation by an element of B. By replacing A by the quotient of A with respect to the maximum Abelian sub-variety contained in B, we can assume without loss of generality that B is a finite group. Since B is a finite group, there exists some point P_0 on D such that the orbit W through P_0 of the vector field $(\sum_{\alpha=1}^{n} c_\alpha \frac{\partial}{\partial z_\alpha})$ on A is not completely contained in D. Integrating the equation

$$(\sum_{\alpha=1}^{n} c_\alpha \frac{\partial}{\partial z_\alpha})^2 \log\theta \equiv \text{constant}$$

on W starting from some point $W - D$ two times and exponenting, we conclude that the function is nowhere zero on W, contradicting the choice of P_0 on the divisor D of θ. □

A6.3 McQuillan's Proof

Faltings [Fal2] proved the analog in number theory of both Bloch's Theorem and Lang's Conjecture (see discussion in section B). Faltings' proof is based on Vojta's idea [Voj2] of the existence of line bundles on the product of copies of a sub-variety of an Abelian variety other than the obvious ones from the Abelian variety if the sub-variety is not a translate of an Abelian sub-variety. McQuillan [McQ2] adapted Faltings' proof to function theory and gave a new proof of Bloch's theorem. We present here McQuillan's proof of Bloch's theorem (Theorem A6.1.1).

Let X^m (respectively A^m) be the product of m copies of X (respectively A). Consider the map $p: X^m \to A^{m(m-1)/2}$ defined by $(x_j)_{1\le j\le m}$ going to $(x_j - x_k)_{1\le j<k\le m}$.

If X is the translate $y + B$ of an Abelian sub-variety B of A, then for $b_j \in B$ $(1 \le j \le m)$ the point $(y + b_j + z)_{1\le j\le m}$ of X^m is mapped to the same point $(b_j - b_k)_{1\le j<k\le m}$ of $A^{m(m-1)/2}$ for any $z \in B$. So in that case every fiber of the map $p: X^m \to A^{m(m-1)/2}$ contains a sub-variety biholomorphic to B and has a positive dimension. However, if X is not a translate of an Abelian variety, then we have the following lemma.

Lemma A6.3.1 *Assume that X is not a translate of an Abelian variety. If $m \ge \dim X + 1$, then the map $p: X^m \to A^{m(m-1)/2}$ defined by $(x_j)_{1\le j\le m}$ going to $(x_j - x_k)_{1\le j<k\le m}$ generically has finite fiber.*

Proof. This is a special case of Lemma 5.1 in [Voj7]. □

For any rational number ϵ let $M(\epsilon)$ be the **Q**-bundle $\epsilon\sum_{j=1}^m pr_j^* L + \sum_{1\le j<k\le m}(pr_j - pr_k)^* L$, where $pr_j : A^m \to A$ is the projection onto the jth factor. Using the property in Lemma A6.3.1 that $p: X^m \to A^{m(m-1)/2}$ generically has finite fiber, we can prove the following lemma.

Lemma A6.3.2 *There exist positive numbers ϵ and c such that for a sufficiently divisible d the dimension of $M(-\epsilon)^{\otimes d}$ over X^m is at least $cd^{m \dim X}$. In particular there exists a non-trivial holomorphic section s of $M(-\epsilon)^{\otimes d}$ over X^m.*

Proof. This is a special case of Proposition 6.1 in [Voj7]. □

The following is an inequality concerning the characteristic function of the difference of two rescaled maps:

Lemma A6.3.3 *Let $f: \mathbf{C} \to A$ be a holomorphic map from $\mathbf{C}$ to an Abelian variety A. Let $f_\lambda(z) = f(\lambda z)$. Then*

$$T_{f_\lambda - f_\mu}(r) \le \frac{|\lambda - \mu| r}{(R - |\lambda| r)(R - |\mu| r)} T_f(R) + O(1)$$

when $\max(|\lambda|, |\mu|)r < R$.

Proof. Let $n = \dim A$. Then the universal covering of A is $\mathbf{C}^n$ with coordinates $w_1, \ldots, w_n$. Since $dd^c \sum_{j=1}^n |w_j|^2$ is a Kähler form on A, it follows that, from Theorem A5.2.8 (Green-Jensen's formula),

$$\begin{aligned} T_f(r) &= \int_0^r \frac{dt}{t} \int_{|z|\le t} f^*(dd^c \sum_{j=1}^n |w_j|^2) = \frac{1}{2}\int_0^{2\pi} \left(\sum_{j=1}^n |w_j \circ f(re^{i\theta})|^2\right) \frac{d\theta}{2\pi} \\ &\quad -\frac{1}{2}\sum_{j=1}^n |w_j \circ f(0)|^2. \end{aligned} \tag{6.27}$$

For any $1 \le j \le n$, by Cauchy's integral formula, we have

$$w_j \circ f(\lambda z) - w_j \circ f(\mu z) = \frac{(\lambda - \mu)z}{2\pi\sqrt{-1}} \int_{|\zeta|=R} \frac{w_j \circ f(\zeta)d\zeta}{(\zeta - \lambda z)(\zeta - \mu z)}.$$

So

$$\sum_{j=1}^n |w_j \circ (f_\lambda(z) - f_\mu(z))|^2 \le \frac{|\lambda - \mu|r}{(R - |\lambda|r)(R - |\mu|r)} \int_0^{2\pi} \left(\sum_{j=1}^n |w_j \circ f(Re^{i\theta})|^2\right) \frac{d\theta}{2\pi}.$$

By (6.27), we have

$$T_{f_\lambda - f_\mu}(r) \le \frac{|\lambda - \mu|r}{(R - |\lambda|r)(R - |\mu|r)} T_f(R) + O(1).$$

□

Proof of Bloch's theorem.

Proof. Suppose Bloch's theorem is false, i.e., there is a non-constant holomorphic map f from $\mathbf{C}$ to A such that the Zariski closure of the image of f is X. We are going to derive a contradiction. Let $m > \dim X$. Let $F(z_1, \ldots, z_m) = (f(z_1), \ldots, f(z_m)) : \mathbf{C}^m \to X^m$. Assume without loss of generalization that $F(0) = 0$. By Lemma A6.3.2, there exists a non-trivial holomorphic section s of $M(-\epsilon)^{\otimes d}$ over X^m. In a local trivialization of the line bundle $M(-\epsilon)^{\otimes d}$ over X^m, we can expand F^*s into homogeneous components $F^*s = \sum_{\mu=l}^{\infty} s_\mu$ so that s_l is not identical to zero. Choose nonzero numbers such that

$$s_l(1 + \lambda_1^0, \ldots, 1 + \lambda_m^0) \ne 0.$$

We now fix a sufficient large number r. Recall that $F(z_1, \ldots, z_m) = (f(z_1), \ldots, f(z_m)) : \mathbf{C}^m \to X^m$. Since the image of f is Zariski dense in X, the image of F is Zariski dense in X^m. So we may choose numbers $\lambda_j = 1 + \eta_r \lambda_j^0 / r$, where $\frac{1}{T_f(r)^4} \leq \eta_r \leq \frac{1}{\lambda_j^0 T_f(r)^3}$ for $1 \leq j \leq m$, such that the image of the map $(z_1, \ldots, z_m) \to (f(\lambda_1 z_1), \cdots, f(\lambda_m z_m))$ from $\mathbf{C}^m \to X^m$ does not lie in the divisor of s. Denote by $\tilde{F}$ (or $F_{\lambda_1,\ldots,\lambda_m}$) such that

$$\tilde{F}(z_1, \ldots, z_m) = F_{\lambda_1,\ldots,\lambda_m}(z_1, \ldots, z_m) = (f(\lambda_1 z_1), \cdots, f(\lambda_m z_m)).$$

Then the image of $\tilde{F}$ (or $F_{\lambda_1,\ldots,\lambda_m}$) does not lie in the divisor of s.

Let L be a very ample line bundle over A. Choose a metric on L so that its Chern form is positive definite. We let the metric of $M(-\epsilon)^{\otimes d}$ be induced from that of L. Then the Chern form Ω of $M(-\epsilon)^{\otimes d}$ is equal to $d(-\epsilon \sum_{j=1}^m pr_j^* \omega + \sum_{1 \leq j < k \leq m} (pr_j - pr_k)^* \omega)$, where ω is the Chern form of L. Since the vanishing order of $|\tilde{F}^* s|^2$ at $z = 0$ is ℓ. By Theorem A5.2.7 (Poincare-Lelong formula),

$$\frac{\sqrt{-1}}{2\pi} \partial\bar{\partial}[\log |\tilde{F}^* s / z^\ell|^2] = Z(\tilde{F}^* s / z^\ell) - \tilde{F}^* \Omega.$$

The above equality, together with Theorem A5.2.8 (Jensen-Green's formula), implies that

$$\int_{\mathbf{S}_m(r)} \log |\tilde{F}^* s / z^\ell| \sigma_m - \log |\tilde{F}^* s / z^\ell|(0) = -T_{\tilde{F}}(r, \Omega) + N_{\tilde{F}}(r, s/z^\ell), \quad (6.28)$$

where

$$T_{\tilde{F}}(r, \Omega) = \int_0^r \frac{dt}{t^{2m-1}} \int_{\mathbf{C}_m[r]} \tilde{F}^* \Omega \wedge v_m^{m-1}$$

and

$$N_{\tilde{F}}(r, s/z^\ell) = \int_0^r \frac{dt}{t^{2m-1}} \int_{Z(\tilde{F}^* s/z^\ell) \cap \mathbf{C}_m[r]} v_m^{m-1}.$$

Thus (6.28) implies that

$$T_{\tilde{F}}(r, \Omega) = N_{\tilde{F}}(r, s/z^\ell) - \int_{\mathbf{S}_m(r)} \log |\tilde{F}^* s| \sigma_m - \ell \log r + \log |\tilde{F}^* s / z^\ell|(0).$$

We re-scale the metric of the line bundle, so that

$$\int_{\mathbf{S}_m(r)} \log |\tilde{F}^* s| \sigma_m \leq 0.$$

Then

$$T_{\tilde{F}}(r, \Omega) \geq -\ell \log r + \log |\tilde{F}^* s/z^\ell|(0).$$

On the other hand,

$$T_{\tilde{F}}(r, \Omega) = -\epsilon \sum_{j=1}^{m} dT_{f_{\lambda_j}}(r) + \sum_{1 \leq j < k \leq m} d(T_{f_{\lambda_j}}(r) - T_{f_{\lambda_k}}(r)),$$

where $f_{\lambda_j}(z) = f(\lambda_j z)$ and

$$T_{f_{\lambda_j}}(r) = T_{f_{\lambda_j}, \omega}(r) = \int_0^r \frac{dt}{t} \int_{\mathbf{C}[r]} f_{\lambda_j}^* \omega.$$

So

$$\begin{aligned} &-\epsilon \sum_{j=1}^{m} dT_{f_{\lambda_j}}(r) + \sum_{1 \leq j < k \leq m} d(T_{f_{\lambda_j}}(r) - T_{f_{\lambda_k}}(r)) \\ &\geq \quad -\ell \log r + \log |\tilde{F}^* s/z^\ell|(0). \end{aligned} \tag{6.29}$$

Using Lemma A6.3.3, we have

$$(T_{f_{\lambda_j}}(r) - T_{f_{\lambda_k}}(r)) \leq \frac{|\lambda_j - \lambda_k| r}{(R - |\lambda_j| r)(R - |\lambda_k| r)} T_f(R) + O(1).$$

Take $R = r + \frac{1}{T_f(r)}$. Recall that $\lambda_j = 1 + \eta_r \lambda_j^0 / r$, and $\frac{1}{T_f(r)^4} \leq \eta_r \leq \frac{1}{\lambda_j^0 T_f(r)^3}$ for $1 \leq j \leq m$. Then

$$R - |\lambda_j| r \geq R - r - \eta_r |\lambda_j^0| \geq \frac{1}{T_f(r)} - \frac{1}{T_f(r)^3} \geq \frac{1}{2T_f(r)}$$

and $|\lambda_j - \lambda_k| r = |\lambda_j^0 - \lambda_k^0| \eta_r \leq \frac{2}{T_f(r)^3}$. So, using Lemma A1.2.4,

$$\begin{aligned} (T_{f_{\lambda_j}}(r) - T_{f_{\lambda_k}}(r)) &\leq \frac{|\lambda_j - \lambda_k| r}{(R - |\lambda_j| r)(R - |\lambda_k| r)} T_f(R) + O(1) \\ &\leq \frac{8}{T_f(r)} \cdot T_f\left(r + \frac{1}{T_f(r)}\right) + O(1) \leq \frac{8}{T_f(r)} \cdot 2T_f(r) + O(1) = O(1), \end{aligned}$$

where the inequality holds for all r outside a set $E \subset (0, +\infty)$ with finite Lebesgue measure. Similarly

$$(T_{f_{\lambda_j}}(r) - T_f(r)) \leq O(1),$$

where the inequality holds for all r outside a set $E \subset (0, +\infty)$ with finite Lebesgue measure. Thus

$$\begin{aligned} T_{\tilde{F}}(r, \Omega) &= -\epsilon \sum_{j=1}^{m} dT_{f_{\lambda_j}}(r) + \sum_{1 \le j < k \le m} d(T_{f_{\lambda_j}}(r) - T_{f_{\lambda_k}}(r)) \\ &= -\epsilon dm T_f(r) + O(1), \end{aligned} \tag{6.30}$$

where the equation holds for all r outside a set $E \subset (0, +\infty)$ with finite Lebesgue measure. We now compute $(\tilde{F}^*_{\lambda_1, \ldots, \lambda_m} s/z^\ell)(0)$. In fact

$$(\tilde{F}^*_{\lambda_1, \ldots, \lambda_m} s/z^\ell)(0) = s_\ell(\lambda_1, \cdots, \lambda_m) = s_\ell(1 + \eta_r \lambda_1^0/r, \cdots, 1 + \eta_r \lambda_m^0/r).$$

Consider the expansion of the function $s_\ell(1+\lambda_1^0 t, \cdots, 1+\lambda_m^0 t)$ as a function of t at $t = 0$. We get

$$s_\ell(1 + \lambda_1^0 t, \cdots, 1 + \lambda_m^0 t) = \varphi_p t^p + \varphi_{p+1} t^{p+1} + \cdots + \varphi_\ell t^\ell$$

with $0 \ne \varphi_p \in \mathbf{C}$. Then, since $\frac{1}{T_f(r)^4} \le \eta_r \le \frac{1}{\lambda_j^0 T_f(r)^3}$ for $1 \le j \le m$,

$$\begin{aligned} |s_\ell(\lambda_1, \cdots, \lambda_m)| &= |s_\ell(1 + \eta_r \lambda_1^0/r, \cdots, 1 + \eta_r \lambda_m^0/r)| \\ &\ge \frac{1}{2} \varphi_p \left(\frac{1}{r T_f(r)^4}\right)^p. \end{aligned}$$

Finally, from (6.29) and (6.30) and the above inequality, we have

$$-\epsilon dm T_f(r) + O(1) \ge -\ell \log r + \log \frac{1}{2} \varphi_p \left(\frac{1}{r T_f(r)^4}\right)^p,$$

where the inequality holds for all r outside a set $E \subset (0, +\infty)$ with finite Lebesgue measure. This gives the estimate $T_f(r) = O(\log r)$ for all r outside a set $E \subset (0, +\infty)$ with finite Lebesgue measure. However, since

$$f^* \omega = \sum_{\alpha, \beta} h_{\alpha \overline{\beta}} f^{\alpha_z} \overline{f}^{\beta_z} \frac{\sqrt{-1}}{2\pi} dz \wedge d\bar{z}$$

and $\sum_{\alpha, \beta} h_{\alpha \overline{\beta}} f^{\alpha_z} \overline{f}^{\beta_z}$ is plurisubharmonic and not identically zero, we know that $T_f(r) \ge O(r^2)$.Thus we have a contradiction. □

Part B: Diophantine Approximation

B6.1 Faltings' Theorem on Rational Points in Abelian Varieties

In 1990 and 1991, Faltings produced two papers in which he proved the following two fascinating theorems.

Theorem B6.1.1 *Let A be an Abelian variety defined over a number field k. Let X be a sub-variety of A, also defined over F. Then the set $X(k)$ of k-rational points in X is contained in a finite union of translated Abelian sub-variety of X.*

Theorem B6.1.2 *Let A be an Abelian variety defined over a number field k and E a sub-variety, also defined over F. Let h be a height on A and v is a valuation on k. Let $\epsilon > 0$. Then we have*

$$\lambda_{v,E}(P) < \epsilon h(P)$$

for almost every point $P \in A(k) - E$.

We will not prove these two theorems since they involve more sophisticated theory. Instead, we will outline the idea. The proof of Faltings' theorems is a higher dimensional generalization of Vojta's proof of Roth's theorem (see [Fal2]). Basically it can be summarized as follows. Assume that X does not contain translates of Abelian subvariety of A and assume that $X(k)$ is infinite. First of all we fix a very ample symmetric line bundle $\mathcal{L}$ over A, and the norms on $\mathcal{L}$ at the Archimedean places of k. Let m be a sufficiently large integer. There exists $x = (x_1, \ldots, x_m)$ in $X^m(k)$ satisfying certain conditions(e.g., the angles between the x_i with respect to the Néron-Tate height associated to $\mathcal{L}$ should be small, the quotient of the height of x_{i+1} by the height of x_i should be big for $1 \le i < m$ and the height of x_1 should be big). Instead of a polynomial, we then construct a global section f of a certain line bundle $\mathcal{L}(\sigma - \epsilon, s_1, \ldots, s_m)^d$ on a certain model of X^m over the ring of integer R of k. This line bundle is a tensor product of pull-backs of $\mathcal{L}$ along maps $A^m \to A$ depending on $\sigma - \epsilon$, the s_i and on d; in particular it comes with the norms at the Archimedean places. By construction, f has small order vanishing at x and has suitably bounded norms at the Archimedean places of k. Then we consider the Arakelov degree of the metrized line bundle $x^*\mathcal{L}(\sigma - \epsilon, s_1, \ldots, s_m)^d$ on $spec(R)$; the conditions satisfied by the x_i give an upper bound, whereas the bound on the norm of f at the Archimedean places gives a lower bound. It turns out that we can choose the parameter ϵ, σ the s_i and d in such a way that the upper bound is smaller than the lower bound, which gives a contradiction. We note that the construction of f is quite involved. Intersection theory is used to show that under suitable hypotheses, the line bundle $\mathcal{L}(-\epsilon, s_1, \ldots, s_m)^d$ is ample on X^m. A new, basic tool here is the so-called Product theorem, a strong generalization by Faltings of Roth's Lemma. Also Siegel's Lemma is used in the construction of the section f.

The Correspondence Table

Nevanlinna Theory	**Diophantine Approximation**
Theorem A6.1.1	Theorem B6.1.1
Theorem A6.2.1	Theorem B6.1.2

Chapter 7

Complex Hyperbolic Manifolds and Lang's Conjecture

Part A: Nevanlinna Theory

A7.1 Schwarz Lemma

The starting point of the theory of complex hyperbolic manifolds is the so-called Schwarz Lemma. Before stating it, we need some preparations. Let X be a Riemann surface, i.e., a 1-dimensional complex manifold. Let

$$d\sigma^2 = 2a(z)dzd\bar{z}$$

be a Hermitian pseudo-metric on X expressed in terms of a local coordinate z. Here the term **pseudo-metric** means that $d\sigma^2$ is only semidefinite, i.e., $a(z) \geq 0$. Let

$$K = \frac{-1}{a}\frac{\partial^2 \log a}{\partial z \partial \bar{z}}. \tag{7.1}$$

K is called the **Gaussian curvature** of $d\sigma^2$. Note that K is defined whenever a is positive.

Example A7.1.1 *Let $\mathbf{D}(r)$ be the disc of radius r on $\mathbf{C}$. The metric*

$$ds^2 = \frac{4r^2 dzd\bar{z}}{(r^2 - |z|^2)^2} \tag{7.2}$$

is called the **Poincaré metric** *on $\mathbf{D}(r)$. It is easy to check that the Gaussian curvature K of the Poincaré metric is -1.*

We will prove a generalization of the Schwarz-Pick Lemma by Ahlfors.

Theorem A7.1.2 (Ahlfors) *Let ds^2 denote the Poincaré metric on the unit disc $\mathbf{D}$. Let $d\sigma^2$ be any Hermitian pseudo-metric on $\mathbf{D}$ whose Gaussian curvature is bounded above by -1. Then*

$$d\sigma^2 \leq ds^2. \tag{7.3}$$

Proof. Let $\mathbf{D}_r$ be the disc of radius $r < 1$ with the Poincaré metric ds^2 of curvature -1 given by

$$ds^2 = 2a_r(z)dzd\bar{z} \quad \text{where} \quad a_r(z) = \frac{2r^2}{(r^2 - |z|^2)^2}.$$

We compare this metric with $d\sigma^2 = 2b(z)dzd\bar{z}$. Put

$$\mu(z) = \log \frac{b(z)}{a_r(z)}.$$

Since $\mu(z) \to -\infty$ as $z \to \partial\mathbf{D}_r$, there is a point $z_0 \in \mathbf{D}_r$ such that

$$\mu(z_0) = \sup\{\mu(z); z \in \mathbf{D}_r\} > -\infty.$$

Then $b(z_0) > 0$. Since z_0 is a maximal point of $\mu(z)$,

$$0 \geq \frac{\partial^2 \mu}{\partial z \partial \bar{z}}(z_0).$$

On the other hand, since the Gaussian curvature of the Poincaré metric is -1 and the curvature of $d\sigma^2$ is bounded above by -1,

$$\frac{\partial^2 \log a_r}{\partial z \partial \bar{z}} = a_r(z) \text{ and } \frac{\partial^2 \log b}{\partial z \partial \bar{z}}(z) \geq b(z).$$

So

$$0 \geq \frac{\partial^2 \mu}{\partial z \partial \bar{z}}(z_0) = \frac{\partial^2 \log b}{\partial z \partial \bar{z}}(z_0) - \frac{\partial^2 \log a_r}{\partial z \partial \bar{z}}(z_0) \geq b(z_0) - a_r(z_0).$$

Hence $a_r(z_0) \geq b(z_0)$ and so $\mu(z_0) \leq 0$. By the choice of z_0, we have $\mu(z) \leq 0$ on $\mathbf{D}_r$, that is

$$a_r(z) \geq b(z).$$

The Theorem is proven by letting $r \to 1$. □

Note that in the proof of Theorem A7.1.2, we see that the theorem holds if $d\sigma^2$ is only continuous at zero points of $d\sigma^2$ and is twice differentiable at the points where it is positive(and hence the curvature is defined). This allows Ahlfors to extend Theorem A7.1.2 to non-smooth metrics. Let $d\sigma^2$ be an upper semi-continuous Hermitian pseudo-metric on the unit disc $\mathbf{D}$. A pseudo-Hermitian metric $d\sigma_0^2$ is called a **supporting pseudo metric** for $d\sigma^2$ at $z_0 \in \mathbf{D}$ if it is defined and of class C^2 in a neighborhood U of z_0 and satisfies the following condition:

$$d\sigma^2 \geq d\sigma_0^2 \text{ on } U \qquad \text{and} \qquad d\sigma^2 = d\sigma_0^2 \text{ at } z_0.$$

We define

$$K_{d\sigma^2}(z_0) = \inf K_{d\sigma_0^2}(z_0),$$

where the infimum is taken over all supporting pseudo metric $d\sigma_0^2$ for $d\sigma^2$ at z_0. Theorem A7.1.2 is generalized to the following theorem.

Theorem A7.1.3 *Let ds^2 denote the Poincaré metric on the unit disc $\mathbf{D}$. Let $d\sigma^2$ be an upper semi-continuous Hermitian pseudo-metric on $\mathbf{D}$ whose curvature is bounded above by -1. Then*

$$d\sigma^2 \leq ds^2.$$

Corollary A7.1.4 *Let X be a Riemann surface with a Hermitian pseudo-metric ds_X^2 whose curvature (wherever defined) is bounded above by -1. Then every holomorphic map $f : \mathbf{D} \to X$ is distance-decreasing, i.e.,*

$$f^* ds_X^2 \leq ds^2,$$

where ds^2 is the Poincaré metric on the unit disc $\mathbf{D}$.

Proof. Set $d\sigma^2 = f^* ds_X^2$. Then $d\sigma^2$ is a Hermitian pseudo-metric on $\mathbf{D}$. If we denote the curvature of ds_X^2 by K_X, then the curvature of $d\sigma^2$ is given by $f^* K_X$. Now Corollary A7.1.4 follows from Theorem A7.1.2. □

The classical Schwarz-Pick Lemma immediately follows from Corollary A7.1.4.

Corollary A7.1.5 Let $\mathbf{D}$ be the unit disc with the Poincaré metric ds^2. Then every holomorphic map $f : \mathbf{D} \to \mathbf{D}$ is distance-decreasing, i.e.,

$$f^*ds^2 \leq ds^2, \quad \text{or equivalently}$$

$$\frac{|f'(z)|}{1-|f(z)|^2} \leq \frac{1}{1-|z|^2}, \quad \text{for } z \in \mathbf{D}.$$

To generalize Corollary A7.1.4 to higher dimensional manifolds, we introduce the concept of **pseudo-length function**.

Definition A7.1.6 *Let M be a complex manifold. By a* **pseudo-length function** *on M we mean a non-negative (and strictly positive somewhere) function $F(x, \xi)$ defined on the complex tangent bundle TM such that*

$$F(x, \lambda\xi) = |\lambda| F(x, \xi) \quad \text{for all } \lambda \in \mathbf{C}.$$

We normally assume that F is smooth. However, for applications it is sometimes necessary to consider upper-semi-continuous pseudo-length functions. If $F(x, \xi) > 0$ for all nonzero $\xi \in T_xM$ and all $x \in M$, then we call F a length function. If, in addition, $F(x, \xi)$ satisfies the property that $F(x, \xi_1 + \xi_2) \leq F(x, \xi) + F(x, \xi_2)$, then we call F as a **Finsler (pseudo-) metric.**

Given an upper-semi-continuous pseudo-length function F on M, let $p \in M$ and $v \in T_pM$, and let $[v]$ denote the complex line spanned by v. We define the **holomorphic sectional curvature** $K_F([v])$ in the direction of $[v]$ by

$$K_F([v]) = \sup K_{f^*F}(0), \tag{7.4}$$

where the supremum is taken over all holomorphic maps from the unit disc $\mathbf{D}$ into M with $f(0) = p$ and $f'(0) \in [v]$.

Corollary A7.1.4 is easily generalized to the following:

Theorem A7.1.7 *Let F be an upper-semicontinuous pseudo-length function on a complex manifold M. If its holomorphic sectional curvature is bounded above by -1, then*

$$f^*F^2 \leq ds^2 \quad \textit{for } f \in \textit{Hol}(\mathbf{D}, M),$$

where ds^2 is the Poincaré metric of the unit disc $\mathbf{D}$.

We now show that for a Hermitian manifold the definition of the holomorphic sectional curvature given above coincides with the usual one. Given a Hermitian metric

$$ds^2 = 2\sum_{i,j=1}^{n} g_{i\bar{j}} dz^i d\bar{z}^j$$

on a complex manifold M, the components of its curvature tensor are expressed by

$$R_{i\bar{j}k\bar{l}} = -\frac{\partial^2 g_{i\bar{j}}}{\partial z^k \partial \bar{z}^l} + \sum g^{p\bar{q}} \frac{\partial g_{i\bar{q}}}{\partial z^k} \frac{\partial g_{p\bar{j}}}{\partial \bar{z}^l}. \tag{7.5}$$

Then given a unit tangent vector $v = \sum v^i(\partial/\partial z^i)$, the **holomorphic sectional curvature** in the direction of v is defined to be

$$H_{ds^2}(v) = \sum R_{i\bar{j}k\bar{l}} v^i \bar{v}^j v^k \bar{v}^l. \tag{7.6}$$

Proposition A7.1.8 *Let M' be a complex submanifold of a Hermitian manifold M. Then the holomorphic sectional curvature H'_{ds^2} of M' does not exceed the holomorphic sectional curvature H_{ds^2} of M, i.e.,*

$$H'_{ds^2}(v) \le H_{ds^2}(v) \quad \textit{for} \quad v \in TM'.$$

Proof. We choose a local coordinate system $z^1, \ldots, z^n$ in such a way that M' is defined by

$$z^{m+1} = \cdots = z^n = 0$$

so that we may use $z^1, \ldots, z^m$ as a local coordinate system for M'. Then the induced Hermitian metric on M' is given by $2\sum_{i,j=1}^{m} g_{i\bar{j}} dz^i d\bar{z}^j$. We shall compute its curvature tensors $R'_{i\bar{j}k\bar{l}}$. Fixing a point $x \in M'$, by a linear change of coordinates, we may assume that $g_{i\bar{j}} = \delta_{ij}$ at x. Then at the point of x we have the following equation of Gauss:

$$R'_{i\bar{j}k\bar{l}} = R_{i\bar{j}k\bar{l}} - \sum_{p=m+1}^{n} \frac{\partial g_{i\bar{p}}}{\partial z^k} \frac{\partial g_{p\bar{j}}}{\partial \bar{z}^l} \tag{7.7}$$

for $i, j, k, l = 1, \ldots, m$. So, by (7.6) and (7.7), the Proposition holds. □

Theorem A7.1.9 *For Hermitian manifold (M, ds^2) the holomorphic sectional curvature defined by (7.4) coincides with the classical holomorphic sectional curvature defined by (7.6).*

Proof. Fix a point $p \in M$, and let $v \in T_pM$ be a unit tangent vector. If M' is a holomorphic curve in M tangent to v at p, then, by Proposition A7.1.8, its Gaussian curvature at p does not exceed the holomorphic sectional curvature $H_{ds^2}(v)$ of M in the direction v. We shall show that there exists a holomorphic curve tangent to v whose Gaussian curvature at p equals $H_{ds^2}(v)$. We start with any holomorphic curve M' tangent to v at p. We may choose a local coordinate system $z^1, \ldots, z^n$ with origin p such that $g_{i\bar{j}} = \delta_{ij}$ and M' is given by $z^2 = \ldots z^n = 0$. From (7.7) we know that if $\partial g_{1\bar{q}}/\partial z^1 = 0$ at p for $q = 2, \ldots, n$, then M' already has the desired property. If not, we consider the following coordinate transformation:

$$z^1 = w^1, z^q = w^q - \frac{1}{2}a^q(w^1)^2, \qquad (q = 2, \ldots, n),$$

where $a^2, \ldots, a^n$ are constants to be chosen appropriately. Substitute the coordinate transformation above into $ds^2 = 2\sum g_{i\bar{j}} dz^i d\bar{z}^j$ to express it in terms of $w^1, \ldots, w^n$. Then we obtain $ds^2 = 2\sum h_{i\bar{j}} dw^i d\bar{w}^j$, with

$$h_{1\bar{q}} = g_{1\bar{q}} - \sum_{r=2}^{m} g_{r\bar{q}} a^r w^1.$$

It follows that if we set $a^q = (\partial g_{1\bar{q}}/\partial z^1)|_p$, then $(\partial h_{1\bar{q}}/\partial w^1)|_p = 0$, so that the holomorphic curve defined by $w^2 = \cdots = w^n = 0$ has the desired property. This finishes the proof. □

Theorem A7.1.9, together with Theorem A7.1.7, implies the following theorem.

Theorem A7.1.10 *Let (M, ds_M^2) be a Hermitian manifold whose holomorphic sectional curvature (in the classical sense) is bounded above by -1, then*

$$f^* ds_M^2 \le ds^2 \quad \text{for} \quad f \in Hol(\mathbf{D}, M),$$

where ds^2 is the Poincaré metric of the unit disc $\mathbf{D}$.

A7.2 Kobayashi Hyperbolicity

Recall that on the unit disc **D** the Poincaré metric is given by $ds^2 = \dfrac{4dzd\bar{z}}{(1-|z|^2)^2}$. The Gaussian curvature K of the Poincaré metric is -1. Let ρ denote the **Poincaré distance** defined by the Poincaré metric ds^2. We first find an explicit formula for ρ.

Theorem A7.2.1

$$\rho(a,b) = \log \frac{|a-a\bar{b}| + |a-b|}{|a-a\bar{b}| - |a-b|} = 2\tanh^{-1}\left|\frac{a-b}{1-a\bar{b}}\right| \quad \textit{for} \ \ a,b \in \mathbf{D}. \quad (7.8)$$

Proof. Let $0 < a < 1$. If $z(t) = x(t) + iy(t), 0 \leq t \leq 1$, is a curve in **D** joining the origin $0 \in \mathbf{D}$ to $a \in \mathbf{D}$, its arc length l with respect to ds^2 satisfies

$$\begin{aligned} l &= \int_0^1 \frac{2(x'(t)^2 + y'(t)^2)^{1/2}}{1 - x(t)^2 - y(t)^2} dt \\ &\geq \int_0^1 \frac{2|x'(t)|}{1-x(t)^2} dt \geq \int_0^a \frac{2dx}{1-x^2} = \log \frac{1+a}{1-a}. \end{aligned}$$

This shows that the ordinary line segment from 0 to a is the shortest path and that

$$\rho(0,a) = \log \frac{1+a}{1-a}.$$

Since the Poincaré metric is invariant under the rotations, we have

$$\rho(0,a) = \log \frac{1+a}{1-a} = 2\tanh^{-1}|a| \quad \text{for} \ \ a \in \mathbf{D}.$$

Given two points a and b in **D**, the transformation

$$w = \frac{z-b}{1-\bar{b}z}$$

is an automorphism of **D** that sends b to 0 and a to $(a-b)/(1-\bar{a}b)$. From the invariance of ρ we obtain

$$\rho(a,b) = \log \frac{|a-a\bar{b}| + |a-b|}{|a-a\bar{b}| - |a-b|} = 2\tanh^{-1}\left|\frac{a-b}{1-a\bar{b}}\right| \quad \text{for} \ \ a,b \in \mathbf{D}.$$

□

Given a complex manifold M, we shall define the **Kobayashi pseudo-distance** on M. Given two points $p, q \in M$, we consider a **Kobayashi chain of discs** from p to q, that is, a chain of points $p = p_0, p_1, \ldots, p_k = q$ of M, pairs of points $a_1, b_1, \ldots, a_k, b_k$ of $\mathbf{D}$ and holomorphic maps $f_1, \ldots, f_k \in \mathrm{Hol}(\mathbf{D}, M)$ such that $f_i(a_i) = p_{i-1}$ and $f_i(b_i) = p_i$ for $i = 1, \ldots, k$.

Definition A7.2.2 *Let M be a complex manifold. The* **Kobayashi pseudo-distance** *on M is defined, for any two points $p, q \in M$, by*

$$d_M(p,q) = \inf \sum_{i=1}^{k} \rho(a_i, b_i)$$

where ρ is the Poincaré distance on the unit disc $\mathbf{D}$ given in (7.8) and the inf is taken over all Kobayashi chains of discs from p to q.

From the definition, we see that d_M has an important property that is the distance decreasing under holomorphic mapping. So

Theorem A7.2.3 *(1) Let M, N be two complex manifolds and $f : M \to N$ be holomorphic. Then*

$$d_N(f(p), f(q)) \leq d_M(p,q).$$

(2) For the unit disc $\mathbf{D}$, the Kobayashi pseudo-distance d_D coincides with the Poincaré distance on $\mathbf{D}$.

(3) The Kobayashi pseudo-distance is the largest pseudo-distance on M such that every holomorphic map $f : \mathbf{D} \to M$ is distance decreasing.

Example A7.2.4 *For the complex plane $\mathbf{C}$ and the punctured plane $\mathbf{C}^* = \mathbf{C} - \{0\}$, we have $d_{\mathbf{C}} = 0$ and $d_{\mathbf{C}^*} = 0$.*

Proof. In fact, given two points $p, q \in \mathbf{C}$ and an arbitrarily small positive number ϵ, there is a holomorphic map $f : D \to \mathbf{C}$ such that $f(0) = p$ and $f(\epsilon) = q$. Hence $d_{\mathbf{C}} \leq \epsilon$. Letting $\epsilon \to 0$ implies that $d_{\mathbf{C}} = 0$. $d_{\mathbf{C}^*} = 0$ follows from the distance-decreasing property (Theorem A7.2.3) and the fact that $d_{\mathbf{C}} = 0$. □

Definition A7.2.5 *A complex manifold M is said to be* **Kobayashi hyperbolic** *if d_M is a distance, i.e., if $d_M(p,q) = 0$ then $p = q$. If the* *distance d_M is complete, then we say that M is* **complete Kobayashi hyperbolic**.

Theorem A7.2.6 *If M is Kobayashi hyperbolic, then every holomorphic map $f : \mathbf{C} \to M$ must be constant.*

Proof. Assume that M is Kobayashi hyperbolic then the Kobayashi pseudo-distance d_M, in fact, is a distance. By Theorem A7.2.3, we have, for every $a, b \in \mathbf{C}$,

$$d_M(f(a), f(b)) \leq d_{\mathbf{C}}(a, b) \equiv 0.$$

Hence, since d_M is a distance, $f(a) = f(b)$, which implies that f is constant. This finishes the proof. □

Let M be a complex manifold and M' be a complex subspace of M with compact closure $\overline{M'}$. We call a point $p \in \overline{M'}$ a **hyperbolic point** if every neighborhood U of p contains a smaller neighborhood V of p, $\bar{V} \subset U$, such that

$$d_{M'}(\bar{V} \cap M', M' - U) > 0,$$

where $d_{M'}$ is the Kobayashi pseudo-distance induced from M. We say that M' is **hyperbolically imbedded** in M if every point of $\overline{M'}$ is a hyperbolic point. Clearly, M' is hyperbolically imbedded in M if and only if, for every pair of distinct points p, q in $\overline{M'} \subset M$, there exists neighborhoods U_p and U_q of p and q in M such that $d_{M'}(U_p \cap M', U_q \cap M') > 0$.

It is an important problem to find many examples of Kobayashi hyperbolic complex manifolds besides those that are easily found. S. Kobayashi raised the following conjecture.

Conjecture A7.2.7 (S. Kobayashi) *(I) Is a generic hypersurface X of degree $\geq 2n+1$ in $\mathbf{P}^n(\mathbf{C})$ hyperbolic?*

(II) Is the complement of such a surface is complete hyperbolic and hyperbolically imbedded in $\mathbf{P}^n(\mathbf{C})$?

The classical result in support (II) is that the complement of $2n+1$ hyperplanes in general position in $\mathbf{P}^n(\mathbf{C})$ is hyperbolically imbedded in $\mathbf{P}^n(\mathbf{C})$. Brody-Green [Br-G] showed that the smooth surface

$$(z^0)^d + \cdots + (z^3)^d + s(z^0 z^1)^{d/2} + t(z^0 z^2)^{d/2} = 0$$

in $\mathbf{P}^3(\mathbf{C})$ is hyperbolic for even degree $d \geq 50$ and generic $s, t \in \mathbf{C}^*$. Nadel [Nad] obtained similar examples of hyperbolic surfaces in $\mathbf{P}^3(\mathbf{C})$ and curves in $\mathbf{P}^2(\mathbf{C})$ with hyperbolically imbedded components. K. Masuda and J. Noguchi [Mas-N] constructed example of the hyperbolic hypersurface of $\mathbf{P}^n(\mathbf{C})$. Dethloff-Schumacher-Wong proved that the complement of the union of generic 2 quadrics in $\mathbf{P}^2(\mathbf{C})$ is hyperbolic, and Siu-Yeung [Siu-Y1] proved the Kobayashi conjecture (II) for generic curves in $\mathbf{P}^2(\mathbf{C})$ of degree greater than some large number.

We know that metric induces distance, and in most case metric is more convenient to use. So next we shall describe Royden's pseudo-length function, of which the Kobayashi distance is the integrated form. Let M be a complex manifold M and let $p \in M$ and $v \in T_pM$. We define

$$R_M(p,v) = \inf\left\{\frac{1}{R} \;\middle|\; \text{there exists a holomorphic map } f : \mathbf{D}(R) \to M \text{ with } f(0) = p, f'(0) = v\right\} \tag{7.9}$$

or equivalently

$$R_M(p,v) = \inf\left\{\frac{1}{R} \;\middle|\; \text{there exists a holomorphic map } f : \mathbf{D} \to M \text{ with } f(0) = p, f'(0) = Rv\right\}. \tag{7.10}$$

Royden proved that it is upper semi-continuous. It follows immediately from the definition and the Schwarz Lemma that on the unit disc $\mathbf{D}$ the Royden function $R_M(p, v)$ is the same as the Poincaré metric.

Royden's pseudo-length function induces a pseudo-distance on M and we shall prove that that distance actually is Kobayashi's pseudo-distance. Let $\gamma : [a, b] \to M$ be a C^1 curve in M. The length induced by Royden's pseudo-length function is

$$L_M(\gamma) = \int_a^b R_M(\gamma(t), \gamma'(t))dt$$

and for any two points $p, q \in M$, the pseudo-distance (we call it Royden's pseudo-distance) induced by Royden's pseudo-length function is given by

$$d'_M(p,q) = \inf L_M(\gamma),$$

where the infimum is taken over all C^1 curves connecting p and q.

Theorem A7.2.8 $d'_M = d_M$*, that is Royden's pseudo-distance is equal to Kobayashi's pseudo-distance.*

Proof. The inequality $d'_M \leq d_M$ follows from Theorem A7.2.3 such that the Kobayashi pseudo-distance is the largest pseudo-distance on M hence every holomorphic map $f : \mathbf{D} \to M$ is distance decreasing.

Next we prove that $d_M \leq d'_M$. Given two points $p, q \in M$, for arbitrary $\epsilon > 0$, there is a C^1 curve $\gamma : [0,1] \to M$ connecting p and q such that

$$L_M(\gamma) = \int_0^1 R_M(\gamma(t), \gamma'(t))dt \leq d'_M(p,q) + \epsilon.$$

Since Royden's function is upper semi-continuous there exists a sequence of continuous functions $\{h_n(t)\}$ increasing to the function $R_M(\gamma(t), \gamma'(t))$. Hence there exists a continuous function $h(t)$ such that

$$R_M(\gamma(t), \gamma'(t)) \leq h(t),$$

but

$$\int_0^1 h(t)dt \leq \int_0^1 R_M(\gamma(t), \gamma'(t))dt + \epsilon \leq d'_M(p,q) + 2\epsilon.$$

Also we find a partition $(t_0, \ldots, t_n)$ of the interval $[0,1]$ such that for $s_i \in [t_{i-1}, t_i]$ we have an estimate for the Riemann sum

$$\sum_{i=1}^{m} h(s_i)(t_i - t_{i-1}) \leq d'_M(p,q) + 3\epsilon. \tag{7.11}$$

Next we prove the following claim, which is a mean-value type theorem:

Claim: *Let* $\gamma : [a,b] \to M$ *be a* C^1 *curve. Then given* $s, \epsilon' > 0$ *there exists* δ *such that for all* t *in the interval* $|t-s| < \delta$ *we have*

$$d_M(\gamma(t), \gamma(s)) \leq [R_M(\gamma(s), \gamma'(s)) + \epsilon']|t-s|.$$

To prove the Claim, we note, by the definition of Royden's pseudo-length function, there exists a disk $\mathbf{D}(R)$ and a holomorphic map $f : \mathbf{D}(R) \to M$ such that $f(0) = \gamma(s)$, $f'(0) = \gamma'(s)$, and

$$\frac{1}{R} < R_M(\gamma(s), \gamma'(s)) + \epsilon.$$

Let $W = D_c^n$ be a coordinate polydisc centered at $\gamma(s)$, and let $W_{c/2}$ be the subpolydisc of half the radius, where $n = \dim M$. For $x, y \in W_{c/2}$ we have

$$d_M(x,y) \leq d_W(x,y).$$

We can choose δ_1 such that $|t-s| < \delta_1$ implies $\gamma(t) \in W_{c/2}$.

Since when f is restricted to the real axis, f also becomes a parametrized curve in M, and two parametrized curves γ and f (restricted to the real axis) have the same tangent vector at $\gamma(s) = f(0)$. Hence there exists δ_2 such that for $|t-s| < \delta_2$ we have

$$d_W(f(t), f(t-s)) \leq \epsilon|t-s|,$$

because d_W defines the topology on the polydisc W. Then

$$d_M(f(t), f(t-s)) \leq \epsilon|t-s|. \tag{7.12}$$

On the other hand, composing f with dilation $m_R : \mathbf{D} \to \mathbf{D}(R)$, directly from the definition of Kobayashi pseudo-distance, we have that for some δ and $|t-s| < \delta$:

$$\begin{aligned} d_M(f(t-s), f(0)) &\leq d_{\mathbf{D}(R)}(t-s, 0) \leq \left(\frac{1}{R} + \epsilon\right)|t-s| \\ &\leq (R_M(\gamma(s), \gamma'(s)) + 2\epsilon)|t-s|. \end{aligned} \tag{7.13}$$

Combining (7.12) with (7.13) and using the triangle inequality we find

$$d_M(\gamma(t), \gamma(s)) \leq (R_M(\gamma(s), \gamma'(s)) + 3\epsilon)|t-s|,$$

thereby proving the Claim.

We now return to the proof. Applying the Claim with $\epsilon' = h(s_i) - R_M(\gamma(s_i), \gamma'(s_i))$ and taking the partition sufficiently small, we have, noting

(7.11),

$$\begin{aligned} d_M(p,q) &\leq \sum d_M(\gamma(t_i), \gamma(t_{i-1})) \\ &\leq \sum [h(s_i)+\epsilon)]|t_i - t_{i-1}| \\ &\leq d'_M(p,q) + 4\epsilon. \end{aligned}$$

This proves the Theorem. □

Finally, we present criteria for using the Royden function to determine the Kobayashi hyperbolicity.

Theorem A7.2.9 *Let M be a complex manifold with a metric h on it. If there is a positive $a > 0$ such that for every point $p \in M$ and every $v \in T_p(M)$, $R_M(p,v) > ah(v)$, then M is Kobayashi hyperbolic.*

Proof. In fact, the condition $R_M(p,v) > ah(v)$, together with the fact that h is a metric, implies that the Royden pseudo-distance induced by $R_M(p,v)$ is actually a distance. But Theorem A7.2.8 says that the Royden pseudo-distance is the same as the Kobayashi distance. So M is Kobayashi hyperbolic. □

A7.3 Brody's Hyperbolicity

Intuitively speaking, a complex manifold is Kobayashi hyperbolic if it does not contain disc of arbitrary large radii. For compact manifolds, as it turns out, the characterization takes an exceptionally simple form.

Theorem A7.3.1 (Brody) *A compact complex manifold M is Kobayashi hyperbolic if and only if M contains no non-constant holomorphic curves, i.e., there is no non-constant holomorphic curve $f : \mathbf{C} \to M$.*

The following example by D. Eisenman and L. Taylor shows that Theorem A7.3.1 does not hold for some non-compact manifolds.

Example A7.3.2 *The domain*

$$M = \left\{ (z,w) \in \mathbf{C}^2 \;\middle|\; |z|<1, |zw|<1 \right\} - \left\{ (0,w) \;\middle|\; |w| \geq 1 \right\}$$

is not Kobayashi hyperbolic, but there is no non-constant holomorphic map $f : \mathbf{C} \to M$.

Proof. The map $h : (z, w) \to (z, zw)$ sends M into the unit bidisc and is one-to-one except at $z = 0$. Let $f : \mathbf{C} \to M$ be a holomorphic map, then $h \circ f : \mathbf{C} \to \mathbf{D}^2$ is constant by Liouville's theorem. So either f is constant or f maps $\mathbf{C}$ into the set $\{(0, w) \in M\}$. But this set is equivalent to the unit disc. Hence f is constant in either case. Since h is distance-decreasing we see that $d_M(p, q) > 0$ for $p \neq q$ unless both p and q are in the subset $\{(0, w) \in M\}$. We shall show that if p and q are in the subset $\{(0, w) \in M\}$, then $d_M(p, q) = 0$. Let $p = (0, b)$ with $b \neq 0$ and $q = (0, 0)$. Set $p_n = (1/n, b)$. Then $d_M(p, q) = \lim d_M(p_n, q)$. Let $a_n = \min\{n, \sqrt{n/|b|}\}$. The mapping $t \in \mathbf{D} \to (a_n t/n, a_n bt) \in M$ maps $1/a_n$ into p_n. Hence

$$\lim d_M(p_n, q) \leq \lim d_{\mathbf{D}}(1/a_n, 0) = 0,$$

which shows that M is not Kobayashi hyperbolic. □

To prove Theorem A7.3.1, we need a lemma about the reparametrization.

Lemma A7.3.3 (Brody's reparametrization) *Let M be a compact Hermitian manifold with a Hermitian metric. Let $f : \mathbf{D}(R) \to M$ be a holomorphic map with $|f'(0)| > 0$, where the norm $|f'(0)|$ is taken with respect to the Hermitian metric. Then for any $0 < r < R$, there exists $0 < t < 1$ and a Möbius transformation T of $\mathbf{D}(r)$ such that the function $g(z) = f(tT(z))$ satisfies $|g'(0)| = \frac{1}{2}|f'(0)|$ and*

$$|g'(z)| \leq \frac{r^2}{2(r^2 - |z|^2)}|f'(0)|.$$

Note that the image of g is contained in the image of f.

Proof. For $0 < t < 1$ let $f_t(z) = f(tz)$. Let

$$s_t(z) = \frac{r^2 - |z|^2}{r^2}|f_t'(z)|.$$

Since $|f_t'(z)| = t|f'(tz)|$,

$$s_t(z) = t\frac{r^2 - |z|^2}{r^2}|f'(tz)|.$$

Let

$$s(t) = \sup_{z \in \mathbf{D}(r)} s_t(z).$$

Since $s_t(z)$ vanishes at points $|z| = r$, the value $s(t)$ is achieved by $s_t(z)$ at some point z_t of $\mathbf{D}(r)$. Since

$$s_t(z) = \frac{t}{t'} s_{t'} \left(\frac{t}{t'} z \right)$$

for $0 \le t < t' \le 1$, it follows that $s(t)$ is an increasing function of $0 \le t \le 1$. Since $s(0) = 0$ and $s(1) \ge |f'(0)|$, there exists some $0 < t_0 \le 1$ such that $s(t_0) = \frac{1}{2}|f'(0)|$. Let $z_{t_0} \in \mathbf{D}(r)$ such that $s(t_0) = s_{t_0}(z_{t_0})$. Let T be the Möbius transformation mapping $\mathbf{D}(r)$ to itself so that T maps the origin to z_{t_0}. Let $g = f_{t_0} \circ T$. To estimate $|g'(z)|$, we use the property that the Poincaré metric on the disc $\mathbf{D}(r)$ is invariant under the Möbius transformation by Schwarz's lemma. That is $ds^2 = T^*(ds^2)$ where ds^2 is Poincaré metric on the disc $\mathbf{D}(r)$. So

$$\frac{r}{r^2 - |z|^2} = \frac{r}{r^2 - |T(z)|^2} \cdot |T'(z)|.$$

Thus

$$\begin{aligned}
|g'(z)| &= |f'_{t_0}(T(z))| \cdot |T'(z)| \\
&= \frac{r^2 - |T(z)|^2}{r^2} \cdot |f'_{t_0}(T(z))| \cdot \frac{r^2}{r^2 - |T(z)|^2} \cdot |T'(z)| \\
&= s_{t_0}(T(z)) \cdot \frac{r^2}{r^2 - |z|^2} \le s(t_0) \cdot \frac{r^2}{r^2 - |z|^2} \\
&= \frac{1}{2}|f'(0)| \cdot \frac{r^2}{r^2 - |z|^2}.
\end{aligned}$$

We also have $|g'(0)| = |f'_{t_0}(T(0))| \cdot |T'(0)| = s_{t_0}(z_{t_0}) = s(t_0) = \frac{1}{2}|f'(0)|$. $\square$

Proof of Theorem A7.3.1:

Proof. Theorem A7.2.6 gives one direction. To prove the other direction, assume that M is not Kobayashi hyperbolic. Let h be a metric on M, then Theorem A7.2.9 implies that there does not exist a positive number a such that for every $p \in M$ and $v \in T_p(M)$, $R_M(p, v) > ah(v)$, where R_M is the Royden pseudo-length function. Hence there is a sequence of points $p_n \in M$ and tangent vectors $v_n \in T_{p_n}(M)$ such that $|v_n| = 1$ and $R_M(p_n, v_n) < 1/n$, where the norm of $|v_n|$ is taken with respect to the metric h. By the definition of R_M given by (7.9), we find an increasing sequence of discs $\mathbf{D}(R_n)$ with radius R_n, $\lim_{n\to} R_n = \infty$, and a sequence of holomorphic maps $f_n : \mathbf{D}(R_n) \to M$ with $f_n(0) = p_n$, $f_n'(0) = v_n$. By applying Lemma A7.3.3, we obtain a sequence of maps $g_n = f_n \circ T_n$ where T_n is Möbius transformation of $\mathbf{D}(R_n)$ such that the function satisfies $|g_n'(0)| = \frac{1}{2}|f_n'(0)| = 1/2$ and

$$|g_n'(z)| \leq \frac{R_n^2}{2(R_n^2 - |z|^2)}.$$

This implies that the family $\mathcal{F}_1 = \{g_n|_{\mathbf{D}(R_{1/2})}\}$ is equicontinuous, the Arzela-Ascoli theorem implies that we can extract a subsequence which converges to a map $h_1 \in \mathrm{Hol}(\mathbf{D}(R_{1/2}), M)$. (We note that this is the place where we use the compactness of M.) Applying the same theorem to $\mathcal{F}_2$, we extract a subsequence from the above subsequence which converges to $h_2 \in \mathrm{Hol}(\mathbf{D}(R_{2/2}), M)$. In this way, we obtain maps $h_k \in \mathrm{Hol}(\mathbf{D}(R_{k/2}), M)$, $k = 1, 2, \ldots$, such that each h_k is an extension of h_{k-1}. Hence we have a map $h \in \mathrm{Hol}(\mathbf{C}, M)$. Moreover, since $|g_n'(0)| = 1/2$, $|h'(0)| = 1/2$. This implies that h is not a constant. This is in contradiction with our assumption that M contains no non-constant holomorphic curves. □

Motivated by Brody's theorem, we provide the following definition.

Definition A7.3.4 *A complex manifold M is said to be* **Brody hyperbolic** *if every holomorphic map $f : \mathbf{C} \to M$ is constant.*

In the previous chapters we show various manifolds which are Brody hyperbolic using Nevanlinna' theory. These manifolds are: $\mathbf{P}^1$ minus 3 distinct points (Corollary A1.3.3); $\mathbf{P}^n$ minus $2n+1$ hyperplanes in general position (Corollary A3.4.5); and an Abelian variety minus an ample divisor (Theorem A6.2.1). The surface $M = \{[x_0 : x_1 : x_2 : x_3] \in \mathbf{P}^4 \mid x_0^m + x_1^m + x_2^m + x_3^{m-2}P(x_0, x_1, x_1, x_3) = 0\}$, where P is a generic form of degree two, is Brody hyperbolic if $m \geq 11$ (Corollary A3.2.10).

By Theorem A7.2.6, we arrive at the following theorem.

Theorem A7.3.5 *If a complex manifold M is Kobayashi hyperbolic then it is Brody hyperbolic.*

A7.4 Differential Geometric Criteria for Hyperbolicity

The results in Section A7.1 about the Schwarz Lemma yield differential geometric criteria for hyperbolicity.

Theorem A7.4.1 *Let M be a complex manifold. If there is a length function F with holomorphic sectional curvature K_F bounded above by a negative constant, then M is hyperbolic. If, moreover, F defines a complete distance on M, then M is complete hyperbolic.*

Proof. By normalizing F, we may assume that $K_F \leq -1$. Let $p \in M$ and $v \in T_pM$. Let $f : D_R \to M$ be a holomorphic map with $f(0) = p, f'(0) = v$. Recall that the Poincaré metric on $\mathbf{D}(R)$ is

$$\frac{4R^2}{(R^2 - |z|^2)^2} dz d\bar{z}.$$

By the Schwarz Lemma, $F(p, v) \leq \frac{2}{R}$. This implies that the Royden pseudo-length function $R_M(p, v)$ satisfies

$$\frac{1}{2} F(p, v) \leq R_M(p, v).$$

By Theorem A7.2.9, M is Kobayashi hyperbolic. □

We note that although M admits a Hermitian metric with negative holomorphic sectional curvature, this is not sufficient to conclude that it is Kobayashi hyperbolic. For example, the Hermitian metric $2(1 + |z|)^2 dz d\bar{z}$ on $\mathbf{C}$ has Gaussian curvature

$$K = -\frac{1}{(1 + |z|^2)^3} < 0,$$

however, $\mathbf{C}$ is not hyperbolic.

Example A7.4.2 *Every compact Riemann surface of genus ≥ 2 is complete Kobayashi hyperbolic.*

Proof. In fact, the universal cover of such a Riemann surface is the upper half-plane. So the Poincaré metric on the upper half-plane induces a complete metric on the Riemann surface with Gaussian curvature as -1. So Theorem A7.4.1 implies that it is complete Kobayashi hyperbolic. □

Example A7.4.3 *The Riemann sphere* $\mathbf{P}^1(\mathbf{C})$ *minus at least three points is complete Kobayashi hyperbolic.*

Proof. In fact, let $M = \mathbf{P}^1(\mathbf{C}) - \{a_i\}_{i=1}^q$ and let $||z, a||$ denote the spherical distance of $\mathbf{P}^1(\mathbf{C})$. Define a hermitian metric $d\sigma^2$ on M by

$$d\sigma^2 = \frac{1}{\prod_{i=1}^q ||z, a_i||^2 (\log c||z, a_i||^2)^2} \cdot \frac{4}{(1+|z|^2)^2} dz d\bar{z}$$

where $c > 0$ is a constant. Taking small $c > 0$, one finds that the Gaussian curvature $K_{d\sigma^2} \leq -k < 0$ with a constant $k > 0$. So M is Kobayashi hyperbolic. For every a_i, we take a local coordinate w around a_i with $w(a_i) = 0$. Then there is a constant $A > 0$ such that

$$d\sigma^2 \sim \frac{A}{|w|^2 (\log |w|^2)^2} dw d\bar{w}.$$

Therefore $d\sigma^2$ is complete. So M is complete Kobayashi hyperbolic. □

Next, we consider $\mathbf{P}^n - \cup_{j=1}^q H_j$, where H_j are hyperplanes in general position and $q \geq 2n + 1$. Assume that there exists a non-constant holomorphic map $f : \mathbf{C} \to \mathbf{P}^n - \cup_{j=1}^q H_j$. We shall construct a pseudo-metric on $\mathbf{C}$ with curvature bounded above by a negative constant, so it derives a contradiction by using Schwarz's Lemma. The procedure basically repeats Ahlfors' method, presented in section A3.5. We recall definitions in section A3.5. Let f be a non-degenerated holomorphic map from $\mathbf{C}$ to $\mathbf{P}^m$. Let F_k be the k-th associated map of f. Let $\phi_k(H) = ||F_k; H||$. Then, we have a Lemma similar to Lemma A3.5.10.

Theorem A7.4.4 *For every* $\epsilon > 0$ *there exists a* $\mu_0(\epsilon) \geq 1$ *such that for all* $\mu \geq \mu_0(\epsilon)$ *and for any hyperplane* $H \subset \mathbf{P}^m$ *we have*

$$dd^c \log \frac{1}{\log^2(\mu/\phi_k(H))} \geq \frac{2\phi_{k+1}(H)}{\phi_k(H) \log^2(\mu/\phi_k(H))} \Omega_k - \epsilon \Omega_k.$$

Proof. First of all,

$$dd^c \log \frac{1}{\log^2(\mu/\phi_k(H))} = 2\frac{dd^c \log \phi_k(H)}{\log(\mu/\phi_k(H))} + 2\frac{d\phi_k(H) \wedge d^c \phi_k(H)}{\phi_k^2(H) \log^2(\mu/\phi_k(H))}. \quad (7.14)$$

Recall that, by Lemma A3.5.8,

$$dd^c \log \phi_k(H) = \frac{\phi_{k-1}(H)\phi_{k+1}(H) - \phi_k^2(H)}{\phi_k^2(H)}\Omega_k$$

and

$$d\phi_k(H) \wedge d^c\phi_k(H) = (\phi_{k+1}(H) - \phi_k(H))(\phi_k(H) - \phi_{k-1}(H))\Omega_k.$$

Thus, we can rewrite (7.14) in the form

$$\begin{aligned}
dd^c \log \frac{1}{\log^2(\mu/\phi_k(H))} &= \frac{2\phi_{k+1}(H)}{\phi_k(H)\log^2(\mu/\phi_k(H))}\Omega_k \\
&+2\frac{\phi_{k-1}(H)\phi_{k+1}(H)}{\phi_k^2(H)}\left(\frac{1}{\log(\mu/\phi_k(H))} - \frac{1}{\log^2(\mu/\phi_k(H))}\right)\Omega_k \\
&+2\frac{\phi_{k-1}(H)}{\phi_k(H)\log^2(\mu/\phi_k(H))}\Omega_k - 2\left(\frac{1}{\log(\mu/\phi_k(H))} + \frac{1}{\log^2(\mu/\phi_k(H))}\right)\Omega_k.
\end{aligned}$$

Then using the fact that $\phi_k(H) \leq 1$, we choose a number μ so large that $\frac{1}{\log(\mu/\phi_k(H))} > 1$ and

$$\frac{1}{\log(\mu/\phi_k(H))} + \frac{1}{\log^2(\mu/\phi_k(H))} < \frac{\epsilon}{2}$$

for the given $\epsilon > 0$ and

$$\frac{1}{\log(\mu/\phi_k(H))} - \frac{1}{\log^2(\mu/\phi_k(H))} > 0.$$

Then

$$dd^c \log \frac{1}{\log^2(\mu/\phi_k(H))} \geq \frac{2\phi_{k+1}(H)}{\phi_k(H)\log^2(\mu/\phi_k(H))}\Omega_k - \epsilon\Omega_k.$$

□

Now let $f : \mathbf{C} \to \mathbf{P}^n - \cup_{j=1}^q H_j$ be a non-constant holomorphic map, where $H_j, 1 \leq j \leq q$ are hyperplanes in general position. Assume that f is m-linearly non-degenerate, i.e., $f(\mathbf{C})$ is contained in a subspace of dimension $m \leq n$, but not any subspace of lower dimension. Without loss of generality, we assume that $f : \mathbf{C} \to \mathbf{P}^m$. Then f is linearly non-degenerate. Furthermore, the hyperplanes $H_j \cap \mathbf{P}^m, 1 \leq j \leq q$ are in m-subgeneral position. Let $\omega(j)$ be the Nochka Weights associated with $\tilde{H}_j = H_j \cap \mathbf{P}^m$. Then, similar to Theorem A3.5.7, we have the following product-to-sum estimate.

Lemma A7.4.5 *For any constant $N \geq 1$ and $1/q \leq \lambda_k \leq 1/(m-k)$, there exists a positive constant $C_k > 0$ which depends only on k and the given hyperplanes such that*

$$C_p \left(\prod_{j=1}^{q} \left(\frac{\phi_{k+1}(H_j)}{\phi_k(H_j)} \right)^{\omega(j)} \frac{1}{(N - \log \phi_k(H_j))^2} \right)^{\lambda_k}$$
$$\leq \sum_{j=1}^{q} \frac{\phi_{k+1}(H_j)}{\phi_k(H_j)(N - \log \phi_k(H_j))^2}$$

on $D_R - \cup_{j=1}^{q} \{\phi_k(H_j) = 0\}$.

To construct the pseudo-metric on $\mathbf{D}(R)$, we write

$$\Omega_k = \frac{\sqrt{-1}}{2\pi} a_k(z) dz \wedge d\bar{z}$$

and

$$\sigma_k = C_k \prod_{j=1}^{q} \left[\left(\frac{\phi_{k+1}(H_j)}{\phi_k(H_j)} \right)^{\omega(j)} \cdot \frac{1}{(N - \log \phi_k(H_j))^2} \right]^{\lambda_k} \cdot a_k,$$

where C_k is the positive constant in the product-to-sum estimate above, $\lambda_k = 1/[m - k + 2q(m-k)^2/N]$ and $N \geq 1$. We take the geometric mean of σ_k and define

$$\Gamma = 2c \prod_{k=0}^{m-1} \sigma_k^{\beta_m/\lambda_k} dz d\bar{z},$$

where $\beta_m = 1/\sum_{k=0}^{m-1} \lambda_k^{-1}$ and $c = 2(\prod_{k=0}^{m-1} \lambda_k^{\lambda_k^{-1}})^{\beta_m}$. Let

$$\Gamma = 2h(z) dz d\bar{z},$$

then

$$h(z) = c \prod_{j=1}^{q} \left(\frac{1}{\phi_0(H_j)^{\omega(j)}} \right)^{\beta_m} \prod_{j=1}^{q} \left[\prod_{k=0}^{m-1} \frac{a_k^{\beta_m/\lambda_k}}{(N - \log \phi_k(H_j))^{2\beta_m}} \right]. \quad (7.15)$$

Theorem A7.4.6 *For $q \geq 2n - m + 2$, and*

$$2q/N < \frac{\sum_{j=1}^{q} \omega(j) - (m+1)}{m(m+2)},$$

we have

$$dd^c \log h(z) \geq \frac{\sqrt{-1}}{2\pi} h(z) dz \wedge d\bar{z}.$$

Proof. From (7.15) it follows that

$$\begin{aligned} &dd^c \log h(z) \\ &= -\beta_m \sum_{j=1}^{q} \omega(j) dd^c \log \phi_0(H_j) + \beta_m \sum_{j=1}^{q} \sum_{k=0}^{m-1} dd^c \log \left(\frac{1}{N - \log \phi_k(H_j)} \right)^2 \\ &\quad + \beta_m \sum_{k=1}^{m-1} (1/\lambda_k) dd^c \log a_k. \end{aligned}$$

By Lemma A3.5.1, $dd^c \log a_k = \Omega_{k+1} - 2\Omega_k + \Omega_{k-1}$ and $dd^c \log \phi_0(H_j) = -\Omega_0$. These, together with Lemma A7.4.4, imply that

$$\begin{aligned} dd^c \log h(z) \geq \beta_m \Bigg(&\sum_{j=1}^{q} \omega(j) \Omega_0 + 2 \sum_{j=1}^{q} \sum_{k=0}^{m-1} \frac{\phi_{k+1}(H_j)}{\phi_k(H_j)(N - \log \phi_k(H))^2} \Omega_k \\ &- \frac{2q}{N} \sum_{k=0}^{m-1} \Omega_k + \sum_{k=0}^{m-1} [(m-k) + (m-k)^2 \frac{2q}{N}] \{\Omega_{k+1} - 2\Omega_k + \Omega_{k-1}\} \Bigg). \end{aligned}$$

Using Lemma A7.4.5, it follows that

$$\begin{aligned} &\sum_{j=1}^{q} \frac{\phi_{k+1}(H_j)}{\phi_k(H_j)(N - \log \phi_k(H_j))^2} \Omega_k \\ &\geq C_k \left(\prod_{j=1}^{q} \left(\frac{\phi_{k+1}(H_j)}{\phi_k(H_j)} \right)^{\omega(j)} \frac{1}{(N - \log \phi_k(H_j))^2} \right)^{\lambda_k} \Omega_k \\ &= \frac{\sqrt{-1}}{2\pi} \sigma_k(z) dz \wedge d\bar{z}. \end{aligned}$$

Notice that

$$\Omega_m = 0,$$

so that

$$\sum_{k=0}^{m-1} (m-k)(\Omega_{k+1} - 2\Omega_k + \Omega_{k-1}) = -(m+1)\Omega_0,$$

and therefore

$$
\begin{aligned}
&dd^c \log h(z) \geq \\
&\quad \beta_m \Bigg(\sum_{j=1}^{q} \omega(j)\Omega_0 + 2\frac{\sqrt{-1}}{2\pi} \sum_{k=0}^{m-1} \sigma_k(z) dz \wedge d\bar{z} - (m+1)\Omega_0 - (m^2+2m)\frac{2q}{N}\Omega_0 \\
&\quad + \sum_{k=1}^{m-2} [(m-k+1)^2 - 2(m-k)^2 + (m-k-1)^2 - 1]\frac{2q}{N}\Omega_k \\
&\quad + \frac{2q}{N}\Omega_{m-1} \Bigg).
\end{aligned}
$$

We use the following elementary inequality: For all positive numbers $x_1, \ldots, x_n$ and $a_1, \ldots, a_q$,

$$a_1 x_1 + \cdots a_n x_n \geq (a_1 + \cdots + a_n)(x_1^{a_1} \cdots x_n^{a_n})^{1/(a_1 + \cdots + a_n)}.$$

Letting $a_k = \lambda_k^{-1}$ we have

$$\sum_{k=1}^{m-1} \sigma_k \geq \frac{c}{2\beta_m} \prod_{k=0}^{m-1} \sigma_k^{\beta_m/\lambda_k} = \frac{h(z)}{2\beta_m}$$

and therefore

$$
\begin{aligned}
dd^c \log h(z) \quad \geq \quad & \beta_m \Bigg[\left(\sum_{j=1}^{q} \omega(j) - (m+1) - (m^2+2m)\frac{2q}{N} \right) \Omega_0 + \sum_{k=1}^{m-2} \frac{2q}{N}\Omega_k \\
& + \frac{2q}{N}\Omega_{m-1} \Bigg] + \frac{\sqrt{-1}}{2\pi} h(z) dz \wedge d\bar{z}.
\end{aligned}
$$

By Theorem A3.4.3, we have

$$\theta(\sum_{j=1}^{q} \omega(j) - (m+1)) = q - 2n + m - 1 > 0,$$

and $\theta > 0$, so $(\sum_{j=1}^{q} \omega(j) - (m+1)) > 0$. Using this and the choice of N gives us

$$dd^c \log h(z) \geq \frac{\sqrt{-1}}{2\pi} h(z) dz \wedge d\bar{z}.$$

□

By Theorem A7.4.6, the Gauss curvature of the pseudo-metric Γ on $\mathbf{D}(R)$ is bounded above by -1. So, using Theorem A7.1.2, we have

$$h(z) \leq \left(\frac{2R}{R^2 - |z|^2}\right)^2.$$

Letting $R \to \infty$, we have $h(z) \equiv 0$ on $\mathbf{C}$, which gives a contradiction. So we again derive the following theorem. (Note, in A5.3, we use Nevanlinna's second main theorem for the derivation.)

Theorem A7.4.7 *$\mathbf{P}^n - \cup_{j=1}^q H_j$ is Brody hyperbolic if $H_j, 1 \leq j \leq q$, are hyperplanes in general position and $q \geq 2n+1$.*

Note that M. Green actually showed that $\mathbf{P}^n - \cup_{j=1}^q H_j$ is Kobayashi hyperbolic and hyperbolically embedded in $\mathbf{P}^n$ if H_j are hyperplanes in general position and $q \geq 2n+1$.

A7.5 Metrics on Jet Bundles

Let M be a compact complex manifold. Let $D = D_1 + \cdots + D_q$ be a divisor on M where $\{D_i \mid i = 1, \ldots, q\}$ are the irreducible components of D. Furthermore we assume that D is of simple normal crossing, i.e., each D_i is smooth; D_i intersects D_j transversally for all $i \neq j$ and each point x is contained in at most n components. For simplicity we also assume that D_i are linearly equivalent, i.e., the line bundle $L = \mathcal{O}(D_i)$ associated to the divisor D_i is the same for all $i = 1, \ldots, q$. Denote by $T_D^*M = \Omega_M^1(\log D)$ the bundle over M of meromorphic 1-forms which are holomorphic except for logarithmic poles at D. T_D^*M is called the logarithmic cotangent bundle of M. The dual T_DM shall be referred to as the logarithmic tangent bundle. The following theorem is due to S. Lu ([Lu]). The proof presented below is provided by P.M. Wong.

Theorem A7.5.1 *Let A be a very ample divisor and D be a divisor of simple normal crossing on a compact complex manifold M or let D be a trivial divisor. Let $f : \mathbf{C} \to M - D$ be a holomorphic curve, then*

$$f^*\omega \equiv 0$$

for all non-trial $\omega \in H^0(M, S^mT_D^*M \otimes \mathcal{O}(-A))$, where $S^mT_D^*M$ is the m-fold symmetric product of T_D^*M.

Proof. Let $\rho_1, \ldots, \rho_N$ be a basis of sections of $H^0(M, \mathcal{O}(A))$ providing an embedding $[\rho_0, \ldots, \rho_N] : M \to \mathbf{P}^N$. Then $\rho = \sum_{i=0}^N \rho_i \otimes \bar{\rho}_i$ may be considered as a Hermitian metric along the fiber of $\mathcal{O}(-A)$. Indeed it is the pull-back of the standard Hermitian metric of the tautological line bundle $\mathcal{O}_{\mathbf{P}^N}(-1)$. Denote by ρ^* the dual metric on $\mathcal{O}(A)$, then

$$c_1(\mathcal{O}(A), \rho^*) > 0.$$

We first deal with the case when D is a trivial divisor. Define a Finsler metric on $T(M)$ by, for every $\xi \in T(M)$,

$$F(\xi) = ||\xi||_{\omega \otimes \rho} = (|\omega(\otimes^m \xi)|^2)^{1/m}.$$

Differentiating the above equation in the sense of the differential form yields:

$$dd^c \log F(\xi) = c_1(\mathcal{O}(A), \rho^*). \tag{7.16}$$

Since M is compact and $c_1(\mathcal{O}(A), \rho^*) > 0$,

$$c_1(\mathcal{O}(A), \rho^*)(\xi, \bar{\xi}) \geq aF(\xi), \tag{7.17}$$

where a is a positive constant, which we may assume to be $1/2\pi$ after normalizing the metric. Combining (7.16) and (7.17) gives, for any holomorphic curve $f : \mathbf{C} \to M$,

$$\begin{aligned} dd^c \log(F \circ f)(\partial/\partial\zeta, \partial/\partial\bar{\zeta}) &= dd^c \log F(f_*\partial/\partial\zeta, f_*\partial/\partial\bar{\zeta}) \\ \geq \quad F(f_*\partial/\partial\zeta, f_*\partial/\partial\bar{\zeta}) &= (F \circ f)(\partial/\partial\zeta, \partial/\partial\bar{\zeta}). \end{aligned}$$

This means that $f^*F = F \circ f$ is a pseudo-length function on $\mathbf{C}$ whose Gaussian curvature is bounded above by -1. So by the Schwarz Lemma, we have

$$f^*F \leq ds_R^2$$

where ds_R^2 is the Poincaré metric on the disc $\mathbf{D}(R)$. Let $R \to \infty$, we have $f^*F \equiv 0$, so $f^*\omega \equiv 0$. This proves the first case.

Now let D be an ample divisor. We have to modify the Finsler metric defined above. Define a Finsler metric on $T_D M$ by, for every $\xi \in T_D M$,

$$||\xi||^2 = ||\xi||_{\omega \otimes \rho} = (|\omega(\otimes^m \xi)|^2)^{1/m},$$

where $T_D M$ is the logarithmic tangent bundle of M. Let $L = \mathcal{O}(D)$ and let $s_i \in H^0(M, L)$ be a holomorphic section such that $D_i = \{s_i = 0\}$ and consider the function

$$\phi(\xi) = \frac{||\xi||_{\omega\otimes\rho}}{\prod_{i=1}^{q}(\log\mu/||s_i||^2)^2}, \quad \xi \in T_D X \tag{7.18}$$

where μ is a constant to be determined and where $||s_i||$ is the norm of a Hermitian metric h along the fiber of L.

Differentiating (7.18) in the sense of the differential form yields:

$$\begin{aligned} dd^c \log\phi &= \frac{1}{m}c_1(\mathcal{O}(A), \rho^*) + \sum_{j=1}^{q} dd^c \log\frac{1}{(\log\mu/||s_i||^2)^2} \\ &= \frac{1}{m}c_1(\mathcal{O}(A), \rho^*) + \sum_{i=1}^{q}\frac{c_1(L, h)}{\log\mu/||s_i||^2} + \sum_{i=1}^{q}\frac{\log||s_i||^2 \wedge d^c\log||s_i||^2}{(\log\mu/||s_i||^2)^2}. \end{aligned}$$

Given any $\epsilon > 0$ we may choose μ so that

$$\frac{1}{\log\mu/||s_i||^2} \geq -\epsilon$$

and for ϵ sufficiently small enough we have (as $c_1(A, \rho^*)$ is positive definite)

$$\frac{1}{3m}c_1(\mathcal{O}(A), \rho^*) - q\epsilon c_1(L, h) \geq 0 \tag{7.19}$$

and so

$$dd^c \log\phi \geq \frac{2}{3m}c_1(\mathcal{O}(A), \rho^*) + \sum_{i=1}^{q}\frac{d\log||s_i||^2 \wedge d^c\log||s_i||^2}{(\log\mu/||s_i||^2)^2}.$$

Since $D = D_1 + \cdots + D_q$ is of simple normal crossing, at any point $x \in M$ there exists an open neighborhood U' and local coordinates $z = (z_1, \ldots z_n)$ such that $D \cap U' = \{z | z_1 \cdots z_k = 0\}$ for some $1 \leq k \leq n$. Assume without loss of generality that $s_i = z_i, i = 1, \ldots, k$ on U' and $s_{k+1}, \ldots, s_q$ are non-vanishing on U'. Since M is compact it is covered by a finite number of such open neighborhoods. In fact we can choose $U \subset \bar{U} \subset U'$ so that a finite number of U covers M.

Each term $d\log||s_i||^2 \wedge d^c\log||s_i||^2/(\log\mu/||s_i||^2)^2$ is regular and on the neighborhood U' is smooth and hence bounded on U for $k+1 \leq i \leq q$. On the other hand, for $1 \le i \le k$, write $\|s_i\| = |z_i|^2 h_i$ where $h_i > 0$ is smooth on U'. Then

$$\begin{aligned}
&d\log\|s_i\|^2 \wedge d^c \log\|s_i\|^2 \\
&= \frac{\sqrt{-1}}{2\pi}\left(\frac{dz_i \wedge d\bar z_i}{|z_i|^2} + \frac{dz_i \wedge \bar\partial \log h_i}{z_i} + \frac{\partial \log h_i \wedge d\bar z_i}{\bar z_i} + \partial \log h_i \wedge \bar\partial \log h_i\right) \\
&= \frac{\sqrt{-1}}{2\pi}\left(\frac{1}{4}\frac{dz_i \wedge d\bar z_i}{|z_i|^2} + \frac{dz_i \wedge \bar\partial \log h_i}{z_i} + \frac{\partial \log h_i \wedge d\bar z_i}{\bar z_i} + 4\partial \log h_i \wedge \bar\partial \log h_i\right) \\
&\quad + \frac{\sqrt{-1}}{2\pi}\frac{3}{4}\frac{dz_i \wedge d\bar z_i}{|z_i|^2} - \frac{\sqrt{-1}}{2\pi} 3\partial \log h_i \wedge \bar\partial \log h_i \\
&\ge \frac{\sqrt{-1}}{2\pi}\left(\frac{3}{4}\frac{dz_i \wedge d\bar z_i}{|z_i|^2} - 3\partial \log h_i \wedge \bar\partial \log h_i\right)
\end{aligned}$$

where we have used the elementary identity $\frac{1}{4}|a|^2 + a\bar b + \bar a b + 4|b| = |\frac{1}{2}a + 2b|$. Thus we have

$$\begin{aligned}
dd^c \log\phi &\ge \frac{2}{3m} c_1(\mathcal{O}(A), \rho^*) \\
&\quad + \frac{3}{4}\frac{\sqrt{-1}}{2\pi}\sum_{i=1}^k \frac{dz_i \wedge d\bar z_i}{|z_i|^2(\log|z_i|^2)^2} - 3\frac{\sqrt{-1}}{2\pi}\sum_{i=1}^k \frac{\partial \log h_i \wedge \bar\partial \log h_i}{(\log\|s_i\|^2)^2} \\
&\ge \frac{2}{3m} c_1(\mathcal{O}(A), \rho^*) + \frac{3}{4}\frac{\sqrt{-1}}{2\pi}\sum_{i=1}^k \frac{dz_i \wedge d\bar z_i}{|z_i|^2(\log|z_i|^2)^2} - 3\epsilon^2 \sum_{i=1}^k d\log h_i \wedge d^c \log h_i.
\end{aligned}$$

Since g_i is smooth on $\bar U$ we have

$$\frac{1}{6m} c_1(\mathcal{O}(A), \rho^*) \ge 3\epsilon^2 \sum_{i=1}^k d\log h_i \wedge d^c \log h_i \tag{7.20}$$

on $\bar U$ for ϵ sufficiently small. Similarly, there exists $\lambda_U > 0$ such that

$$\frac{1}{6m} c_1(\mathcal{O}(A), \rho^*) \ge \lambda_U \sum_{i=k+1}^n dz_i \wedge d\bar z_i. \tag{7.21}$$

Putting all these together we get

$$dd^c \log\phi \ge \frac{1}{3m} c_1(\mathcal{O}(A), \rho^*) + a_U\left(\sum_{i=1}^k \frac{dz_i \wedge d\bar z_i}{|z_i|^2(\log|z_i|^2)^2} + \sum_{i=k+1}^n dz_i \wedge d\bar z_i\right) \tag{7.22}$$

where a_U is a positive constant. Consider

$$\xi = a_1 z_1 \frac{\partial}{\partial z_1} + \ldots + a_k z_k \frac{\partial}{\partial z_k} + \ldots + a_n z_n \frac{\partial}{\partial z_n} \in T_D X|_U$$

with $\sum_{i=1}^n |a_i|^2 \neq 0$. Then on U there exists constant $b_U > 0$ such that

$$\begin{aligned} L_U(\xi) &= \sum_{i=1}^k \frac{|a|_i^2}{(\log \mu/|z_i|^2)^2} + \sum_{i=k+1}^n |a|_i^2 \geq \frac{b_U}{\prod_{i=1}^k (\log \mu/|z_i|^2)^2} \sum_{i=1}^n |a_i|^2 \\ &\geq \frac{b'_U}{\prod_{i=1}^k (\log \mu/|z_i|^2)^2} \sum_{i=1}^n |a_i|^2. \end{aligned}$$

On the other hand

$$||\xi||^2_{\omega\otimes\rho} \leq c_U \sum_{i=1}^n |a_i|^2$$

for some constant $c_U > 0$ on U. Thus there exists a positive constant $\alpha_U > 0$ such that $L_U(\xi) \geq \alpha_U \phi(\xi)$ on U. Let $\alpha = \min \alpha_U > 0$ where the minimum is taken over the finite covering U's. Then

$$dd^c \log \phi(\xi, \bar{\xi}) \geq \frac{1}{3m} c_1(\mathcal{O}(A), \rho^*)(\xi, \bar{\xi}) + \alpha \frac{1}{2\pi} \phi(\xi), \xi \in T_D M. \qquad (7.23)$$

By rearranging the constant to the norm, we may assume that $\alpha = 1$. Let $f : \mathbf{C} \to M - D$ be a holomorphic curve. Then (7.23) implies that

$$\begin{aligned} dd^c \log(\phi \circ f)(\partial/\partial\zeta, \partial/\partial\bar{\zeta}) &= dd^c \log \phi(f_*\partial/\partial\zeta, f_*\partial/\partial\bar{\zeta}) \\ &\geq \frac{1}{2\pi}\phi(f_*\partial/\partial\zeta, f_*\partial/\partial\bar{\zeta}) = \frac{1}{2\pi}(\phi \circ f)(\partial/\partial\zeta, \partial/\partial\bar{\zeta}). \end{aligned}$$

This means that $f^*\phi = \phi \circ f$ is a pseudo-length function on $\mathbf{C}$ whose Gaussian curvature is bounded above by -1. So by the Schwarz Lemma, we have

$$f^*\phi \leq ds_R^2$$

where ds_R^2 is the Poincaré metric on the disc $\mathbf{D}(R)$. Letting $R \to \infty$, we have $f^*\phi \equiv 0$, so $f^*\omega \equiv 0$. □

To apply Theorem A7.5.1, we introduce some notation. Let E be a holomorphic vector bundle of rank r over M. At each point $x \in M$, let $\mathbf{P}(E_x)$ be the $(r-1)$-dimensional projective lines through the origin in the fiber E_x. Let $\mathbf{P}(E)$ be the fiber bundle over M whose fiber at x is $\mathbf{P}(E_x)$. In other words,

$$\mathbf{P}(E) = \cup_{x \in M} \mathbf{P}(E_x) = (E - \text{zero section})/\mathbf{C}^*.$$

Using the projection $p : \mathbf{P}(E) \to M$, we pull back the bundle E to obtain the vector bundle p^*E of rank r over $\mathbf{P}(E)$. We define the **tautological bundle** $\mathcal{O}_{\mathbf{P}(E)}(-1)$ over $\mathbf{P}(E)$ as the subbundle of p^*E whose fiber at any point $(x, v) \in \mathbf{P}(E)$ is the line in E_x represented by v. A vector bundle E over M is said to be **big** (resp. **ample**) if the line bundle $\mathcal{O}_{\mathbf{P}(E^*)}(1)$ over $\mathbf{P}(E^*)$ is big (resp. ample), where E^* is the dual of E.

We now consider the case that the cotangent bundle T^*M is big. By Grothendieck's well-known result ([Gro], p. 162-163),

$$H^0(M, S^m T^*M \otimes \mathcal{O}(-A)) \cong H^0(\mathbf{P}(TM), \mathcal{O}_{\mathbf{P}(TM)}(m) \otimes (-p^*A)),$$

where TM is the tangent bundle over M. In the case that the cotangent bundle T^*M is big, i.e., the line bundle $\mathcal{O}_{\mathbf{P}(TM)}(1)$ is big, there exists a global section in $H^0(\mathbf{P}(TM), \mathcal{O}_{\mathbf{P}(TM)}(m) \otimes (-p^*A))$ for m large enough. So there exists an $\omega \in H^0(M, S^m T^*M \otimes \mathcal{O}(-A))$. Applying Theorem A7.5.1 with D as a trivial divisor, $f^*\omega \equiv 0$. In general, it is very hard to go from $f^*\omega \equiv 0$ to the conclusion that f is constant. In the case that the cotangent bundle T^*M is ample, then the base locus is empty, so we conclude that f is constant. This proves Kobayashi's result that *For a compact complex manifold M, if its cotangent bundle Ω_M is ample, then M is hyperbolic.*

Next we consider the surfaces of general type. Let M be a compact complex surface with $c_1^2 > c_2$, where c_1, c_2 are the first and second Chern forms of M. To produce a meromorphic differential

$$\omega \in H^0(M, S^m T^*M \otimes \mathcal{O}(-A)),$$

we use Hirzebruch's Riemann-Roch formula (see [Hi]). By Hirzebruch's Riemann-Roch formula,

$$\chi(M, S^m T^*M) = \frac{m^3}{6}(c_1^2 - c_2) + O(m^2).$$

On the other hand, Serre duality implies that

$$h^2(M, S^m T^*M) = h^0(M, S^m T^*M \otimes K_M)$$

where K_M is the canonical bundle over M. A vanishing theorem due to Bogomolov [Bo] implies that, on a surface M of general type,

$$h^0(M, S^mT^*M \otimes K_M^{\otimes q}) = 0$$

for all m, q such that $m - 2q > 0$. In particular, $h^0(M, S^mT^*M \otimes K_M) = 0$ whenever $m \geq 3$ and we get, by combining Grothendieck's result

$$\begin{aligned} h^0(\mathbf{P}(TM), \mathcal{O}_{\mathbf{P}(TM)}(m)) &= h^0(M, S^mT^*M) \\ &\geq \chi(M, S^mT^*M) = \frac{m^3}{6}(c_1^2 - c_2) + O(m^2). \end{aligned}$$

As a consequence, the line bundle $\mathcal{O}_{\mathbf{P}(TM)}(1)$ is big when $c_1^2 > c_2$. So there exists a nontrivial section $\omega \in H^0(\mathbf{P}(TM), \mathcal{O}_{\mathbf{P}(TM)}(m) \otimes (-p^*A)) \cong H^0(M, S^mT^*M \otimes \mathcal{O}(-A))$ for m large enough. If $f : \mathbf{C} \to M$ is a holomorphic curve, then, by Theorem A7.5.1 (with $D = \emptyset$), we have $f^*\omega \equiv 0$. This means that f satisfies a differential equation. However, to conclude hyperbolicity, i.e., f is constant, one needs some new techniques. We present here two results which give some conclusions on complex hyperbolicity under various conditions. The first one is due to Lu-Yau.

Theorem A7.5.2(Lu-Yau) *Let M be a surface of general type and $c_1^2(M) > 2c_2(M)$, where $c_1(M), c_2(M)$ are the first and second Chern forms of M, then M is Brody hyperbolic.*

Proof. By the above discussion, we see that there exists a nontrivial section $\omega \in H^0(M, S^mT^*M \otimes \mathcal{O}(-A))$ such that $f^*\omega \equiv 0$ for m large enough. That is $f(\mathbf{C})$ is contained in a horizontal component of the zero set Z^ω of ω. Let Y be such a horizontal component of the zero set Z^ω. By the condition $c_1^2(M) > 2c_2(M)$, one can prove (see [Lu-Y]), using Riemann-Roch again, that

$$\dim H^0(Y, S^mT^*Y \otimes \mathcal{O}(-A)|_Y) > bm^2$$

for m big enough. Thus, applying Theorem A7.5.1 again, we conclude that the image of f is contained in the zero set of some nontrivial section in $H^0(Y, S^mT^*Y \otimes \mathcal{O}(-A)|_Y)$. By the dimension reason, this zero set has dimension zero. Hence f is constant. □

The second result is due to Dethloff-Schumacher-Wong [DSW1]. They used meromorphic 1-jet differentials with only logarithmic pole singularities to show the hyperbolicity of the complement of 3 generic nonsingular plane curves with very mild conditions on their degrees. One new ingredient they introduced is the use of a Borel Lemma-type argument after the use of Theorem A7.5.1 for meromorphic differentials with only logarithmic pole singularities.

Theorem A7.5.3 *Let C_j $(1 \leq j \leq 3)$ be three nonsingular curves in $\mathbf{P}^2$ such that the degree of C_1 is at least 3 and the degrees of C_2 and C_3 are at least 2. Suppose that the three nonsingular curves intersect only in normal crossing and if one curve C_j is a quadric then there does not exist a line which is tangential to C_k for $k \neq j$ and which intersects C_j at precisely the two points of tangency with C_k $(k \neq j)$. Then there is no non-constant holomorphic map from $\mathbf{C}$ to $\mathbf{P}^2 - \cup_{j=1}^3 C_j$.*

Proof. First we use the theorem of Riemann-Roch to produce a meromorphic 1-jet differential which is holomorphic except for logarithmic poles along $\cup_j C_j$. Let $C = \cup_{j=1}^3 C_j$ and the degree of C_j be b_j. Let $\Omega^1_{\mathbf{P}^2}(\log C)$ be the bundle over $\mathbf{P}^2$ of meromorphic 1-forms which are holomorphic except for logarithmic poles at C. By using the exact sequence $0 \to \Omega^1_{\mathbf{P}^2} \to \Omega^1_{\mathbf{P}^2}(\log C) \to \Omega^1_{\mathbf{P}^2}(\log C)/\Omega^1_{\mathbf{P}^2} \to 0$, where $\Omega^1_{\mathbf{P}^2}$ is the cotangent bundle of $\mathbf{P}^2$, one computes $c_1^2(\Omega^1_{\mathbf{P}^2}(\log C)) - c_2(\Omega^1_{\mathbf{P}^2}(\log C))$ to be $-3(\sum_{j=1}^3 b_j - 4) - 6 + \sum_{i<j} b_i b_j$ which is positive when $b_1 \geq 3$ and $b_2 \geq 2$ and $b_3 \geq 2$. Similar to the above discussion, using the theorem of Riemann-Roch and the result of Bogomolov [Bog], we can find a nontrivial section

$$\omega \in H^0(\mathbf{P}^2, S^m \Omega^1_{\mathbf{P}^2}(\log C) \otimes \mathcal{O}(-A))$$

for some ample curve A. Let $f : \mathbf{C} \to \mathbf{P}^2 - C$ be a non-constant holomorphic map. Theorem A7.5.1 implies that the pull-back $f^*\omega$ of ω by f is identically zero. The paper of Dethloff-Schumacher-Wong [DSW94] introduced a new technique to show that the image of f is contained in an algebraic curve. It uses a refinement of the Borel Lemma argument. First of all, since there cannot exist any non-constant holomorphic map from $\mathbf{C}$ to C, by using Brody's reparametrization, according to Lemma A7.3.3, we can assume without loss of generality that the pointwise norm of df is uniformly bounded on $\mathbf{C}$. Let a_j be positive integers such that $a_j b_j$ is the same for $1 \leq j \leq 3$. Let σ_j be the homogeneous polynomial of degree b_j defining C_j. Consider the map $\Phi : \mathbf{P}^2 \to \mathbf{P}^2$ with finite fibers defined by $[\sigma_1^{a_1}, \sigma_2^{a_2}, \sigma_3^{a_3}]$. Let $\tilde{\omega}$ be the meromorphic 1-jet differential on $\mathbf{P}^2$ which is obtained by locally taking the product of the $a_1 b_1$ values of ω on the $a_1 b_1$ sheets of Φ. We claim that the three homogeneous components of $\Phi \circ f$ are given by $\exp(p_j(z))$ for some polynomial $p_j(z)$ of degree at most 2 ($0 \leq j \leq 2$). In fact, by Lemma A7.3.3 (Brody's reparametrization Lemma) we can assume that $|d\Phi \circ f|$ is uniformly bounded on $\mathbf{C}$. It follows that $T_{\Phi\circ f}(r) = O(r^2)$. Let $g_0, \ldots, g_2$ be any entire function on $\mathbf{C}$ without common zeros such that $[g_0 : \ldots : g_2]$ defines $\Phi \circ f$. For any $1 \leq j \leq 2$, by Theorem A3.1.2,

$$N_{g_j/g_0}(r, \infty) \leq T_{g_j/g_0}(r) \leq T_{\Phi\circ f}(r) = O(r^2).$$

Let B_j be the pole divisor of g_j/g_0. Let Z be the zero-set of g_0. Since $g_0, \ldots, g_2$ have no common zeros, $Z \leq \sum_{j=1}^n B_j$ as divisors. Write $Z = \sum_{\nu=1}^\infty a_\nu$ and let W_Z be the Weierstrass canonical product $\prod_{\nu=1}^\infty E(\frac{z}{a_\nu}, 2)$ where $E(\frac{z}{a_\nu}, 2) = (1-z)exp(\sum_{\nu=1}^\infty \frac{z^\nu}{\nu})$. Then $T_{W_Z}(r) = O(r^2)$. Let

$$\exp(p_0(z)) = W_Z \quad \text{and} \quad \exp(p_j(z)) = W_Z \frac{g_j}{g_0} \text{ for } 1 \leq j \leq 2.$$

Then we have $T_{\exp(p_j(z))}(r) \leq O(r^2)$ for $0 \leq j \leq 2$. So $p_j(z)$ are polynomials of degree at most 2, $0 \leq j \leq 2$. So the claim holds. We now continue our proof. The identical vanishing of $f^*\omega$ from Theorem A7.5.1 implies that there exist rational functions $R_j(\xi_1, \xi_2)$ of ξ_1, ξ_2 such that

$$\sum_{j=1}^k R_i(\xi_1, \xi_2)(d\xi_1)^j (d\xi_2)^{k-j} \equiv 0$$

when $\xi_j = \exp(p_j(z) - p_0(z))$ ($j = 1, 2$). This means that we have an equation of the type

$$\sum_{j=1}^k \exp(P_j(z))Q_j(z) \equiv 0, \tag{7.24}$$

where P_j and Q_j are polynomials such that $Q_1, \cdots, Q_k$ have no common zeros. Consider the map ψ from $\mathbf{C}$ to $\mathbf{P}^{k-1}$ defined by

$$[\exp(P_1(z))Q_1(z), \cdots, \exp(P_k(z))Q_k(z)].$$

By (7.24) the image of ψ lies in the hyperplane $H := \{\sum_{j=1}^k \zeta_j = 0\}$, where $[\zeta_1, \cdots, \zeta_k]$ is the homogeneous coordinate of $\mathbf{P}^{k-1}$. The algebraic degeneracy of $\Phi \circ f$ (and therefore of f) comes from applying the Second Main Theorem to the map ψ from $\mathbf{C}$ to H and the hyperplanes $H \cap \{\zeta_j = 0\}$ in H $(1 \leq j \leq k)$, because, though $\exp(P_j(z))Q_j(z)$ may be zero, the number of its zeros is finite and does not make the defect for $H \cap \{\zeta_j = 0\}$ smaller than 1. The condition on C_j $(1 \leq j \leq 3)$ rules out the existence of such an algebraic curve. □

The case of logarithmic jet differentials are defined analogously. Let M be a compact complex manifold. Let $D = D_1 + \ldots + D_q$ be a divisor on M where $\{D_i \mid i = 1, \ldots, q\}$ are the irreducible components of D. Furthermore we assume that D is of simple normal crossing, i.e., each D_i is smooth; D_i intersects D_j transversally for all $i \neq j$ and each point x is contained in at most n components. For simplicity we also assume that D_i' are linearly equivalent, i.e., the line bundle $L = \mathcal{O}(D_i)$ associated to the divisor D_i is the same for all $i = 1, \ldots, q$. For every $P \in M$, since D is of simple normal crossing, D is locally defined by $\prod_{j=1}^q z_j = 0$ for some local coordinate system $z_1, \ldots, z_n$ of P. A meromorphic k-jet differential ω on M of total weight m with at most logarithmic pole singularities along D is expressed in the form

$$\begin{aligned} \omega \quad = & \\ \sum_\nu \quad & \omega_{\nu_{1,1}\cdots\nu_{1,k}\cdots\nu_{n,1}\cdots\nu_{n,k}}(\frac{dz_1}{z_1})^{\nu_{1,1}}\cdots(\frac{d^kz_1}{z_1})^{\nu_{1,k}}\cdots(\frac{dz_q}{z_q})^{\nu_{q,1}}\cdots(\frac{d^kz_q}{z_q})^{\nu_{1,k}} \\ & \cdot(dz_{q+1})^{\nu_{q,1}}\cdots(d^kz_{q+1})^{\nu_{1,k}}\cdots(dz_n)^{\nu_{n,1}}\cdots(d^kz_n)^{\nu_{n,k}} \end{aligned} \tag{7.25}$$

where the summation is over the kn-tuple

$$\nu = (\nu_{1,1}\cdots\nu_{1,k}\cdots\nu_{n,1}\cdots\nu_{n,k})$$

of non-negative integers with

$$(\nu_{1,1} + 2\nu_{1,2} + \cdots + k\nu_{1,k}) + \cdots + (\nu_{n,1} + 2\nu_{n,2} + \cdots + k\nu_{n,k}) = m$$

and $\omega_{\nu_{1,1}\cdots\nu_{1,k}\cdots\nu_{n,1}\cdots\nu_{n,k}}$ is a meromorphic function locally defined.

Let M be a projective compact complex manifold. Fix a positive $(1,1)$ form ω on M and write $T_f(r) = T_{f,\omega}(r)$. The following theorem is due to P.M. Wong [Wong 6].

Theorem A7.5.4 (Logarithmic derivative Lemma for Jet Differentials with possible log poles) *Let k be a positive integer. Let M be a* *projective compact complex manifold of dimension n. Let D either be an effective divisor in M with simple normal crossings or just a trivial divisor. Let ω be a k-jet meromorphic differential form of total weight m with at most logarithmic pole singularities along D. Let $f : \mathbf{C} \to M$ be a holomorphic map such that $f(\mathbf{C})$ is not contained in D. Assume that $f^*\omega = \xi(d\zeta)^m$ on $\mathbf{C}$. Then either $f^*\omega \equiv 0$ or*

$$\frac{1}{2\pi}\int_0^{2\pi} \log^+ |\xi(re^{i\theta})| d\theta \leq O(\log^+ T_f(r)),$$

where the inequality holds for all r outside a set $E \subset (0, +\infty)$ with finite Lebesgue measure.

Proof. We claim that there exists a finite set $\mathcal{H}$ of rational functions on M with the property that for each point $x \in M$ there exists a subset $\mathcal{H}_x$ of $\mathcal{H}$ such that $\{(d^{(\alpha)}h/h)^{m/\alpha} \mid h \in \mathcal{H}_x, 1 \leq \alpha \leq k\}$ span the fiber of $\mathcal{J}_k^m M(\log D)$ over $x \in M$, where $\mathcal{J}_k^m M(\log D)$ is the sheaf of germs of meromorphic k-jet differentials on M of total weight m with at most logarithmic pole singularities along D.

We now construct $\mathcal{H}$. We may assume without loss of generality that D is irreducible, since otherwise we will just consider each component D_i because D has only finitely many components and a finite union of a finite number of rational functions is still finite. We also assume that D is ample; otherwise we may replace D by $D + D'$ so that $D + D'$ is ample. Observe that if s is a function holomorphic on a neighborhood U such that $[s = 0] = D \cap U$ then $[s^\tau = 0] = \tau D \cap U$ where τ is a real constant. Thus $d^{(j)}(\log s^\tau) = \tau d^{(j)}(\log s)$. This means that we may assume that D is very ample by replacing D with τD for some τ so that τD is very ample.

Let $u \in H^0(M, \mathcal{O}(D))$ be a section such that $D = [u = 0]$. At a point $x \in D$ choose a section $v_1 \in H^0(M, \mathcal{O}(D))$ so that $E_1 = [v_1 = 0]$ is smooth, $D + E_1$ is of simple normal crossings and v_1 is non-vanishing at x. The rational function $t_1 = u/v_1$ is regular on the affine open neighborhood $M - E_1$ of x and $(M - E_1) \cap [t_1 = 0] = (M - E_1) \cap D$. Choose rational functions $t_2 = u_2/v_2, ..., t_n = u_n/v_n$ where u_i and v_i are sections of a very ample bundle L on M so that $t_2, ..., t_n$ are regular at x, the divisors $D_i = [u_i = 0], E_i = [v_i = 0]$ are smooth and $D + D_2 + \cdots + D_n + E_1 + \cdots + E_n$ is of simple normal crossings. Moreover, since the bundles involved are very ample the sections can be chosen so that $dt_1 \wedge \cdots \wedge dt_n$ is non-vanishing at x; the complete system of sections provides an embedding, hence at each point there are $n+1$ sections with the property that n of the quotients of these $n+1$ sections form a local coordinate system on some open neighborhood U_x of x. Then the set $\{(d^{(\alpha)}t_j/t_j)^{m/\alpha} \mid 1 \le j \le n, 1 \le \alpha \le k\}$ generates $\mathcal{J}_k^m M(\log D)$ over U_x. However, since D is compact it is covered by a finite number of such open neighborhoods, say $U_1, ..., U_m$ and we obtain a finite number of rational functions. Moreover, there exist relatively compact open subsets U_i' of U_i $(1 \le i \le m)$ such that $\cap_{1\le i\le m} U_i'$ still covers D. Next we consider a point x in the compact set $M - \cap_{1\le i\le m} U_i'$. Repeating the procedure as above we can find rational functions $s_1 = a_1/b_1, ..., s_n = a_n/b_n$ where a_i and b_i are sections of some very ample line bundle L so that $s_1, ..., s_n$ form a holomorphic local coordinate on some open neighborhood V_x of x. Then the set $\{(d^{(\alpha)}s_j/s_j)^{m/\alpha} \mid 1 \le j \le n, 1 \le \alpha \le k\}$ generates $\mathcal{J}_k^m M$ over V_x. We must also choose these sections so that the divisor $H = [s_1 \cdots s_n = 0]$ together with those divisors (finite in number), which had been already constructed, is still a divisor with simple normal crossings(this is possible by the very ampleness of the line bundle L). Since $M - \cap_{1\le i\le m} U_i'$ is compact, it is covered by a finite of such coordinate neighborhoods. So the number of rational functions obtained is still finite. Let $\mathcal{H}$ be the collection of all the rational functions obtained above, then the claim holds. Note if D is the trivial divisor, then it is enough to use only the second part of the construction.

We can also describe the above argument in the following simple way. Let $\mathcal{L}_0$ be a very ample line sheaf on M. Let $E_{00}, \dots, E_{0,2n}$ be effective divisors corresponding to $\mathcal{L}_0$ such that any $n+1$ have empty intersection, and for integer i, j with $0 \le i < j \le 2n$ choose a rational function f_{ij} on M such that $(f_{ij}) = E_{0i} - E_{0j}$. Then the set $\{(d^{(\alpha)}f_{ij}/f_{ij})^{m/\alpha} \mid 0 \le i < j \le 2n, 1 \le \alpha \le k\}$ has the property that, for each point $x \in M$, some subset of this is regular at x and generates $\mathcal{J}_k^m M$ there. Next, let $\mathcal{L}_1$ and $\mathcal{L}_1'$ be very ample line sheaves on M such that $\mathcal{L}_1 \cong \mathcal{L}_1'(D)$. For each $j = 1, \dots, n$, choose divisors E_{1j} and E_{1j}' associated to $\mathcal{L}_1$ and $\mathcal{L}_1'$ respectively, such that $\cap_j(\mathrm{supp}E_{1j} \cup \mathrm{supp}E_{1j}')$ is disjoint from $\mathrm{supp}D$. Choose a rational function g_j on M with $(g_j) = D + E_{1j}' - E_{1j}$. Then the set $\{(d^{(\alpha)}g_j/g_j)^{m/\alpha} \mid 1 \le j \le n, 1 \le \alpha \le k\}$ has the property that, for each point $x \in M$, some subset of this is regular at x and generates $\mathcal{J}_k^m M(\log D)$ there. Let $\mathcal{H} = \{f_{ij} \mid 0 \le i < j \le 2n\} \cup \{g_j \mid 1 \le j \le n\}$. By compactness of M, $\mathcal{H}$ satisfies the desired condition.

Since $f^*\omega = \xi(d\zeta)^m$, by the claim, using the compactness of M and using also the inequality $\log^+(a+b) \le \log^+ a + \log^+ b$, we have

$$\frac{1}{2\pi}\int_0^{2\pi} \log^+ |\xi(re^{i\theta})| d\theta \le C \sum_{h\in\mathcal{H}} \sum_{1\le\alpha\le k} \frac{1}{2\pi}\int_0^{2\pi} \log^+ \left| \frac{(h\circ f)^{(\alpha)}}{h\circ f} \right| d\theta$$

where $C > 0$ is a constant. Now by Theorem A1.2.5, the inequality

$$\int_0^{2\pi} \log^+ |(h\circ f)^{(\alpha)}/h\circ f| \frac{d\theta}{2\pi} \le O(\log^+ T_{h\circ f}(r))$$

holds for all r outside a set $E \subset (0,+\infty)$ with finite Lebesgue measure. Since h is a rational function, $\log^+ T_{h\circ f}(r) \le O(\log^+ T_f(r))$ and we arrive at the estimate. □

Theorem A7.5.5 (Wong) *Let M be a projective compact complex manifold. Let D either be an effective divisor in M with simple normal crossings or just a trivial divisor. Let ω be a meromorphic k-jet differential of weight m with logarithmic singularity along D that vanishes along an ample divisor A. Then for every holomorphic map $f : \mathbf{C} \to M - D$ (resp. $f : \mathbf{C} \to M$ in the second case), we have $f^*\omega \equiv 0$.*

Proof. Let ω be a meromorphic k-jet differential of weight m with logarithmic singularity along D vanishing along an ample divisor A. Let $f : \mathbf{C} \to M - D$ (resp. $f : \mathbf{C} \to M$ in the second case) be a holomorphic map. Assume that $f^*\omega \not\equiv 0$. We will derive a contradiction. Choose a positive integer ℓ such that ℓA is very ample. Then a basis of holomorphic sections $H^0(M, \mathcal{O}(\ell A))$ embeds M into the projective space $\mathbf{P}^N(\mathbf{C})$ with homogeneous coordinates $[w_0, \ldots, w_N]$. By Cartan's Second Main Theorem, we conclude that for any $1 > \epsilon > 0$, there exists a hyperplane $H = \{[w_0 : \cdots : w_N] \mid \sum_{i=0}^N a_i w_i = 0\}$ such that $N_f(r,H) \ge (1-\epsilon)T_f(r)$. Let s_A be the holomorphic section of the line bundle associated with the given ample divisor A so that the divisor of s_A is A. By replacing ω by $\left(\frac{\omega}{s_A}\right)^\ell (\sum_{i=0}^N a_i w_i)$ we can assume without loss of generality that $\ell = 1$ and $A = \{[w_0 : \cdots : w_N] \mid \sum_{i=0}^N a_i w_i = 0\} \cap M$. So we have

$$N_f(r,A) \ge (1-\epsilon)T_f(r).$$

Since ω vanishes along an ample divisor A, by the First Main Theorem,

$$T_{\omega\circ j^k f}(r) \ge N_f(r,A) \ge (1-\epsilon)T_f(r).$$

On the other hand, by Theorem A7.5.4,

$$T_{\omega \circ j^k f}(r) \;\le\; O(\log^+ T_f(r))$$

where the inequality holds for all r outside a set $E \subset (0, +\infty)$ with finite Lebesgue measure. This gives a contradiction. □

Part B: Diophantine Approximation

B7.1 Lang's Conjecture

The following Conjecture is due to Lang.

Conjecture B7.1.1 (Lang) *Let $M \subset \mathbf{P}^N$ be a complex submanifold which is defined over a number field k. If M is Kobayashi hyperbolic as a complex manifold, then there exist only finitely many k-rational points of M, where the k-rational points of M are the points whose homogeneous coordinates are represented by values in k.*

It is Lang's optimistic view that there would be no way to obtain infinitely many k-rational points of M other than to construct them from projective space $\mathbf{P}^1_k$ (rational curves) or from Abelian varieties into M. Of course, in such case, M is not hyperbolic. In the case of $\dim M = 1$, M is Kobayashi hyperbolic if and only if the genus of $M \geq 2$. Hence the Conjecture in this case is equivalent to Mordell's conjecture, which was solved by Faltings.

The Correspondence Table

Nevanlinna Theory	**Diophantine Approximation**
Theorem A7.3.1	Conjecture B7.1.1

Bibliography

[Ad-S] Adachi, Y. and Suzuki, M.(1990) "On the family of holomorphic mappings into projective space with lacunary hypersurfaces", *J. Math. Kyoto Univ.* **30**, 451–459.

[Ahl] Ahlfors, L.(1941) "The theory of meromorphic curves", *Acta Soc. Sci. Fenn. Nova* Ser. A **3**(4), 171–183.

[Ash] Ashline, G.L.(1999) *The defect relation of meromorphic maps on parabolic manifolds*, Memoirs of AMS, Vol. 139, Num. 665.

[Ax] Ax, J.(1972) "Some topics in differential algebraic geometry II", *Amer. J. Math.* **94**, 1205–1213.

[Az-S] Azukawa, K. and Suzuki, M.(1980) "Some examples of algebraic degeneracy and hyperbolic manifolds", *Rocky Mountain J. Math.* **10**, 655–659.

[Bab] Babets, V.A.(1984) "Theorems of Picard type for holomorphic mappings", *Siberian Math. J.* **25**, 195–200.

[Bak] Baker, A.(1990) *Transcendental number theory (3rd edn.)*, Cambridge University Press.

[Bia] Biancofiore, A.(1982) "A hypersurface defect relation for a class of meromorphic maps", *Trans. A.M.S.* **270**, 47–60.

[Bia-S] Biancofiore, A. and Stoll, W.(1981) "Another proof of the lemma of the logarithmic derivative in several complex variables", In: J. Fornaess (ed.) Recent developments in several complex variables, pp. 29–45.

[Blo] Bloch, A.(1926) "Sur les systèmes de fonctions uniformes satisfaisant à l'équation d'une variété algébrique dont l'irrégularité dépasse la dimension", *J. de Math.* **5**, 19–66.

[Bog] Bogomolov, F.A.(1979) "Holomorphic tensors and vector bundles on projective varieties", *Math. USSR, Izv.* **13**, 499–555.

[Bom] Bombieri, E.(1991) "The Mordell conjecture revisited", *Ann. Scu. Norm. Sup. Pisa.* **17**, 615–640, and *Errata-Corrige, ibid* **18**, 473.

[Bo-V] Bombieri, E. and Vaaler, J.(1983) "On Siegel's lemma", *Invent. Math.* **73**, 11–32.

[Bo-C] Bott, R. and Chern, S.S.(1965) "Hermitian vector bundles and the equidistribution of the zeros of their holomorphic sections", *Acta Math.* **114**, 71–112.

[Bro] Brody, R.(1978) "Compact manifolds and hyperbolicity", *Trans. Amer. Math. Soc.* **235**, 213–219.

[Br-G] Brody, R. and Green, M.(1977) "A family of smooth hyperbolic hypersurfaces of $\mathbf{P}_3$", *Duke Math. J.* **44**, 873-874.

[Bro-M] Brownawell, W.D. and Masser, D.W.(1986) "Vanishing sums in function fields", *Math. Proc. Comb. Phil. Soc.* **100**, 427-434.

[Ca-G] Carlson, J. and Green, M.(1976) "Holomorphic curves in the plane", *Duke Math. J.* **43**, 1–9.

[Ca-G] Carlson, J. and Griffiths, P.(1972) "A defect relation for equidimensional holomorphic mappings between algebraic varieties", *Ann. of Math.* **95**, 557–584.

[Car1] Cartan, H.(1928) "Sur les systeme de fonctions holomorphes a variétés linéaires lacunaires et leurs applications", *Ann. Sci. Ecole Norm. Sup (3)* **45**, 255–346.

[Car2] Cartan, H.(1933) "Sur les zeros des combinaisons linéaires de p fonctions holomorphes données", *Mathematica (Cluf)* **7**, 5–31.

[Che] Chen, W.(1990) "Defect relations for degenerate meromorphic maps", *Trans. Amer. Math. Soc.* **319**, 499–515.

[Chern] Chern, S.S.(1968) "On holomorphic mappings of Hermitian manifolds of the same dimension", *Proc. Symp. Pur. Math. Amer. Math. Soc.* **11**, 157–170.

[Ch-Y] Cherry, W. and Ye, Z.(1999) *Navanlinna's theory of value distribution with an emphasis on error terms*, preprint.

[Chu] Chuang, C.T.(1964) "Une Generalisation d'une Ineaires de Nevanlinna", *Scientia Sinica* **13**, 887–895.

[Co-G] Cowen, M. and Griffith, Ph.(1976) "Holomorphic curves and metrics of nonnegative curvature", *J. Analyse Math.* **29**, 93–153.

[Del1] Deligne, L.(1970) *Equations différentielles à points singuliers réguliers*, Lecture Notes in Math. **163**, Springer, Berlin.

[Dem2] Demailly, J.P.(1993) "A numerical criterion for very ample line bundles", *J. Diff. Geom.*, **37**, 323–374.

[Dem3] Demailly, J.P.(1995) "Algebraic criteria for Kobayashi hyperbolic varieties and jet differentials", *Proc. Symp. Pur. Math. Amer. Math. Soc.* **62**, Part 2, 285–360.

[Dem4] Demailly, J.P.(1997) "Variétés hyperboliques et q́uations différentielles algébriques", *Gazette Math,* **73**, 3-23.

[Dem-G1] Demailly, J.P. and Goul J. El(1997) "Connexions méromorphes projectives et variétés hyperboliques", *C.R. Acad. Sci. Paris* **324**, 1385–1390.

[Dem-G2] Demailly, J.P. and Goul J. El(2000) "Hyperbolicity of generic surfaces of high degree in projective 3-space", *American J. Math.* **122**, 515–546.

[DSW1] Dethloff, G., Schumacher, G. and Wong, P.M.(1995) " Hyperbolicity of the complements of plane algebraic curves", *Amer. J. Math.* **117**, 573–599.

[DSW2] Dethloff, G., Schumacher, G. and Wong, P.M.(1995) "On the hyperbolicity of the complements of curves in algebraic surfaces: the three component case", *Duke Math. J.* **78**, 193–212.

[Ere1] Eremenko, A.E.(1999) "A Picard theorem for meromorphic curves in projective space", *Period. Math. Hungar.* **38**, 39–42.

[Ere2] Eremenko, A.E.(1996) "Holomorphic curves omitting five planes in projective spaces", *Amer. J. Math.* **118**, 1141–1151.

[Ere3] Eremenko, A.E.(1996) "A counter example to Cartan's conjecture on holomorphic curves omitting hyperplanes", *Proc. Amer. Math. Soc.* **124**, 3097–3100.

[Er-S] Eremenko, A.E. and Sodin, M.L.(1992) "The value distribution of meromorphic functions and meromorphic curves from the point of view of potential theory", *St. Petersburg Math. J.* **3**, 109–136.

[Eve1] Evertse, J.-H.(1984) "On equations in S-units and the Thue-Mahler equation", *Invent. Math.* **75**, 561–584.

[Eve2] Evertse, J.-H.(1984) "On sums of S-units and linear recurrences", *Compos. Math.* **53**, 225–244.

[Eve3] Evertse, J.-H.(1995) "The number of solutions of decomposable form equations", *Invent. Math.* **122**, 559–601.

[Eve4] Evertse, J.-H.(1995) "An explicit version of Faltings' Product Theorem and an improvement of Roth's lemma", *Acta Arith.* **73**, 215–248.

[Eve5] Evertse, J.-H.(1996) "An improvement of the quantitative subspace theorem", *Compos. Math.* **101**, 225–311.

[Eve6] Evertse, J.-H.(1997) "The number of solutions of the Thue–Mahler equation", *J. Reine Angew. Math.* **482**, 121–149.

[Ev-G1] Evertse, J.-H. and Győry, K.(1985) "On unit equations and decomposable form equations" *J. Reine Angew. Math.* **358**, 6–19.

[Ev-G2] Evertse, J.-H. and Győry, K.(1988) "Finiteness criteria for decomposable form equations", *Acta Arith.* **50**, 357–379.

[Ev-G3] Evertse, J.-H. and Győry, K.(1988) "Decomposable form equations" In: New Advances in Transcendence Theory (ed. by A. Baker). Cambridge Univ. Press, pp. 175–202.

[Ev-G4] Evertse, J.-H. and Győry, K.(1997) "The number of families of solutions of decomposable form equations", *Acta Arith.* **80**, 367–394.

[Ev-G-S-T] Evertse, J.-H., Győry, K., Stewart, C.L. and Tijdeman, R.(1988) "S-unit equations and their applications", In: New Advances in Transcendence Theory, Proc. Conf. Durham, UK, (ed. by A. Baker), 110–174. Cambridge Univ. Press.

[Ev-S] Evertse, J.-H. and Schlickewei, H.P.(2000) "A quantitative version of the absolute subspace theorem", preprint.

[Fal1] Faltings, G.(1983) "Endlichkeitssätze für abelsche Varietäten über Zahlkörpern", *Invent. Math.* **73** (1983), 349–366.

[Fal2] Faltings, G.(1991) "Diophantine approximations on abelian varieties", *Ann.*

of Math. **129**, 549–576.

[Fuj1] Fujimoto, H.(1974) "On meromorphic maps into the complex projective space", *J. Math. Sco. Japan* **26**, 272–288.

[Fuj2] Fujimoto, H.(1975) "The uniqueness problem of meromorphic maps into the complex projective space", *Nogoya Math. J.* **58**, 1–23.

[Fuj3] Fujimoto, H.(1993) *Value Distribution Theory on the Gauss Map of Minimal Surfaes in* $\mathbf{R}^m$, Aspect of Mathematics, E21, Vieweg.

[Ful] W. Fulton, W.(1976) "Ample vector bundles, Chern classes, and numerical criteria", *Invent. Math.* **32**, 171–178.

[Gra] Grauert, H.(1989) "Jetmetriken und hyperbolische Geometrie", *Math. Z.* **200**, 149–168.

[Gr-P] Grauert, H. and Peternell, U(1985). "Hyperbolicity of the complement of plane curves", *Manuscripta Math.* **50**, 429–441.

[Gre1] Green, M.(1975) "Some Picard theorems for holomorphic maps to algebraic varieties," *Amer. J. Math.* **97**, 43–75.

[Gre2] Green, M.(1977) "The hyperbolicity of the complement of $2n+1$ hyperplanes in general position in $\mathbf{P}_n$ and related results," *Proc. Amer. Math. Soc.* **66**, 109–113.

[Gre3] Green, M.(1978) "Holomorphic maps to complex tori", *Amer. J. Math.*, **100**, 615–620.

[Gre-G] Green, M. and Griffiths,P.(1980) "Two applications of algebraic geometry to entire holomorphic mappings", The Chern Symposium 1979, Proc. Internat. Sympos., Berkeley, 1979, Springer-Verlag.

[Gri-K] Griffiths, P. and King, J.(1973) "Nevanlinna theory and holomorphic mappings between algebraic varieties", *Acta Math.* **130**, 145–220.

[Gro-O] Gross, F. and Osgood, C.F. "Recent developments in the application of Diophantine Approximation to the Value Distribution Theory of meromorphic Functions", *J. of Complex Variables*, **13**, 143–154.

[Gro] Grothendieck, A.(1958) "La theorie des classes de Chern", *Bull. Soc. Math. France*, **86**, 137–154.

[Győ1] Győry, K.(1993) "Some applications of decomposable form equations to resultant equations", *Colloq. Math.*, **65**, 267–275.

[Győ2] Győry, K.(1994) "On the irreducibility of neighbouring polynomials", *Acta Arith.* **67**, 283–296.

[Győ-R] Győry, K. and Ru, M.(1998) "Integer solutions of a sequence of decomposable form inequalities", *Acta Arith.*, **86**, 227–237.

[Hay] Hayman, W.(1964) *Meromorphic Functions*, Oxford University Press.

[H-S] Hindry, M. and Silverman, J.(2000) *Diophantine Geometry: an Introduction*, Graduate Texts in Mathematics **201**, Springer-Verlag.

[Hi] Hirzebruch, F.(1966) *Topological methods in Algebraic Geometry, Grundlehren Math. Wiss.*, **131**, Springer-Verlag.

[Iit] Iitaka, S.(1976) "Logarithmic forms of algebraic varieties", *J. Fac. Univ. Tokyo Sect. IA Math.* **23**, 525–544.

[Ji] Ji, S.(1988) "Uniqueness problem without multiplicity in value distribution theory", *Pacific J. of Math.* **135**, 323–348.

[Ka] Kawamata, Y.(1980) "On Bloch's conjecture", *Invent. Math.* **57**, 97–100.

[Ko] Kobayashi, R.(1991) "Holomorphic curves into algebraic subvarieties of an abelian variety", *Int. J. Math.* **2**, 711–724.

[Kob] Kobayashi, S.(1998) *Hyperbolic Complex Spaces*, Springer-Verlag.

[Lang1] Lang, S.(1983) *Fundamentals of Diophantine Geometry*, Springer-Verlag.

[Lang2] Lang, S. (1987) *Introduction to Complex Hyperbolic Spaces*, Springer-Verlag, New York-Berlin-Heidelberg.

[Lang3] Lang, S.(1988) "The error term in Nevanlinna theory", *Duke Math. Journal*, **56**, 193–218.

[Lang-C] Lang, S. and Cherry, W.(1990) *Topic in Nevanlinna theory*, Lecture Notes in Math. 1433, Springer-Verlag.

[Lau] Laurent, M.(1984) "Equations diophantiennes exponentielles", *Invent. Math.* **78**, 299–327.

[Lu] Lu, S.(1991) "On meromorphic maps into varieties of log-general type", *Proc. Symp. Amer. Math. Soc.* **52**, 305–333.

[Lu-Y] Lu, S., Yau, S. T.(1990) "Holomorphic curves in surfaces of general type", *Proc. Natl. Acad. Sci. USA* **87**, 80–82.

[Mas-N] Masuda, K. and Noguchi, J.(1996) "A construction of hyperbolic hypersurfaces of $\mathbf{P}^n(\mathbf{C})$", *Math. Ann.*, **304**, 339–362.

[McQ1] McQuillan, M.(1994) "Quelques compléments á une démonstration de Faltings", *C. R. Acad. Sci. Paris, Séris I*, **319**, 649–652.

[McQ2] McQuillan, M.(1996) "A new proof of the Bloch conjecture", *J. of Algebraic Geom.* **5**, 107–117.

[McQ3] McQuillan, M.(1997) "A dynamical counterpart to Faltings' diophantine approximation on abelian varieties", Preprint.

[McQ4] McQuillan, M.(1998) "Diophantine approximations and foliations", *Publ. Math. IHES* **87**, 121–174.

[Nad] Nadel, A.(1989) "Hyperbolic surfaces in $\mathbf{P}^3$", *Duke Math. J.* **58**, 749–771.

[Nev1] Nevanlinna, R.(1925) "Zur Theorie der meromorphen Funktionen", *Acta Math.* **46**, 1–99.

[Nev2] Nevanlinna, R.(1929) *Le théorème de Picard-Borel et la théorie des fonctions meromorphes*, Gauthiers-Villars, Paris, 1929.

[Nev3] Nevanlinna, R.(1970) *Analytic Functions*, Springer-Verlag 1970.

[Noc1] Nochka, E.I.(1982) "On a theorem from linear algebra", *Izv. Akad. Nauk. Modav. SSR Ser. Fiz. Teklam Mat. Nauk*, 29–33.

[Noc2] Nochka, E.I.(1983) "On the theory of meromorphic curves", *Dokl. Akad. Nauk. SSR*, 377–381.

[Nog1] Noguchi, J.(1977) "Holomorphic curves in algebraic varieties", *Hiroshima Math. J.* **7**, 833–853.

[Nog2] Noguchi, J.(1981) "Lemma on logarithmic derivatives and holomorphic curves in algebraic varieties", *Nagoya Math. J.* **83**, 213–233.

[Nog3] Noguchi, J.(1981) "A higher dimensional analogue of Mordell's conjecture over function fields", *Math. Ann.* **258**, 207–212.

[Nog4] Noguchi, J.(1988) "Moduli spaces of holomorphic mappings into hyperbolically imbedded complex spaces and locally symmetric spaces", *Invent. Math.* **93**, 14–34.

[Nog5] Noguchi, J.(1999) "On holomorphic curves in semi-abelian varieties", *Math. Z.* **228**, 713–721.

[No-O] Noguchi, J. and Ochiai, T.(1990) *Geometric Function Theory in Several Complex Variables*, Transl. Math. Mon. **80**, Amer. Math. Soc., Providence, R.I..

[No-S] Noguchi, J. and Sunada, T.(1982) "Finiteness of the family of rational and meromorphic mappings into algebraic varieties", *Amer. J. Math.* **104**, 887–900.

[Och] Ochiai, T.(1977) "On holomorphic curves in algebraic varieties with ample irregularity", *Invent. Math.* **443**, 83–96.

[Osg] Osgood, C.F.(1985) "Sometimes effective Thue-Siegel-Roth-Schmidt Nevanlinna bounds or better", *J. Number Theory* **21**, 1–15.

[OR] Osserman, R. and Ru, M.(1997) "An estimate for the Gauss curvature on minimal surfaces in $\mathbf{R}^m$ whose Gauss map omits a set of hyperplanes", *J. Diff. Geom.*, **46**, 578–593.

[Po-S] Poorten, A.J. van der and Schlickewei, H.P.(1991) "Additive relations in fields", *J. Austral. Math. Soc.* (Ser. A) **51**, 154–170.

[Rot] Roth, K.F.(1955), "Rational approximations to algebraic numbers", *Mathematika* **2** 1–20; Corrigendum, ibid., 168.

[Ru1] Ru, M.(1993), "Integral points and the hyperbolicity of the complement of hypersurfaces", *J. Reine Angew. Math.* **442**, 163–176.

[Ru2] Ru, M.(1995) "Geometric and arithmetic aspects of $\mathbf{P}^n$ minus hyperplanes", *Amer. J. Math.* **117**, 307–321.

[Ru3] Ru, M.(1997) "On a general form of the second main theorem", *Trans. of A.M.S.* **349**, 5093–5105.

[Ru4] Ru, M.(1997) "The second main theorem with moving targets on parabolic manifolds", *Indiana Univ. Math. J.*, **46**, 299–318.

[Ru5] Ru, M.(2000) "Algebroid functions, Wirsings theorem and their relations", *Math. Zeitschrift*, **233**, 137–148.

[Ru6] Ru, M.(2000) "An uniqueness theorem for moving targets without counting multiplicities", to appear in *Proc. Amer. Math. Soc.*

[Ru-S1] Ru, M. and Stoll, W.(1990) "Courbes holomorphes evitant des hyperplans mobiles", *C. R. Acad. Sci. Paris* **310**, Serie I, 45–48.

[Ru-S2] Ru, M. and Stoll, W.(1991) "The second main theorem for moving targets," *J. Geom. Anal.* **1**, 99–138.

[Ru-S3] Ru, M. and Stoll, W.(1991) "The Cartan's conjecture for moving targets", *Proceedings of Symposia in Pure Mathematics, American Math. Soc.*, **52**, Part 2, 477–508.

[Ru-V] Ru, M. and Vojta, P.(1997) "Schmidt's subspace theorem with moving targets", *Invent. Math.* **127**, 51–65.

[Ru-W] Ru, M. and Wong, P.M.(1991) "Integral points of $\mathbf{P}^n - \{2n+1$ hyperplanes in general position}", *Invent. Math.* **106**, 195–216.

[Sak1] Sakai, F.(1978) "Symmetric powers of the cotangent bundle and classification of algebraic varieties", in *Proceedings of Copenhagen Summer Meeting in Algebraic Geometry.*

[Sak2] Sakai, F.(1980) "Semi-stable curves on algebraic surfaces and logarithmic pluricanonical maps", *Math. Ann.* **254**, 98–120.

[Sch1] Schlickewei, H.P.(1977) "The $\wp$-adic Thue-Siegel-Roth-Schmidt theorem", *Arch. Math.* **29**, 267–270.

[Sch2] Schlickewei, H.P.(1990) "S-unit equations over number fields," *Invent. Math.* **102**, 95–107.

[Sch3] Schlickewei, H.P.(1992) "The quantitative Subspace Theorem for number fields", *Compos. Math.* **82**, 245–274.

[Schm1] Schmidt, W.M.(1971), "Linearformen mit algebraischen Koeffizienten II", *Math. Ann.* **191**, 1–20.

[Schm2] Schmidt, W.M.(1972) "Norm form equations", *Ann. of Math.* **96**, 526–551.

[Schm3] Schmidt, W.M.(1980) *Diophantine approximation*, Lecture Notes in Math. 785, Springer-Verlag.

[Schm4] Schmidt, W.M.(1989) "The subspace theorem in diophantine approximations", *Compositio Math.* **69**, 121–173.

[Schm5] Schmidt, W.M.(1990) "The number of solutions of norm form equations", *Trans. Amer. Math. Soc.* **317**, 197–227.

[Schm6] Schmidt, W.M.(1991) *Diophantine approximations and diophantine equations*, Lecture Notes in Math. 1467. Springer-Verlag.

[Sha] Shabat, B.V.(1985) *Distribution of values of holomorphic mappings*, Translations of Mathematical Monographs Vol. **61**, American Mathematical Society.

[Shi1] Shiffman, B.(1975) "Nevanlinna defect relations for singular divisors", *Invent. Math.* **31**, 155–182.

[Shi2] Shiffman, B.(1977) "Holomorphic curves in algebraic manifolds", *Bull. Amer. Math. Soc.* **83**, 553–568.

[Shi3] Shiffman, B.(1979) "On holomorphic curves and meromorphic maps in projective space", *Indiana Univ. Math. J.* **28**, 627–641.

[Shi4] Shiffman, B.(1982) "New defect relations for meromorphic functions on $\mathbf{C}^n$", *Bull. Amer. Math. Soc.* **7**, 594–601.

[Shi5] Shiffman, B.(1984) "A general second main theorem for meromorphic functions on $\mathbf{C}^n$", *Amer. J. Math.* **106**, 509–531.

[Shir] Shirosaki, M.(1991) "Another proof of the defect relation for moving targets", *Tohoku Math. J.*, **43**, 355–360.

[Sil1] Silverman, J.H.(1984) "Elliptic Curves", *Duke Math. J.* **51**, 395–403.

[Sil2] Silverman, J.H.(1986) *The arithmetic of elliptic curves*, Graduate Texts in Mathematics, Springer-Verlag.

[Siu1] Siu, Y.T.(1987) "Defect relations for holomorphic maps between spaces of different dimensions", *Duke Math. J.* **55**, 213–251.

[Siu2] Siu, Y.T.(1990) "Nonequidimensional value distribution theory and meromorphic connections", *Duke Math. J.* **61**, 341–367.

[Siu3] Siu, Y.T.(1995) "Hyperbolicity Problems in Function Theory", in *Five Decades as a Mathematician and Educator - on the 80th birthday of Professor Yung-Chow Wong*, ed. Kai-Yuen Chan and Ming-Chit Liu, World Scientific: Singapore, New Jersey, London, Hong Kong, pp.409–514.

[Siu-Y1] Siu, Y.T. and Yeung, S.K.(1996) "Hyperbolicity of the complement of a generic smooth curve of high degree in the complex projective plane", *Invent. Math.* **124**, 573–618.

[Siu-Y2] Siu, Y.T. and Yeung, S.K.(1996) "A generalized Bloch's theorem and the hyperbolicity of the complement of an ample divisor in an abelian variety", *Math. Ann.* **306**, 743–758.

[Siu-Y3] Siu, Y.T. and Yeung, S.K.(1997) "Defects for ample divisors of abelian varieties, Schwarz lemma, and hyperbolic hypersurfaces of low degree", *Amer. J. Math.* **119**, 1139–1172.

[Ste] Steinmetz, N.(1986) "Eine Verallgemeinerung des zweiten Nevanlinnasche Hauptsatzes", *J. Reine Angew. Math.* **368**, 134–141.

[Sto1] Stoll, W.(1953)(1954) "Die beiden Hauptsätze der Wertverteilungstheorie bei Funktionen mehrerer komplexen Veränderlichen. I", *Acta Math.* **90**, 1–115, II. Acta Math. **92**, 55–169.

[Sto2] Stoll, W.(1970) *Value distribution of holomorphic maps into compact, complex manifolds*, Lecture Notes in Mathematics **135**, Springer-Verlag.

[Sto3] Stoll, W.(1977) *Value distribution on parabolic spaces*, Lecture Notes in Mathematics **600**, Springer-Verlag.

[Sto4] Stoll, W.(1983) *The Ahlfors-Weyl theory of meromorphic maps on parabolic manifolds*, Lecture Notes in Mathematics **981**, 101–219.

[Sto5] Stoll, W.(1985) *Value distribution theory for meromorphic maps*, Aspects of Math., E7.

[Sto6] Stoll, W.(1988) "An extension of the theorem of Steinmetz-Nevanlinna to holomorphic curves", *Math. Ann.* **282**, 185–222.

[Sto7] Stoll, W.(1989) "On the propagation of dependences", *Pacific J. of Math.* **139**, 311–337.

[Sto8] Stoll, W.(1995) *Value Distribution Theory in Several Complex Variables*, Shandong Science and Technology Press.

[St-W] Stoll, W. and Wong P.M.(1994) "Second Main Theorem of Nevanlinna Theory for non- equidimensional Meromorphic Maps", *Amer. J. Math.* **116**.

[Thu] Thue, A.(1909) "Über Annäherungswerte algebraischer Zahlen", *J. Reine Angew. Math.* **135**, 184–305.

[Tod1] Toda, N.(1971) "Sur quelques combinaisons linéaires exceptionelles au sens de Nevanlinna", *Tohoku Math. J.* **23**, 67–95.

[Tod2] Toda, N.(1971) "On the functional equation $\sum_{i=0}^{p} a_i f_i^{n_i} = 1$", *Tohoku Math. J.* **23**, 289–299.

[Tod3] Toda, N.(1981) "Sur l'inégalité fondamental de H. Cartan pour les systemes de fonctions entières", *Nagoya Math. J.* **83**, 5–14.

[Val] Valiron, G.(1931) "Sur la dérivée des fonctions algébroides", *Bull. Soc. Math. France* **59**, 17–39.

[Vit] Vitter, Al.(1977) "The lemma of the logarithmic derivative in several complex variables", *Duke Math. J.* **44**, 89–104.

[Voj1] Vojta, P.(1987) *Diophantine approximations and value distribution theory*, Lecture Notes in Math. **1239**, Springer-Verlag.

[Voj2] Vojta, P.(1991) "Siegel's theorem in the compact case", *Ann. Math.* **133**, 509–548.

[Voj3] Vojta, P.(1992) "A generalization of theorems of Faltings and Thue-Siegel-Roth-Wirsing", *J. Amer. Math. Soc.* **5**, 762–804.

[Voj4] Vojta, P.(1992) "A quantitative proof of Roth's theorem with moving targets", preprint, 27 pp.

[Voj5] Vojta, P.(1993) "Arithmetic of subvarieties of abelian and semiabelian varieties", Advances in number theory (Kingston, ON, 1991), 233–238, Oxford Sci. Publ., Oxford Univ. Press, New York.

[Voj6] Vojta, P.(1996) "Roth's theorem with moving targets", *International Mathematics Research Notices*, 109–114.

[Voj7] Vojta, P.(1996) "Integral points on subvarieties of semiabelian varieties I.", *Invent. Math.* **126**, no. 1, 133–181

[Voj8] Vojta, P.(1997) "On Cartan's theorem and Cartan's conjecture", *Amer. J. Math.* **119**, 1–17.

[Voj9] Vojta, P.(1998) "A more general *abc* conjecture", *Internat. Math. Res. Notices*, **21**, 1103–1116.

[Voj10] Vojta, P.(1999) "Integral points on subvarieties of semiabelian varieties. II", *Amer. J. Math.* **121**, No. 2, 283–313.

[Voj11] Vojta, P.(1999) "Nevanlinna theory and Diophantine approximation", Several complex variables (Berkeley, CA, 1995–1996), 535–564, Math. Sci. Res. Inst. Publ., 37, Cambridge Univ. Press, Cambridge.

[Voj12] Vojta, P.(2000) "On the ABC conjecture and Diophantine approximation by rational points", *Amer. J. Math*, **122**, 843–872.

[Wa1] Wang, J. T-Y.(1996) "The truncated second main theorem of function fields", *J. Number Theory* **58**, 137–159.

[Wa2] Wang, J. T-Y.(1996) "An effective Roth's theorem of function fields" *Rocky Mt. J. Math.* **26**, 1225–1234.

[Wa3] Wang, J. T-Y.(1998) "Integral points of $\mathbf{P}^n - \{2n+1$ hyperplanes in general position$\}$ over number fields and function fields", *Trans. Amer. Math. Soc.*, **348**, 3379–3389.

[We-We] Weyl, H. and Weyl, F.J.(1943) *Meromorphic Functions and Analytic Curves*, Princeton University Press, Princeton.

[Wir] Wirsing, E.(1971) "On approximations of algebraic numbers by algebraic numbers of bounded degree", *Proc. of Symp. in Pure Math.*, 213–247.

[Wong1] Wong, P.M.(1976) *Defect relations for maps on parabolic spaces and Kobayashi metrics on projective spaces omitting hyperplanes*, Thesis, University of Notre Dame, 231 pp.

[Wong2] Wong, P.M.(1980) "Holomorphic mappings into Abelian varieties", *Amer. J. Math.* **102**, 493–501.

[Wong3] Wong, P.M.(1989) "On the second main theorem of Nevanlinna theory", *Amer. J. Math.* 549–583.

[Wong4] Wong, P.M.(1992) Recent results in hyperbolic geometry and diophantine geometry, in Proc. International Symp. in Holomorphic Mappings, Diophantine Geometry and Related Topics, RIMS, Kyoto University, 120-135(1992) .

[Wong5] Wong, P.M.(1993) "Holomorphic curves in spaces of constant curvature. In "Complex Geometry" edited by G. Komatsu and Y. Sakane, 201–223. Lecture Notes in Pure and Applied Mathematics, vol 143, Marcel Dekker.

[Wong6] Wong, P.M.(1999) "Nevanlinna theory for holomorphic curves in projective varieties", Preprint.

[Wu] Wu, H. *The equidistribution theory of holomorphic curves.* Ann. of Math. Studies **64**, Princeton University Press, Princeton, N.J. 1970.

[Ya] Yamanoi, K. "Algebro-geometric version of Nevanlinna's lemma on logarithmic derivative and applications", Preprint.

[Ye] Ye, Z.(1995) "On Nevanlinna's second main theorem in projective space", *Invent. Math.*, **122**, 475–507.

[Zai1] Zaidenberg, M.(1983) "Picard's theorem and hyperbolicity", *Siberian Math. J.* **24**, 858–867.

[Zai2] Zaidenberg, M.(1990) "A function-field analog of the Mordell conjecture: A noncompact version", *Math. USSR Izevestiya* **35**, 61–81.

[Zai3] Zaidenberg, M.(1992) "Shottky-Laundau growth estimates for s-normal families of holomorphic mappings", *Math. Ann.* **293**, 123–141.

[Zai4] Zaidenberg, M.(1989) "Stability of hyperbolic embeddedness and construction of examples", *Math. USSR Sbornik* **63**, 351–361.

[Zha] Zhang, S.(1992) "Positive line bundles on arithmetic surfaces", *Ann. of Math.* **136**, 569–587.

List of frequently used notation

$\mathbf{C}$	the set of complex numbers
$\mathbf{C}^n$	the n-th Cartesian products of $\mathbf{C}$
$\mathbf{C}^n[r]$	$= \{z \in \mathbf{C}^n \mid \|z\| \leq r\}$
$\mathbf{C}^n(r)$	$= \{z \in \mathbf{C}^n \mid \|z\| < r\}$
$< \, , \, >$	the standard inner product on $\mathbf{C}^n$
$c_1(L)$	Chern form of the metrized line bundle L, 55
d^c	$= \frac{\sqrt{-1}}{4\pi}(\bar{\partial} - \partial)$, 50
$\mathbf{D}$	the unit disc on $\mathbf{C}$
$\mathbf{D}(r)$	the open disc of radius r on $\mathbf{C}$
$\overline{\mathbf{D}}(r)$	the closure of $\mathbf{D}(r)$ in $\mathbf{C}$
$\mathcal{O}(D)$	the line bundle associated to the divisor D, 54
$\mathrm{Im} F$	the imaginary part of F, 8
$[D]$	the current associated to the divisor D, 50, 226
$d^n f$	the k-jet lifting of f, 240
$\mathbf{f}$	a reduced representation of $f : \mathbf{C} \to \mathbf{P}(\mathbf{C}^n)$, 101
$f^*\omega$	the pull-back of ω by f
$G_{L/k}$	Galois group of L over k, 77
k	a number field, 26
$\bar{k}$	algebraic closure of k, 67
$k(C)$	function field of the algebraic curve C, 77
$h(x)$	absolute logarithmic height of x, 29, 70
$h_D(x)$	absolute logarithmic height of x with respect to D, 70
$h(\mathbf{x})$	logarithmic height of $\mathbf{x}$ for $\mathbf{x} = [x_0 : \cdots : x_n] \in \mathbf{P}^n$, 160
$J_k(M)$	k-jet bundle, 241

$\lambda_{v,D}$	Weil function with respect to the divisor D, 71
$\lambda_H(f(z))$	Weil function of f with respect to the hyperplane H, 102
$\lambda_{v,H}$	Weil function with respect to the hyperplane H, 160
$m_f(r,a)$	proximity function of f with respect to a, 5
$m_f(r,H)$	proximity function of f with respect to the hyperplane H, 102
$m(x,a)$	proximity function of x with respect to a, 29
$m(x,D)$	proximity function of x with respect to the divisor D, 71
$m(x,H)$	proximity function of x with respect to the hyperplane H, 160
M_k	canonical set of places in the number field k, 28
M_k^∞	Set of Archimedean places in M_k, 28
M_k^0	Set of non-Archimedean places in M_k, 28
$N_f(r,a)$	counting function of f with respect a, 4
$N_f(r,H)$	counting function of f with respect to the hyperplane H, 102
$N_f^{(1)}(r,a)$	truncated counting function of f with respect a, 21
$N_f^{(n)}(r,H)$	truncated counting function of f with respect the hyperplane H, 102
$N_{\mathrm{ram},f}(r)$	ramification term of f, 18
$N(x,a)$	counting function of x with respect to a, 29
$N(x,D)$	counting function of x with respect to the divisor D, 71
$N(x,H)$	counting function of x with respect to the hyperplane H, 160
$O(1)$	a constant
$f(r) \le g(r) + O(h(r))$	$\limsup_{r\to+\infty} \frac{f(r)-g(r)}{h(r)} < +\infty$
$f(r) \le g(r) + o(h(r))$	$\limsup_{r\to+\infty} \frac{f(r)-g(r)}{h(r)} \le 0$
$\mathrm{ord}_{z_0} f$	order of f at the point z_0, 4
$\mathrm{ord}_z^+ f$	the multiplicity of the zero of f at the point z, 4
$\mathcal{O}_S$	the ring of S-integers, 159
$\mathcal{O}_S^*$	the multiplicative group of S-units, 159
$\omega_n(z)$	$= dd^c \log \lvert z\rvert^2$, 214

$\Omega_M(log D)$	logarithmic cotangent bundle of M with at worst logarithmic simple poles along D, 289
$\mid\ \mid_p$	p-adic absolute value, 25
$\mid\ \mid_{\mathcal{P}}$	p-adic absolute value associated to a prime ideal, 27
$\|\ \|_{\mathcal{P}}$	normalized p-adic absolute value associated to a prime ideal, 27
$\mathbf{P}^n$	n-th dimensional projective space
$\mathbf{P}^n(\mathbf{C})$	the complex projective space of dimension n
$\mathbf{Q}$	the set of rational numbers
$\overline{\mathbf{Q}}$	algebraic closure of $\mathbf{Q}$
$\mathbf{Q}_p$	p-adic closure of $\mathbf{Q}$, 26
$\mathbf{R}$	the set of real numbers
$\mathrm{Re}F$	real part of F
$\mathrm{Ric}(\omega)$	Ricci form of ω, 58
$R_M(p, v)$	Royden's pseudo-distance, 276
$\mathbf{R}_k$	the ring of algebraic integers of k, 27, 81
$\mathbf{R}_S$	the ring of S-integers, 84
ρ_n	$= (dd^c\|z\|^2)^n$, 214
S	a finite subset of M_k containing all Archimedean places, 29
$S_n(r)$	$= \{z \in \mathbf{C}^n \mid \|z\| = r\}$, 214
σ_n	$= d^c \log\|z\|^2 \wedge (dd^c \log\|z\|^2)^{n-1}$, 214
$T_f(r)$	characteristic function of f
$T_{f,\omega}(r)$	characteristic function of f with respect to ω, 57, 228
$T_f(r, L)$	characteristic function of f with respect to a line bundle L, 229
$T_f(r, \mathcal{O}(D))$	characteristic function of f with respect to the divisor D, 58, 229
v	a place on the number field k
$\|\ \|_v$	normalized almost-absolute value associated to a place $v \in M_k$, 28
$v(x)$	$= -\log\|x\|_v$, 79
$\|\mathbf{x}\|_v$	$= \max_{0\le i\le n}\|x_i\|_v$ for $\mathbf{x} = (x_0 : \cdots : x_n)$, 159
$v_n(z)$	$= dd^c\|z\|^2$, 214
$\wedge$	the wedge product
$W(f_0, \ldots, f_n)$	the Wronskian of $f_0, \ldots, f_n$, 104
$X(k)$	k-rational points on X, 67
$X[m]$	m-torsion subgroup of the elliptic curve X, 76
$\mathbf{Z}$	the set of integers

Index